Confocal Microscopy

Confocal Microscopy

Edited by

T. Wilson

Department of Engineering Science
University of Oxford, Oxford, UK

ACADEMIC PRESS
London   San Diego   New York   Boston
Sydney   Tokyo   Toronto

ACADEMIC PRESS LIMITED
24/28 Oval Road
London NW1 7DX

United States Edition published by
ACADEMIC PRESS INC.
San Diego
CA 92101

Copyright © 1990 by
ACADEMIC PRESS LIMITED

This book is printed on acid-free paper

All Rights Reserved

No part of this book may be reproduced in any form by photostat, microfilm, or any other means, without written permission from the publishers.

British Library Cataloguing in Publication Data
Wilson, T.
   Confocal microscopy.
   1. Microscopy
   I. Title
   502.82

ISBN 0-12-757270-8

Printed in Great Britain by
St Edmundsbury Press Ltd, Bury St Edmunds,
Suffolk

Preface

It is now over five years since *Theory and Practice of Scanning Optical Microscopy* was published, and the subject has advanced considerably during that time. We have seen conferences and special issues of journals devoted entirely to the technique, but, perhaps more telling, a host of commercial instruments have become available. It therefore seemed appropriate to bring out a new book on the subject, as a good many of the three-dimensional processing techniques, especially in biology, had been developed since 1984, as had semiconductor line-width measuring machines.

The question was how to do it. I decided that the way to get the best coverage was to invite specialists in the various branches of the technology to contribute chapters on areas about which they had an intimate knowledge. I thank all of those who responded to my invitation. I think that they have all done a splendid job, and I hope that the result will be found to be a useful book.

We have concentrated on the science and applications of these instruments, referring to commercial instruments only when a particular microscope was used to obtain an image. We have made no attempt to compare products. Such comparisons are invidious, and indeed are likely to change with time as new models supersede old. We have tried to present a unified account of the current position of confocal microscopy, laying out the physical basis of the imaging as well as the practical methods by which images are obtained, in both laser-based and tandem scanning confocal microscopes. Specific applications areas are also discussed.

These instruments are already finding a great many applications in biology and the other life sciences, ophthalmology, industrial inspection and semiconductor line-width metrology. As time goes on, we can be sure that new applications will emerge, based on the unique three-dimensional imaging capability of these instruments.

Finally, for their help in editing the text, thanks are due to Catherine Wilson and to Simon Hewlett. Without Simon's expert knowledge of confocal microscopy and of LaTeX, the preparation of the text would have been a vastly more difficult undertaking. For any errors which may remain, I, alas, must take responsibility.

Tony Wilson
Oxford, 1990

Contributors

A. Boyde, Department of Anatomy and Developmental Biology, University College London.
G. J. Brakenhoff, Department of Electron Microscopy and Molecular Cytology, University of Amsterdam.
C. J. Cogswell, Department of Biology, University of Oregon.
P.-O. Forsgren, Department of Physics IV, The Royal Institute of Technology, Stockholm.
O. Franksson, Department of Physics IV, The Royal Institute of Technology, Stockholm.
M. Hadravský, Department of Biophysics, Faculty of Medicine, Charles University, Plzen.
Z. Hegedus, CSIRO, Division of Applied Physics, Lindfield, New South Wales.
V. Howard, Department of Human Anatomy and Cell Biology, University of Liverpool.
G. S. Kino, Edward L. Ginzton Laboratory, Stanford University.
A. Liljeborg, Department of Physics IV, The Royal Institute of Technology, Stockholm.
B. R. Masters, School of Electrical Engineering, Georgia Institute of Technology.
J. L. Oud, Department of Electron Microscopy and Molecular Cytology, University of Amsterdam.
M. Petráň, Department of Biophysics, Faculty of Medicine, Charles University, Plzen.
V. Sarafis, University of Western Sydney.
C. J. R. Sheppard, School of Physics, University of Sydney.
E. H. K. Stelzer, European Molecular Biology Laboratory, Heidelberg.
H. T. M. van der Voort, Department of Electron Microscopy and Molecular Cytology, University of Amsterdam.
R. W. Wijnaendts-van-Resandt, Heidelberg Instruments, Heidelberg.
T. Wilson, Department of Engineering Science, University of Oxford.
G. Q. Xiao, Edward L. Ginzton Laboratory, Stanford University.

Table of Contents

Preface

Contributors

1. **Confocal Microscopy**
   by T. Wilson.

I.	Introduction	1
II.	Image formation in scanning microscopes	7
III.	Applications of depth discrimination	13
IV.	Theory of imaging in scanning microscopes	22
	IV.1. Fundamentals of optical image formation	22
	IV.2. The imaging of simple objects	25
	IV.3. Fourier imaging	31
	IV.4. Optical sectioning	38
	IV.5. Flare and scattered light	39
V.	Fluorescence microscopy	41
VI.	The tandem scanning microscope	45
VII.	Non-confocal imaging in scanning microscopes	47
	VII.1. Differential contrast	49
	VII.2. Differential amplitude contrast	50
	VII.3. Differential phase contrast	53
VIII.	Conclusions	58
	References	60

2. **Software and Electronics for a Digital 3-D Microscope**
   by P-O. Forsgren, O. Franksson and A. Liljeborg.

I.	Introduction	65
II.	System concepts	68
III.	Display techniques	71
	III.1. Display principles	71
	III.2. Look-through and depth-coded projections	72
	III.3. Surface shading	73
	III.4. Rotations	76
IV.	Filtering of volumes	78
V.	Data storage	79
VI.	Instrument control and computer requirements	81
	VI.1. Control processor selection	81
	VI.2. Communication bus selection	82
	VI.3. Computer considerations for data-processing	83
	VI.4. Instrument control	85
VII.	Electronics	87

	VII.1. Microprocessor system	87
	VII.2. Position encoding of the scanning mirror	87
	VII.3. Detector signal enhancement	90
	References	91

3. Optical Aspects of Confocal Microscopy
by T. Wilson.

I.	Introduction	93
II.	Brightfield image formation with finite-sized detectors	94
	II.1. The point object and the planar reflector	94
	II.2. Images with a finite-sized detector	99
	II.3. The role of aberrations	102
	II.4. Alternative detector geometries	108
III.	Fluorescence imaging	114
	III.1. Fluorescence imaging with a slit-shaped detector	115
IV.	Coherent detectors in scanning microscopy	118
V.	Pupil function filters	120
VI.	Enhancing the optical sectioning	125
VII.	The direct view microscope	130
VIII.	Scanning mechanisms	135
IX.	Conclusions	139
	References	139

4. Three-dimensional Imaging in Confocal Microscopy
by C. J. R. Sheppard and C. J. Cogswell.

I.	Introduction	143
II.	Imaging in confocal systems: a single point object	147
III.	Three-dimensional image formation	155
IV.	Axial response	161
V.	The axial response for a low aperture system	164
VI.	Conclusions	167
	References	168

5. Pupil Filters in Confocal Imaging
by Z. S. Hegedus.

I.	Introduction	171
II.	Composite point spread and transfer functions	171
III.	Imaging with modified pupils	173
IV.	Pupils in confocal imaging	174
	IV.1. Amplitude point spread functions	174
	IV.2. Transfer functions	176
	IV.3. General remarks and some experimental results	179
V.	Pupil generation	182
	Acknowledgements	183
	References	183

6. Three-dimensional Image Representation in Confocal Microscopy
by G. J. Brakenhoff, H. T. M. van der Voort and J. L. Oud.

I.	Introduction	185
II.	Aspects of confocal image formation	186
III.	Confocal data collection	188
IV.	Three-dimensional data sets and representation techniques	190
	IV.1. Sections and cross-sections	190
	IV.2. Superposition and stereoscopic images	192
V.	Representation by simulated fluorescence processing (SFP)	193
VI.	Graphic techniques	194
VII.	Conclusions and comments	196
	Acknowledgements	196
	References	196

7. Optical Cell Splicing with the Confocal Fluorescence Microscope: Microtomoscopy
by E. H. K. Stelzer and R. W. Wijnaendts-van-Resandt.

I.	Issues in confocal fluorescence microscopy	199
II.	Depth discrimination in a confocal fluorescence microscope	201
	II.1. Generating x-z images	202
III.	Biological applications of confocal fluorescence microscopy	205
	III.1. Consequences for biological applications	205
	III.2. Applying confocal fluorescence microscopes	206
IV.	Conclusions	210
V.	Acknowledgements	211
	References	211

8. Confocal Brightfield Imaging Techniques Using an On-Axis Scanning Optical Microscope
by C. J. Cogswell and C. J. R. Sheppard.

I.	Introduction	213
II.	Microscope design considerations	214
	II.1. Choosing the optimum design configuration for brightfield optics	214
	II.2. Scanning speed	217
III.	Reflected brightfield	218
	III.1. Sample characteristics	218
	III.2. Improving visual quality and resolution	227
	III.3. Improving visibility of image features by electronic differentiation	233
IV.	Differential phase contrast	234
V.	Confocal transmission	236
VI.	Confocal differential interference contrast (DIC)	237
VII.	Conclusion	241
	Acknowledgements	242
	References	242

9. Direct View Confocal Microscopy
by M. Petráň, A. Boyde and M. Hadravský.

I.	Introduction	245
	I.1. Historical background	245
	I.2. Double scanning: tandem scanning	246
	I.3. Disc scanning	247
II.	Choice of aperture disc for the rapid scanning method	248
III.	Some ways in which disc scanning can be implemented	249
	III.1. Reflection or transmission?	249
	III.2. Reflection: one or two discs?	249
	III.3. Reflection: one disc with central symmetry	250
	III.4. Reflection: the autocollimation case	253
	III.5. Transmission TSM: one disc with central symmetry	254
	III.6. Transmission TSM: two discs	254
IV.	Illumination	254
V.	The aperture disc	255
	V.1. Disc size: why not Nipkow's original design?	255
	V.2. Pinhole layout	256
	V.3. Pinhole size in the disc and spot size at the object	257
	V.4. Disc materials in practice	258
	V.5. Practical design information on discs to date	259
	V.6. Slits	259
	V.7. Driving the disc	259
VI.	Other components	260
	VI.1. Mirrors and beam-splitters	260
	VI.2. Filters	260
	VI.3. Alignment aids	260
	VI.4. Objective lenses	261
	VI.5. Eyepieces and relay optics	263
	VI.6. Cameras, TV cameras and image intensifiers	264
	VI.7. Stands: rigidity versus flexibility	264
	VI.8. Stage xyz automation	265
	VI.9. Simultaneous illumination for non-confocal modes	268
VII.	Image processing and computers	268
VIII.	Applications, specimen preparation and imaging modes	270
	VIII.1. The main areas of application	270
	VIII.2. For which kinds of specimens is TSM most suited?	270
	VIII.3. Specimen preparation	271
	VIII.4. Fluorescence	271
	VIII.5. Reflective stains	271
	VIII.6. Reflective coatings	271
	VIII.7. Chromatic aberration colour-coding for depth	272
	VIII.8. Stereoscopic image recording	272
	VIII.9. Colour-coding without a computer	274
	VIII.10 Depth limitation	275
	VIII.11 Particle counting and stereology	275
	VIII.12 Reflection and interference contrast	275

	IX.	Resolution and contrast	275

IX.   Resolution and contrast 275
       IX.1. Resolution from the standpoint of information theory 275
       IX.2. The equivalence of the TSM and other CSLMs ... 276
       IX.3. Methods which compete with overlapping advantages 276
X.   Summary and conclusions 277
       X.1.   Present standing of the TSM 277
       X.2.   Future directions 277
XI.   Acknowledgements 278
XII.   Applications bibliography 278
       References 281

10. The Confocal Microscope as an Instrument for Measuring Microstructural Geometry
by V. Howard.

I.   Introduction 285
II.   Dimensions and probes 286
III.   Volume estimation 289
IV.   Number estimation 290
V.   Surface and length estimation 292
VI.   Estimation of particle size 295
VII.   Other zero-dimensional quantities 298
VIII. Future developments 299
       References 301

11. Confocal Microscopy of Ocular Tissue
by B. R. Masters.

I.   Introduction 305
II.   The structure and optical properties of the eye 305
III.   Development of optically sectioning microscopes in ophthalmology 306
IV.   Applications of confocal microscopy to ophthalmology ... 309
V.   Development of a real-time confocal system for clinical ophthalmology 320
       Acknowledgements 322
       References 322

12. Biological Perspectives of Confocal Microscopy
by V. Sarafis.

I.   Introduction 325
II.   Applications of reflection microscopy 326
       II.1. Reflection from phase specimens 326
       II.2. Reflection from gold immunolabelled material 327
       II.3. Autoradiography 327
III.   Transmission microscopy 327
       III.1. Transmission in brightfield 327
       III.2. Correcting for spherical aberration 328
IV.   Fluorescence confocal microscopy 328

	IV.1. Fluorescence microscopy	328
	IV.2. Fluorescence microscopy with thick scattering and absorbing specimens	329
	IV.3. Application of confocal scanning light microscopy in fluorescence	329
	IV.4. Advantages of confocal fluorescence microscopy for increasing resolving power	330
V.	Application to stereology	331
VI.	Confocal microscopy applications for real-time imaging	331
VII.	White light confocal microscopy	332
VIII.	Extensions of confocal microscopy methodology	332
	References	333

13. Semiconductor Metrology
by R. W. Wijnaendts-van-Resandt.

I.	Introduction	339
II.	VLSI metrology	341
	II.1. Linewidth measurements	341
	II.2. Registration measurements	342
	II.3. Defect inspection	343
	II.4. Critical dimensions of masks and reticles	343
	II.5. Other optical inspection techniques	344
III.	Quantitative confocal microscopy	345
	III.1. The z-response	346
	III.2. Edge-response, object scan versus beam scan	347
	III.3. Confocal contrast modes	348
	III.4. Data quality and analysis	349
IV.	Technical realizations and results	349
	IV.1. Stage-scanning systems	349
	IV.2. Beam-scanning systems	350
	IV.3. Three-dimensional image generation	351
	IV.4. Data analysis	353
V.	The size measurement of transparent microstructures	355
	V.1. Polarization contrast (QIC)	356
	V.2. Ultraviolet microscopy	358
VI.	Confocal microscopy versus scanning electron microscopy	358
VII.	Conclusions and outlook	359
	References	359

14. Real-time Scanning Optical Microscopes
by G. S. Kino and G. Q. Xiao.

I.	The tandem scanning optical microscope	361
II.	The real-time tandem scanning optical microscope	365
III.	Theory of the real-time scanning optical microscope	367
IV.	The transverse response of a confocal microscope	373
V.	Experimental results	378
VI.	Conclusions	386
	References	386

15. Confocal Interference Microscopy
by C. J. R. Sheppard.

I.	Introduction	389
II.	Surface profiling by confocal interferometry	395
III.	Heterodyne interferometry	398
IV.	Aberration measurement by confocal interferometry	399
V.	Conclusions	409
	References	410

1. Confocal Microscopy

T. WILSON

I. Introduction

There has been a variety of reasons for the growth of interest in scanning *optical* microscopy during the last few years. Amongst these we must certainly include the ready availability of laser light sources and the tremendous increase in powerful low-cost computer image- and data-processing systems. However, these reasons are insufficient in themselves. It is probably fair to say that if these instruments could not operate in the *confocal* mode, they would not attract a fraction of their present popularity. In essence, the confocal system offers, amongst many other things, the chance to image thick biological tissue in three dimensions. These instruments operate well in the brightfield and fluorescence modes. The remaining chapters of this book will discuss many of the features of confocal microscopy in great detail. In this introductory chapter we will merely try to set the scene for the rest of the book. In particular we will develop a very simple theory which will allow us to predict and understand the main aspects of image formation in brightfield and fluorescence imaging. However, we will begin by considering the conventional optical microscope and will show that, by choosing a scanning approach, we have the freedom to modify the optical system in simple ways to permit many novel imaging modes, of which the confocal arrangement is one of the most powerful.

In essence, the conventional microscope is a parallel processing system which images the entire object field simultaneously. This is quite a severe requirement for the optical components. We can relax this requirement if we no longer try to image the whole object at once. The limit of the relaxation is to require an image of only *one* object point at a time. In this case, all that we ask of the optics is that it should provide a good image of one point. The price that we have to pay is that we must *scan* in order to build up an image of the entire field. The answer to the question as to whether this price is worth paying will, to some extent, depend on the application in question. In general, though, the flexibility of modifying the optical system in

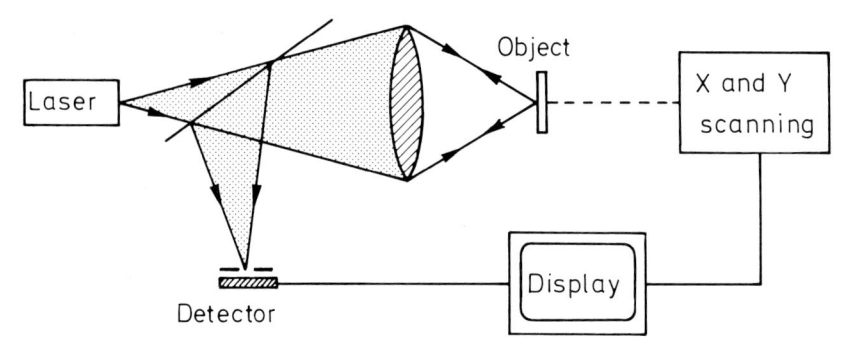

Figure 1.1. Schematic arrangement of a reflection mode scanning optical microscope.

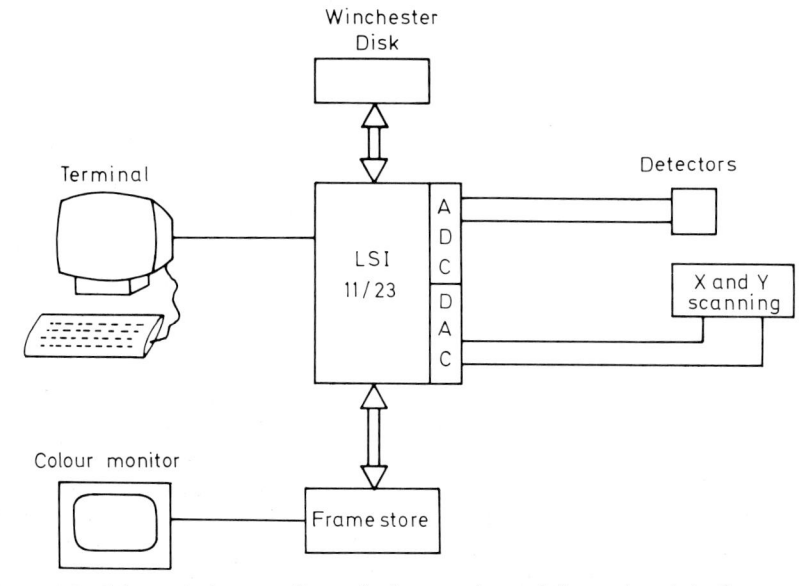

Figure 1.2. Schematic layout of a typical computer and framestore interface.

order to obtain some specific unique imaging mode makes this a small price to pay.

A typical arrangement of a scanning optical microscope is shown in Figure 1.1. It is also possible, in principle, to operate these systems in transmission. The essential components are some form of mechanism for scanning the light beam (usually from a laser) relative to the specimen and appropriate photodetectors to collect the reflected or transmitted light [Wilson and Sheppard, 1984]. Most of the early systems were analogue in nature (see, for example, Montgomery, 1962), but it is now more usual, thanks to the serial nature of the image formation, to use a computer both to drive the microscope and to collect and display the image. It is also common to use a framestore to enable images to be taken, processed and displayed. A typical computer framestore interface is shown in Figure 1.2 and a typical, low-magnification image is shown in Figure 1.3. We deliberately choose low magnification to illustrate that a scanning system - in this case a mechanical object-scanning system - can produce good images without loss of resolution and contrast at the edge of the field.

An inevitable consequence of our approach is that the image is obtained in an electrical form. This permits many forms of image enhancement and processing to be undertaken. The simplest example is probably contrast enhancement. Figure 1.4 shows a low magnification image of pen strokes on a piece of paper. It is clear from the image which was the last line to have been made. Quantitative measurements may also be made to a high degree of accuracy. As an example, Figure 1.5 shows an image of the cleaved end of an optical fibre. The intensity of the light reflected from the end of the fibre depends on the refractive index of the material, which permits changes of about one percent to be measured and hence the refractive index profile of the fibre to be determined [Wilson *et al.*, 1980]. Although the enhancement here is linear, non-linear techniques such as histogram equalisation and homomorphic filtering are also available to allow full use to be made of the dynamic range of display [Gonzales and Wintz, 1987; Cox, 1981]. A further advantage of the scanning approach is that the processing

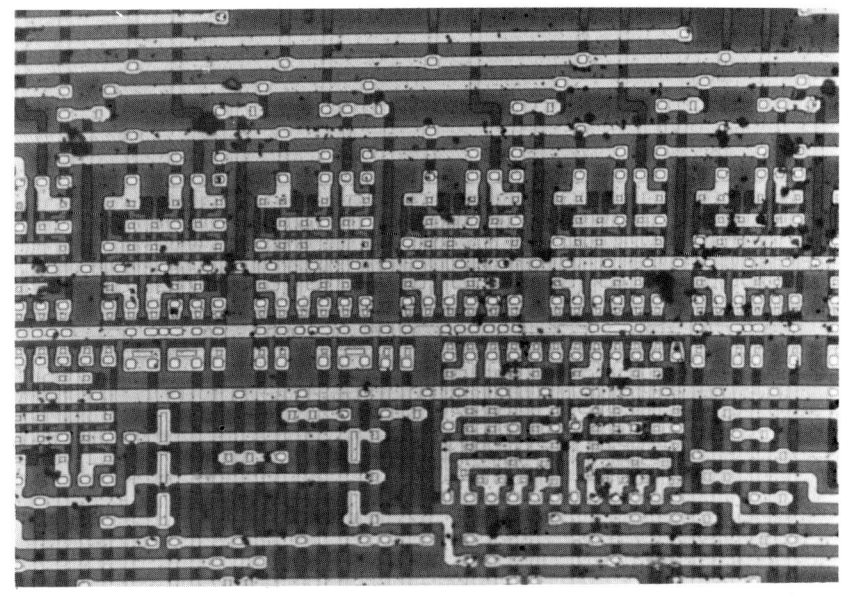

Figure 1.3. A low-magnification image of a portion of a microcircuit.

Figure 1.4. An image of a portion of a character written in ink on a piece of paper. The use of electrical contrast enhancement makes it clear which pen stroke was the last to have been made.

is done on data from an ordinary image. There is no need to modify the optical system to increase contrast by, for example, stopping down the collector lens or using dark-ground techniques. These latter techniques also modify the resolution of the image formation, whereas electrical processing does not. A discussion of other advantages which arise from scanning is given by Sheppard (1987).

The images that we have presented so far have all been taken with microscope systems in which the scanning was achieved by mechanically rastering the specimen relative to a finely focussed spot of light. The advantage here [Wilson and Sheppard, 1984; Brakenhoff et al., 1986] is that the optical system is very simple: we only require the objective lens to focus light to an axial diffraction-limited spot. In principle, this relaxes the requirements made of the objective lens and suggests that specially designed lenses could be used to advantage in these systems. This approach gives space-invariant imaging, which ensures that the resolution and contrast are identical across the entire field of view and are completely decoupled from the magnification. It is thus possible for continuously variable magnification to be obtained with a single objective lens. This, in turn, means that low-magnification images can be taken with high numerical aperture objectives: a particularly important advantage in, for example, fluo-

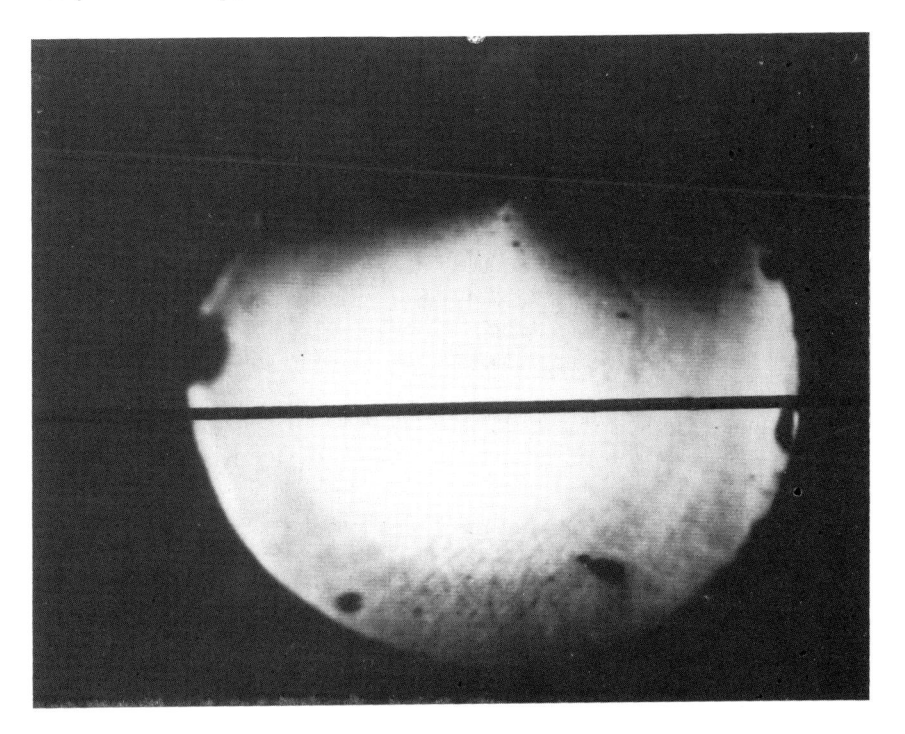

Figure 1.5. The image of the cleaved end of an optical fibre. The difference between peak white and black in this image corresponds to a change in reflectivity of 0.58%.

rescence microscopy, where high aperture also implies greater signal collection level. A possible drawback to this approach concerns the relatively slow speed of image acquisition: typically, a few seconds. However, the use of a framestore makes this, in most cases, a relatively unimportant consideration. Indeed, in the very important case of fluorescence microscopy, the ability to scan slowly may be an advantage.

In some situations, of course, speed is very important, and here we must consider a beam-scanning approach. This may be achieved by using vibrating galvanometer-type mirrors, rotating mirror wheels or acousto-optic beam deflectors. The use of the latter two alternatives gives the possibility of TV-rate scanning, whereas vibrating mirror systems are usually relatively slow. The drawbacks to these systems lie in the increased complexity of the optical system and the non-space-invariant imaging, which may be very important both at the edge of the field and in very highly sensitive image manipulation. The magnification is now coupled to the resolution, because it is usually necessary to change objective lenses in order to cover the entire magnification range. One advantage of these systems, however, is that they are usually built around conventional optical microscopes, Figure 1.6, which permits conventional binocular viewing in order to locate the region of interest. However, we must be very careful with high-speed scanning systems and ensure that the speed of scanning does not influence, and hence modify, the image. This will usually not occur, but it is worth noting that in the optical beam induced current technique of examining semiconductor materials and devices, the scan speed can dramatically modify the images [Wilson and Pester, 1987a, 1987b]. An alternative beam-scanning approach has been proposed recently by Goldstein [Goldstein, 1989] in which the probe beam is scanned with acousto-optic beam deflectors and descanned using an image dissector tube.

A middle-ground alternative scheme is to elect to scan the objective lens itself relative to a stationary optical system and a stationary object [Hamilton and Wilson, 1986]. This overcomes some of the drawbacks of both object-scanning and beam-scanning systems. It permits

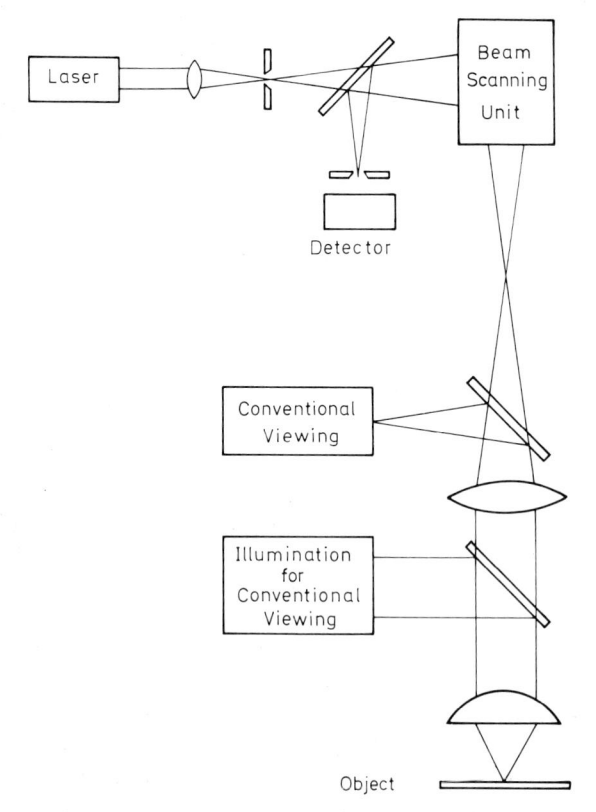

Figure 1.6. Schematic diagram illustrating the combination of a scanning microscope with a conventional microscope.

Confocal Microscopy 7

the use of an effectively on-axis system while relaxing the restrictions on object size and environment imposed by object-scanning systems. It may well also prove to be an attractive approach to scanning fluorescence imaging in situations where the signal levels are extremely weak, because it employs a minimum of optical components.

It might be claimed that a serious disadvantage of these approaches is that one needs to use a television screen in order to see an image. It would be much more convenient in some applications to be able to view a confocal image directly by eye in real time. The microscope which resulted from this line of thought is the tandem scanning microscope [Egger and Petràñ, 1967; Baer, 1970] which will be discussed in detail later in this chapter and also in Chapters 3, 9 and 14. An alternative to these schemes which has been proposed by Lichtman [Lichtman *et al.*, 1989] will be discussed in Chapter 3.

It is clear that there is no absolute answer as to the *best* method of scanning, and that the specific application may well present constraints which limit the choice. However, it is probably fair to say that for the ultimate in imaging combined with a simple optical system, a mechanically-scanned object or objective approach should be seriously considered.

It is appropriate now to pause, and discuss how the imaging in scanning microscopes compares to conventional instruments. We will show that relaxing the requirement to simultaneously image the whole object permits us to modify the optical system in a very simple way, resulting in the confocal arrangement. We will then go on to describe other imaging modes which arise from the use of the confocal approach before presenting a simple theory of the operation of confocal microscopes.

II. Image formation in scanning microscopes

We will base our initial discussion on a very simple model of image formation in conventional microscopes. Using this model, we can derive an equivalent form of scanning microscope which will allow us to suggest the modification which leads to the confocal optical system.

Figure 1.7(a) illustrates the key elements of the optical system of a conventional optical microscope. In essence, the object is illuminated by a patch of light from an extended source via a condenser lens. The illuminated patch of the object is then imaged by the objective lens into the image plane, where it is finally viewed by eye. In this case the resolution results primarily from the objective lens. The condenser lens plays a much less important role: its aberrations are unimportant, and it serves to determine the coherence of the imaging [Hopkins, 1953; Born and Wolf, 1975]. A scanning microscope could be built by

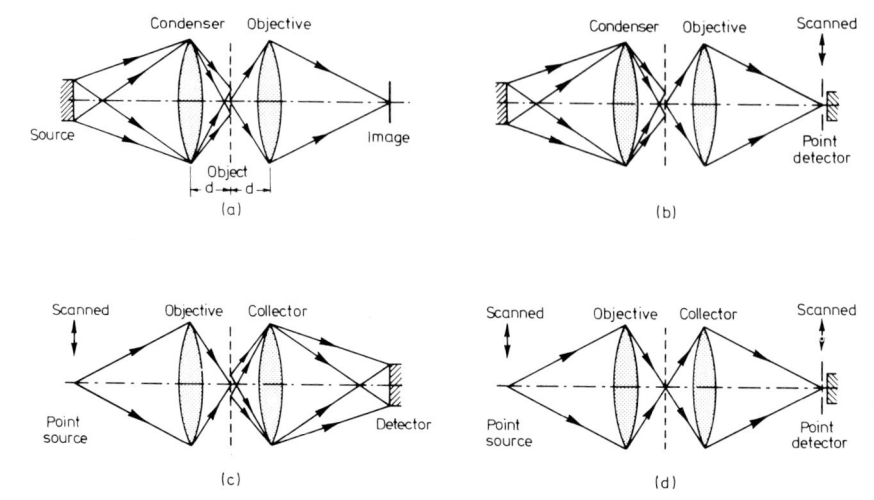

Figure 1.7. The optical arrangement of various forms of scanning optical microscopes. (a) Conventional microscope. (b) A form of conventional scanning microscope. (c) A form of conventional scanning microscope. (d) The confocal optical system.

simply scanning a point detector through the image plane to measure the image intensity of each object point, Figure 1.7(b). In this way we would build up an image point by point, which could be stored in a computer and displayed on a television screen. An analogous optical system is shown in Figure 1.7(c), which uses a point source and a large-area incoherent detector. Here the first lens, the objective lens, probes the specimen, point by point, with a finely focussed spot of light, while the second lens merely serves to collect the transmitted light. This collector lens is analogous to the condenser lens in conventional microscopy, and thus plays a minor role in determining the resolution.

In order to make the second lens contribute significantly to the resolution, we need to make it into an *imaging* lens rather than a *collector* lens. We achieve this by using a point detector rather than a large-area one. This is the form of the *confocal* scanning microscope, Figure 1.7(d). Here light from the point source probes a very small region of the object, and the point detector ensures that only light from that same small area is detected. In the figure, an image is built up by scanning the source and detector in synchronism. In practice, it may prove more convenient to scan the light beam or the object rather than the detector, and also to operate the system in reflection.

In this symmetric confocal configuration, we see that both lenses play equal roles in the imaging. The use of the point detector has caused the second lens to *image* the region of the object probed by the first lens. We might expect, therefore, that as two lenses are employed simultaneously to image the object, the resolution will be improved. This prediction is borne out both in theory and in practice, Figure 1.8, and in the later chapters of this book.

This improvement may at first seem implausible. However, it can be simply explained by a principle given by Lukosz (1966) which states, in essence, that resolution can be increased at the expense of field of view. The field of view can then be increased by scanning. One way of taking advantage of Lukosz's principle is to place an extremely small aperture extremely close to the object [McCutchen, 1967; Ash and Nicholls, 1972]. The resolution is now determined by the size of the hole rather than the radiation. In the confocal microscope we do not use a physical aperture in the focal plane. Instead, we use the back-projected image of a point detector in conjunction with the focussed point source.

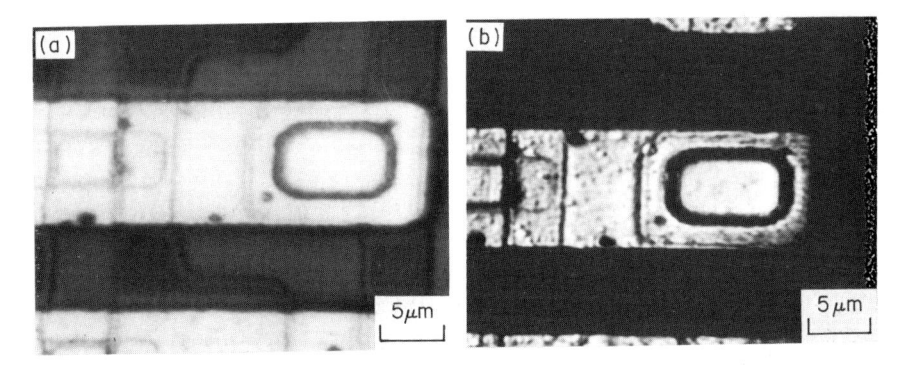

Figure 1.8. A portion of a microcircuit taken in (a) a conventional and (b) a confocal scanning microscope. The enhanced resolution is noticeable, as is the optical sectioning whereby the confocal image has rejected detail below the plane of the metallisation where the microscope is focussed.

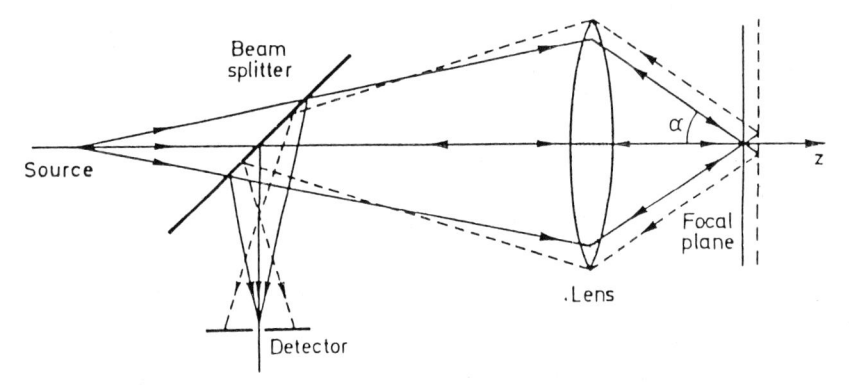

Figure 1.9. The origin of the depth discrimination or optical sectioning property of confocal optical systems.

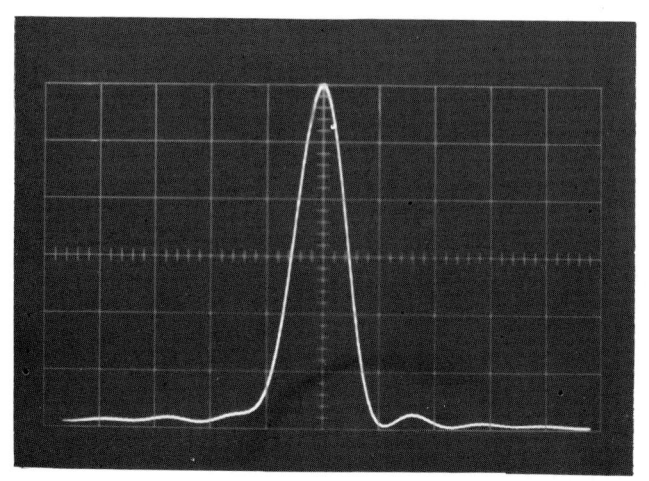

Figure 1.10. The variation in detected signal as a plane mirror is scanned axially through focus. The curve was obtained using red light, wavelength 0.6328μm, and a 0.5 N.A. objective lens. The full axial scan corresponds to 20μm.

The confocal principle was first described by Minsky [Minsky, 1961] and, in the tandem scanning form, by Petráň and Baer. The motivation for all these authors' work was the requirement of obtaining an image from a prescribed section of a thick translucent object, without the presence of out-of-focus information from surrounding planes. The confocal principle fulfils this requirement at the same time as providing enhanced lateral resolution. Light from the specimen is focussed through a small aperture, thus ensuring that information is obtained only from one particular level of the specimen. A comparison of the images in Figure 1.8 shows that the confocal image contains object detail only from the plane of focus – the plane of the metallisation. The rest of the image is much darker than the corresponding regions in the conventional image.

This optical sectioning or depth discrimination property has become the major motivation for using confocal microscopes, and is the basis of many of the novel imaging modes of these instruments. It may be understood very easily from Figure 1.9, where we show a reflection-mode confocal microscope and consider the imaging of a specimen with a rough surface. The full lines show the optical path when an object feature lies in the focal plane of the lens. At a later scan position, the object surface is supposed to be located in the plane of the vertical dashed line. In this case, simple ray tracing shows that the light reflected back to the detector pinhole arrives as a defocussed blur, only the central portion of which is detected and contributes to the image. In this way the system discriminates against features which do not lie within the focal region of the lens. A very simple method of both demonstrating the effect and giving a measure of its strength is to scan a perfect reflector axially through focus and measure the detected signal strength. Figure 1.10 shows a typical response. These responses are frequently termed the V(z), by analogy with a similar technique in scanning acoustic microscopy, and, although the correspondence is not perfect, the name is likely to stick. The simple paraxial theory we will present later in this chapter models this response as:

$$I(u) = \left[\frac{\sin(u/2)}{u/2}\right]^2 \quad (1.1)$$

where u is a normalised axial co-ordinate which is related to real axial distance, z, via:

$$u = \frac{8\pi}{\lambda} z \sin^2(\alpha/2) \quad (1.2)$$

where λ is the wavelength and $\sin\alpha$ the numerical aperture. As a measure of the strength of the sectioning, we can choose the full width at half intensity of the $I(u)$ curves. Figure 1.11 shows this value as a function of numerical aperture for the specific case of imaging with red light from a helium neon laser. These curves were obtained using a high aperture theory [Sheppard and Wilson, 1981], which is more

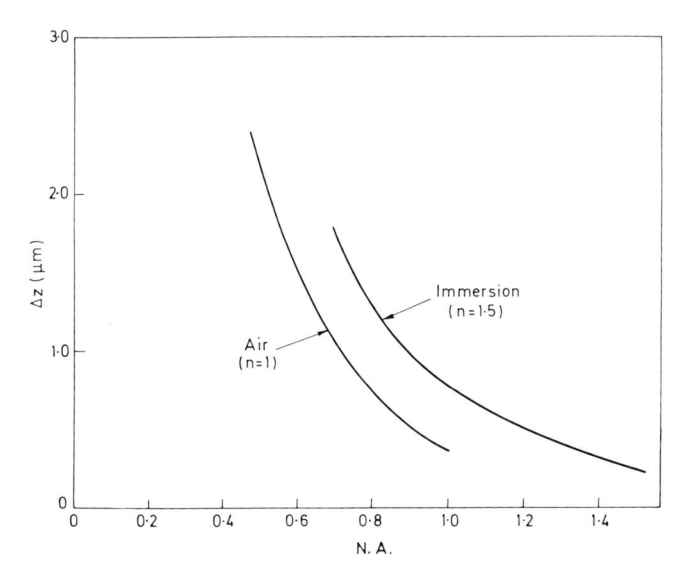

Figure 1.11. The optical sectioning width as a function of numerical aperture. The curves are for red light (0.6328 μm wavelength). Δz is the full width at the half-intensity points of the curves of $I(u)$ against u.

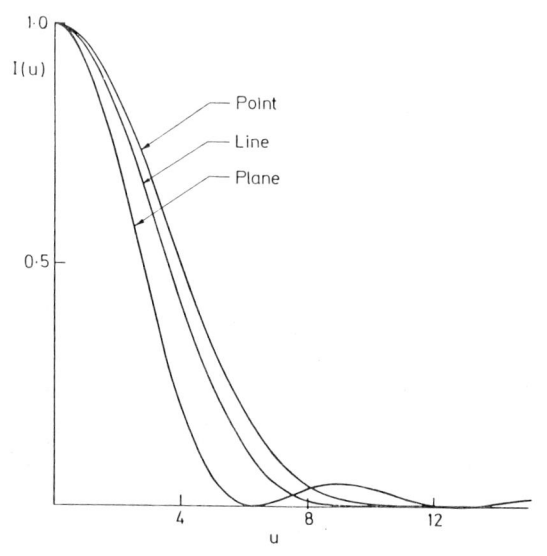

Figure 1.12. The variation in detected signal as a point, line and plane are scanned through focus in a confocal microscope. In the case of the point and line the microscope is focussed on their centres.

reliable than equation (1.1) at the highest values of numerical aperture.

Although Figure 1.11 gives a good indication of the degree of sectioning that we might expect, we should remember that these values refer to the imaging of planes. If we consider other object features such as points and lines, we obtain different degrees of sectioning. Figure 1.12 shows a comparison of $I(u)$ curves in these cases. In particular, the response for a point object is given by:

$$I(u) = \left[\frac{\sin(u/4)}{u/4}\right]^4 \qquad (1.3)$$

which shows that the sectioning is weaker than the case for the plane. It is also worth remembering that these curves refer to brightfield reflection imaging. The image formation in the practically important case of fluorescence imaging is completely different from the brightfield case, and although our remarks apply broadly to both cases, the actual numerical values are different. In the fluorescence case, for example, the sectioning strength is generally weaker.

Although the degree of optical sectioning is object-dependent, the planar reflector proves to be a very useful object in order to test the alignment and performance of a microscope. It is also useful, as we will see later in Chapters 3, 4 and 15, in determining the quality of objective lenses.

III. Applications of depth discrimination

As this property is almost certainly the major reason for the popularity of confocal microscopes, it is worthwhile, at this point, to review some of the novel imaging techniques which have become available with confocal microscopy. The review will be relatively short, because other chapters of this book will consider these and other techniques in greater detail.

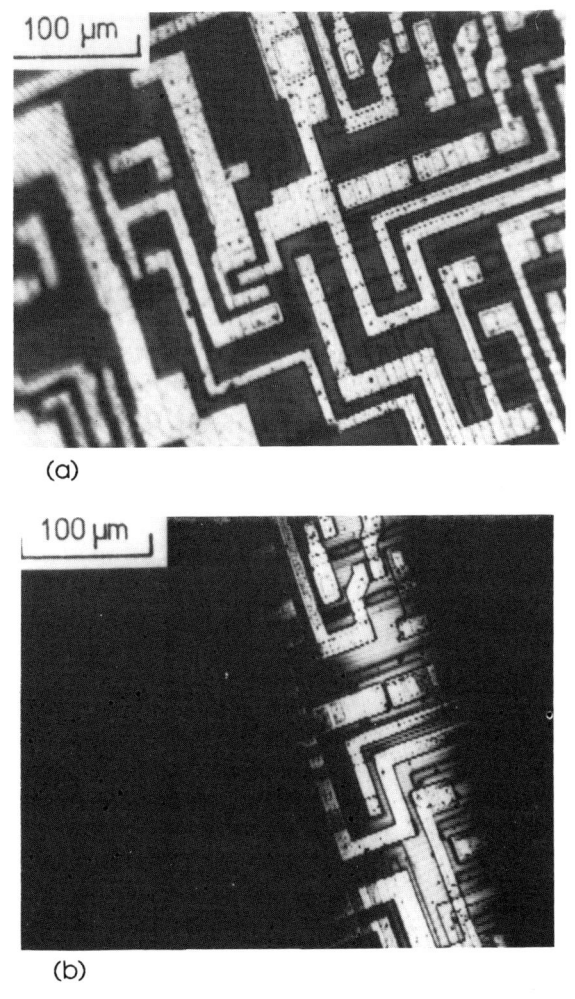

Figure 1.13. (a) Conventional scanning microscope image of a tilted microcircuit: the parts of the object outside the focal plane appear blurred. (b) Confocal image of the same microcircuit: only the part of the specimen within the focal region is imaged strongly.

Figure 1.13 illustrates the essential effect: Figure 1.13(a) shows a conventional image of a planar microcircuit which has deliberately been mounted with its normal at an angle to the optic axis. We see that only one portion of the circuit, running diagonally, is in focus. Figure 1.13(b) shows the corresponding confocal image: here the discrimination against detail outside the focal plane is clear. The areas which were out of focus in Figure 1.13(a) have been rejected. Furthermore, the confocal image appears to be in focus throughout the visible band, which illustrates that the sectioning property is stronger than the depth of focus.

This suggests that if we try to image a thick translucent specimen, we can arrange, by the choice of our focal position, to image detail from one specific volume region. In essence, we can section the specimen optically without having to resort to mechanical means. Figure 1.14 shows an idealised schematic of the process. The portion of the beehive-shaped object that we see is determined by the focus position. In this way it is possible to take a *through-focus* series and obtain data about the three-dimensional structure of the specimen. If we represent the volume image by $I(x, y, z)$, then by focussing at a position $z = z_1$, we obtain, ideally, the image $I(x, y, z_1)$. This, of course, is not true in practice because the optical section is not infinitely thin.

It is clear that the confocal microscope allows us to form high resolution images with a depth of focus sufficiently small that all the detail which is imaged appears in focus. This suggests immediately that we can extend the depth of focus of the microscope by adding to-

Focus position   Extended focus

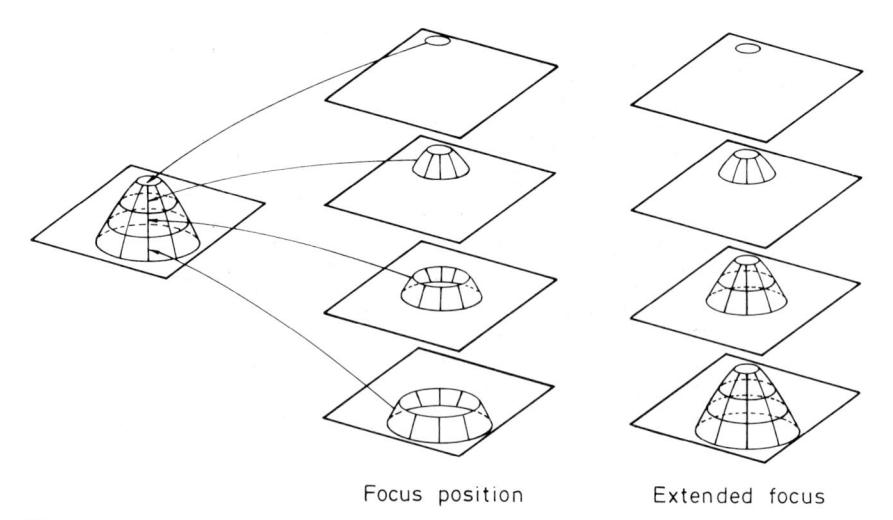

Figure 1.14. An idealisation of the optical sectioning property showing the ability to obtain a through-focus series of images, which may then be used to reconstruct the original volume object at high resolution.

Confocal Microscopy

gether (integrating) the images taken at different focal settings *without* sacrificing the lateral resolution. Mathematically, this *extended-focus* image is given by:

$$I_{EF}(x,y) = \int I(x,y,z)\,dz \qquad (1.4)$$

Figure 1.15(a) shows the result of scanning the specimen of Figure 1.13 axially by an amount such that each portion of its surface passes

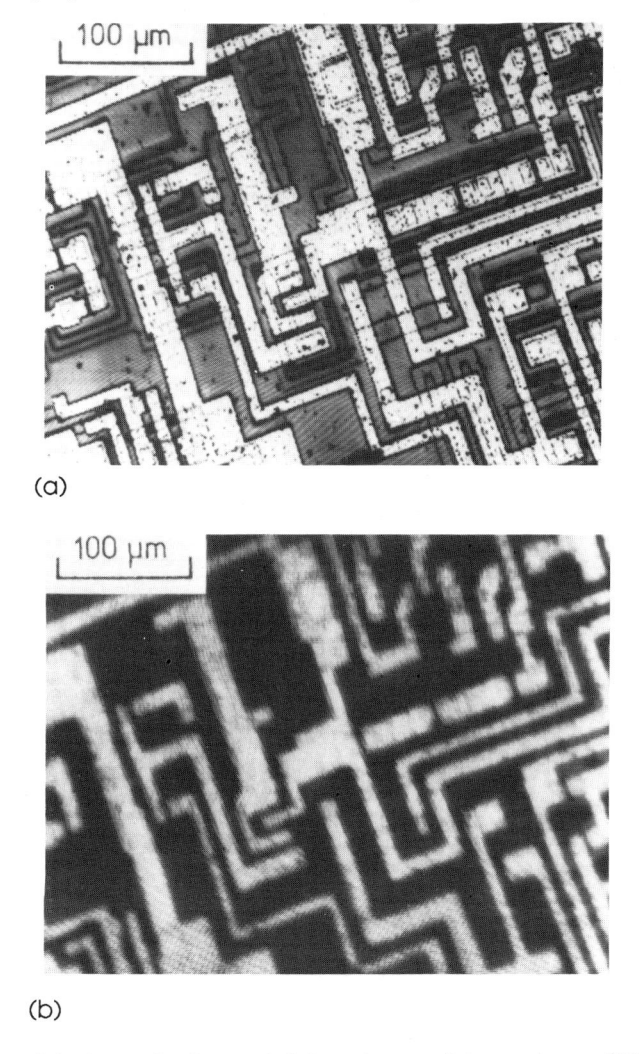

Figure 1.15. (a) The confocal extended-focus image of the specimen of Figure 1.13. (b) The conventional extended-focus image formed by summing the conventional microscope images obtained at different focal settings.

through the focal plane, and adding up the images from each plane [Wilson and Hamilton, 1982]. Figure 1.15(b) is the corresponding image obtained by summing conventional microscope images. This gives the expected blurred out-of-focus image, whereas the confocal extended-focus image is excellent and is in focus across the entire field. This should be compared, in particular, to the conventional image of Figure 1.13(a).

An example of this technique at high resolution is shown in Figure 1.16, in which the hairs on an ant's leg, with two hairs projecting to the left, have been imaged. The axial distance between the tips of these hairs is 30μm, and Figure 1.16(a) is an extended-focus image which shows both excellently resolved along their full length, as well as much detail on the leg itself. Figure 1.16(b), in which the microscope has been focussed on the tip of the projecting hair, shows an attempt to increase the depth of focus in conventional microscopy by using a very low numerical aperture lens, but it is clear that the resolution has suffered dramatically as a result. Even so, nothing approaching the depth of field of the extended-focus image has been achieved.

As an alternative to the extended-focus method, we can form an auto-focus image by scanning the object axially and, instead of integrating, selecting the focus at each picture point by recording the maximum in the detected signal. Mathematically, this might be written:

$$I_{AF}(x,y) = I(x,y,z_{max}) \qquad (1.5)$$

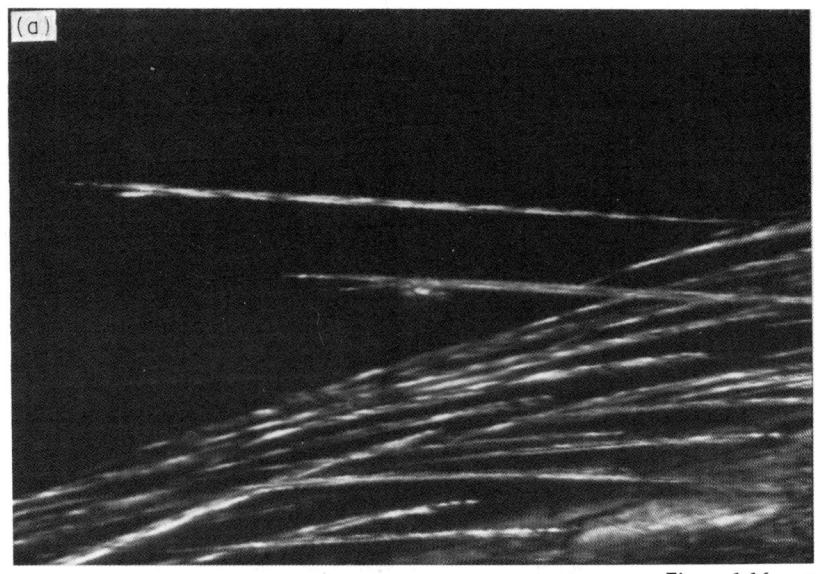

Figure 1.16 cont.

Confocal Microscopy

Figure 1.16. The hairs on an ant's leg: (a) extended focus with a 0.85 N.A. objective and He-Ne laser light and (b) conventional scanning image with reduced (1/32) numerical apertures.

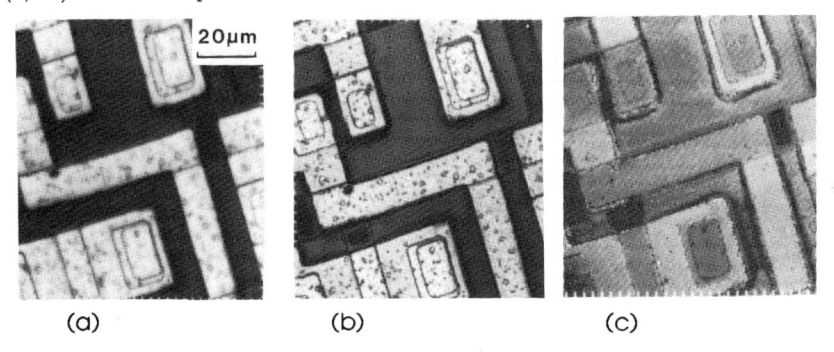

Figure 1.17. Digitally scanned and stored images of an integrated circuit: (a) conventional, (b) auto-focus with edge enhancements and (c) a height image where brightness corresponds to surface heights.

where z_{max} corresponds to the focus setting giving the maximum signal. The images obtained are somewhat similar to the extended focus and, again, substantial increases in depth of focus may be obtained. Figure 1.17 shows a digitally scanned auto-focus image which has been further enhanced by edge enhancement using standard matrix techniques. We can go one step further with the auto-focus method and turn the microscope into a non-contacting surface profilometer. Here we simply display z_{max}. This is illustrated in Figure 1.17(c), where we have mapped surface height to grey level, and Figure 1.18, where we show the profile of a metal strip on a microcircuit. The strip is clearly defined and the difference between the surface texture of the metal and the surrounding semiconductor is apparent. The reflectivities of the metal and semiconductor are quite different, but this does not affect the height-measuring system. These various techniques may be combined into an isometric image, Figure 1.19, where height and surface reflectivity data have been combined.

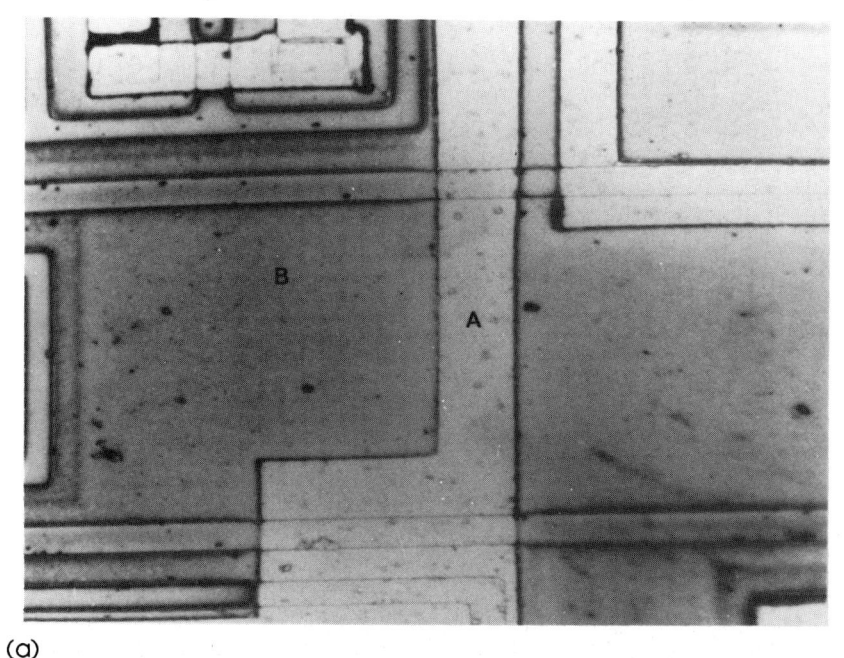

Figure 1.18. (a) A portion of microcircuit: 'A' indicates the metal strip and 'B' the semiconductor. (b) The height profile across the metal strip.

Confocal Microscopy 19

The sensitivity of this profiling technique can be improved by using interference techniques [Matthews, 1987]. The basic idea consists in moving the object axially from point to point such that the interferometer is always locked to one dark fringe. As an alternative to moving the object axially, one may incorporate a phase shifter into the reference arm of the interferometer. This has the advantage of higher speed and a sensitivity of better than 1nm. It does not, however, permit profiling over such a large range. Figure 1.20 shows an example of a profile obtained using this method. This, and other interference techniques, are discussed in more detail in Chapter 15.

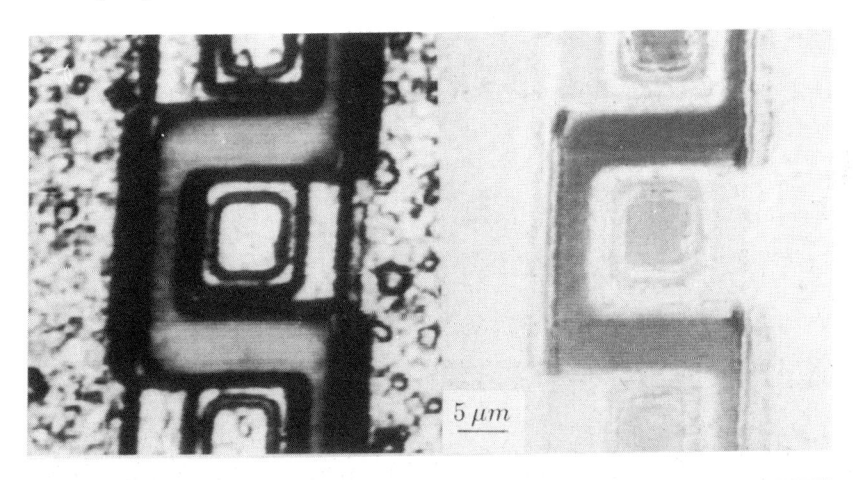

Figure 1.19. The top left-hand image shows an extended focus image of a portion of a particularly deep transistor. The top right-hand image is the corresponding height image where object height has been coded as image brightness. The bottom image is a computer-generated isometric view of the same microcircuit.

It is clear by now that the confocal method gives us a convenient tool for studying three-dimensional structures in general. We essentially record the image as a series of slices and play it back in any desired fashion. Naturally, in practice it is not as simple as this, but we can, for example, display the data as an $x - z$ image rather than an $x - y$ image. This is somewhat similar to viewing the specimen from the side. As another example [van der Voort, 1989], we might choose to recombine the data as stereo pairs by introducing a slight lateral offset to each image slice as we add them up. If we do this twice, with an offset to the left in one case and the right in another, we obtain, very simply, stereo pairs. Mathematically, we form images of the form:

$$\int I(x \pm \gamma z, y, z)\, dz \qquad (1.6)$$

where γ is a constant. In practice it may not be necessary to introduce offsets in both directions to obtain an adequate stereo view.

All that we have said so far on these techniques has been by way of simplified introduction to the following chapters. In particular, we have not presented any fluorescence images, because these will be dealt with adequately later. The key point is that, in both brightfield

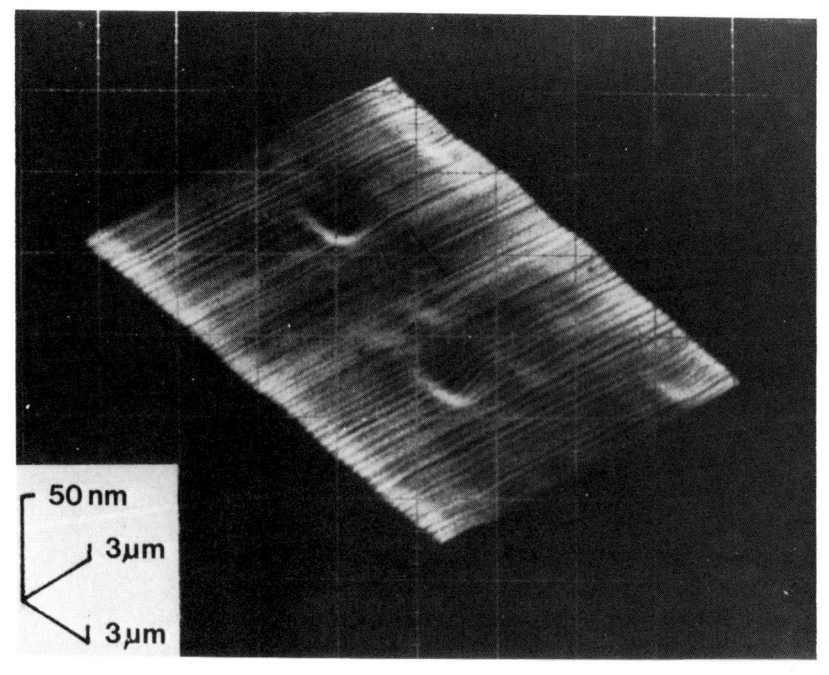

Figure 1.20. A surface profile of a portion of a surface acoustic wave device bonding pad obtained by an interferometric technique. (Courtesy of Dr. D. K. Hamilton.)

and fluorescence modes, the confocal principle permits the imaging of specimens in three dimensions. The situation is, of course, more involved than we have implied. A thorough knowledge of the image formation process, together with the effects of lens aberrations and absorption, is necessary before accurate data manipulation can take place.

We should emphasise that although most of the images which are presented in this book were obtained using visible light, this is not necessary. All that is required is a suitable laser source, objective lens and detector. As an example, we show an advantage of using infrared light in the imaging of silicon semiconductor devices. Figure 1.21(a) shows a conventional red light (0.6328μm wavelength) image of the surface of the microcircuit. Figure 1.21(b) is the conventional infra-red (1.14μm wavelength) image, but now focussed 60μm into the specimen. The detail is clear, but the contrast is fairly low and the defocussed image of the surface metallisation is confusing the image. Figure 1.21(c) is the corresponding confocal image. All the focal-plane-specific detail has been rejected and the contrast of the features of interest is much higher. This technique has implications for the investigation of bonding problems and die-attach difficulties in finished devices [Hamilton and Wilson, 1987].

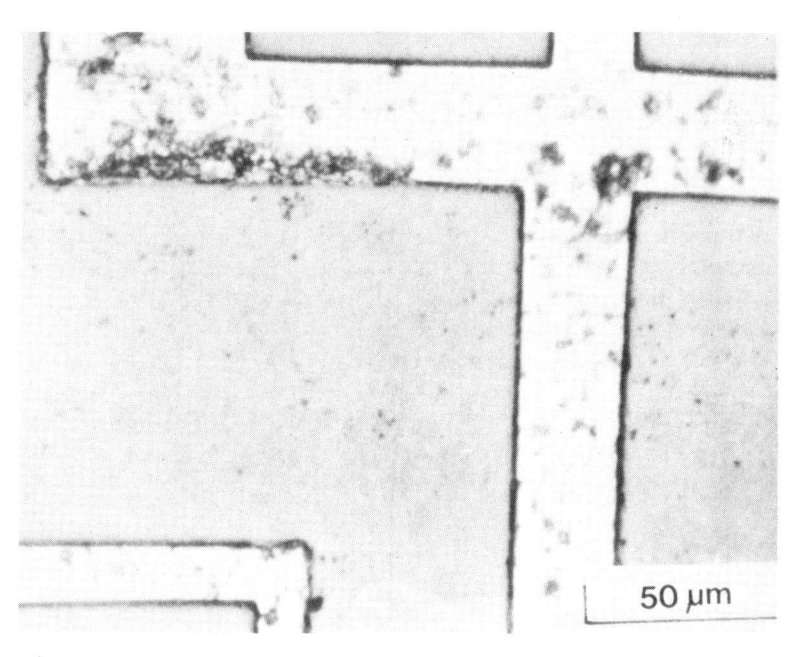

(a)

Figure 1.21 cont.

(b)   (c)

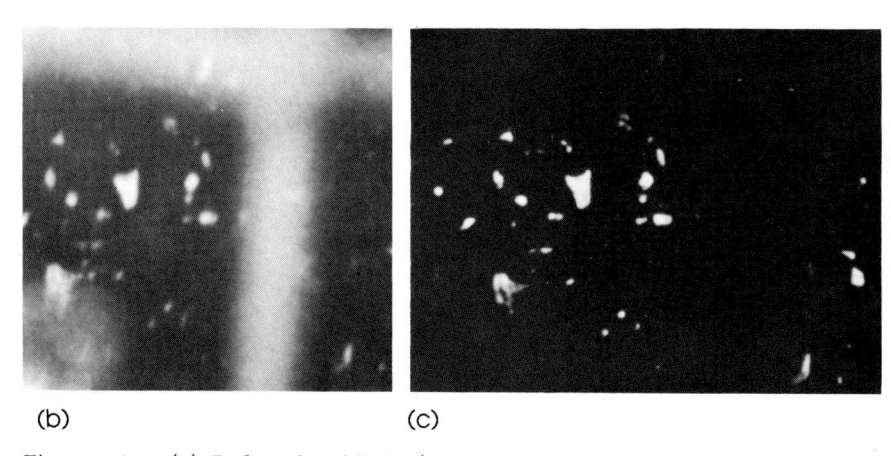

Figure 1.21. (a) Reflected red-light (0.6328μm wavelength) image of a silicon device formed using a 0.5 numerical aperture objective lens. (b) Conventional reflected infrared light (1.15μm wavelength) image of the device when focused on the bottom of the substrate. (c) Confocal, infrared, reflected-light image of the region of Figure 1.21(b).

IV. Theory of imaging in scanning microscopes

IV.1. Fundamentals of optical image formation

It is not appropriate here to go into great detail of the theory of image formation in conventional and confocal microscopes. However, it is important to present a little theory in order to understand better the differences between the various microscope systems. Our theory will be paraxial, and will be simplified by the fact that we will consider only thin specimens which are characterised by a complex amplitude transmittance (or reflection) function, t. The function is complex in order to allow for variations of phase owing to changes of optical thickness or surface roughness.

We will begin by considering the image formed by a simple lens as shown in Figure 1.22. If we imagine that the amplitude of the field at the plane (x_1, y_1) is $U(x_1, y_1)$, then we can write the field at the plane (x_3, y_3) [Wilson and Sheppard, 1984; Goodman, 1968], neglecting premultiplying constants and phase variations, as:

$$U(x_3, y_3) = \int \int U(x_1, y_1) \, h(x_1 + x_3/M, y_1 + y_3/M) \, dx_1 dy_1 \quad (1.7)$$

where M is the magnification, d_2/d_1, and h is the amplitude point spread function (or impulse response) which is given by [Goodman, 1968]:

$$h(x,y) = \int\int P(x_2,y_2)\exp\left[\frac{2\pi j}{\lambda}(x_2 x + y_2 y)\right] dx_2 dy_2 \quad (1.8)$$

where P is the pupil function of the lens. The pupil function serves to describe the physical extent of the lens and its transmissivity. It is, in general, a complex function which describes the aberrations present, including the degree of defocus of the system.

Suppose now that $U(x_1,y_1)$ represents our object and that it consists simply of a single bright point in an opaque background, so that:

$$U(x_1,y_1) = \delta(x_1)\delta(y_1) \quad (1.9)$$

where $\delta(.)$ represents a Dirac delta function. The intensity in the image is now given by equation (1.7):

$$I(x_3,y_3) = |h(x_3/M, y_3/M)|^2 \quad (1.10)$$

If we consider a circularly symmetric pupil function of radius a, we can introduce a pupil function $P(\rho)$, where:

$$\rho = r_2/a \quad (1.11)$$

which is zero for $\rho \geq 1$. It is now straightforward to rewrite equation (1.10) [Born and Wolf, 1975; Wilson and Sheppard, 1984] as:

$$I(v) = \left|2\int_0^1 P(\rho)J_0(v\rho)\rho\, d\rho\right|^2 \quad (1.12)$$

where J_0 is a zero order Bessel function of the first kind and v is a normalised optical co-ordinate related to the real radial co-ordinate, r_3, via:

$$v = \frac{2\pi}{\lambda}r_3 \sin\alpha \quad (1.13)$$

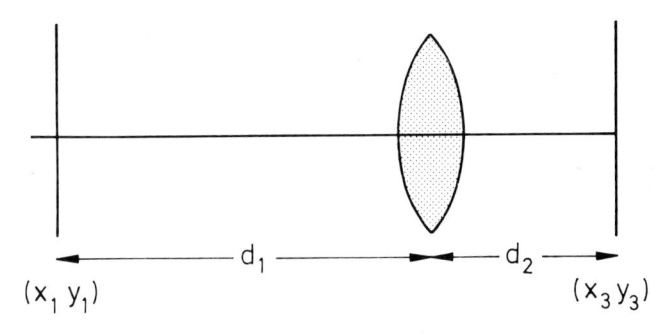

Figure 1.22. The image formation geometry.

where $\sin\alpha$, the numerical aperture, is given by a/d_2. It is clear that the form of the image now depends critically on the form of the pupil function, $P(\rho)$. In the idealised case, when:

$$P(\rho) = \begin{cases} 1 & \rho \leq 1 \\ 0 & \text{otherwise} \end{cases} \quad (1.14)$$

we can write:

$$I(v) = \left(\frac{2J_1(v)}{v}\right)^2 \quad (1.15)$$

which is plotted in Figure 1.23 and is the normal form of the Airy disc. The first zero occurs at $v = 1.22\pi$.

As a dramatic example of the role of the form of the pupil function, we consider an infinitely thin annular pupil such that light is transmitted only at a normalised radius $\rho = 1$. It is clear from equation (1.12) that:

$$I(v) = J_0^2(v). \quad (1.16)$$

This is also plotted in Figure 1.23, where we see that the central spot becomes narrower, but at the expense of higher side lobes.

We should emphasise again that equation (1.7) is approximate in the sense that it has neglected a pre-multiplying phase term. This was unimportant in our discussion of the image of a point object, but could be important in the imaging of extended objects at higher aperture or lower values of Fresnel number [Wilson and Sheppard, 1984]. However, we will happily ignore all these complications in the following.

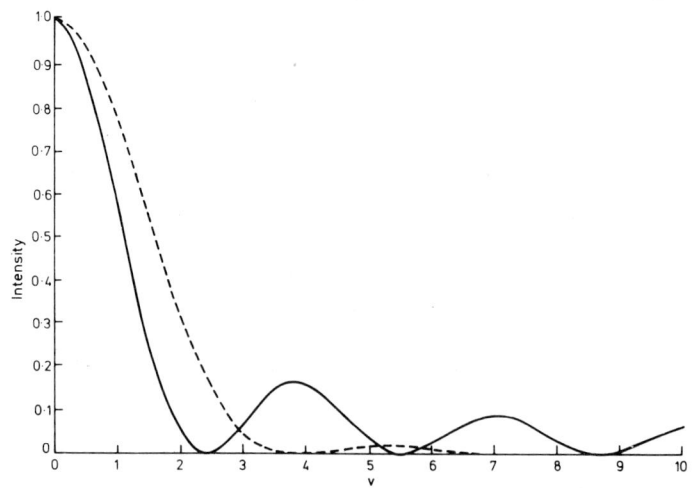

Figure 1.23. The form of the Airy disc for a full lens is shown in dashed lines whereas the full line shows the corresponding result for the use of an annular aperture.

IV.2. The imaging of simple objects

The image formation in a conventional microscope is unfortunately rather complicated. We will therefore discuss it only briefly here and will refer to the main results [Wilson and Sheppard, 1984; Martin, 1966]. The situation in the confocal case is considerably simpler. Figure 1.24 shows the form of a conventional scanning microscope. If we consider the image of a thin object amplitude transmittance, t, we can write the image intensity in the limiting case when the pupil P_2 becomes extremely large as:

$$I = |h_1|^2 \otimes |t|^2 \qquad (1.17)$$

where h_1 is the amplitude point spread function of the first lens and the symbol $\otimes$ denotes the convolution operation. In this case the imaging is referred to as *incoherent*. On the other hand, as P_2 becomes extremely small, the imaging becomes *coherent* and may be written as:

$$I = |h_1 \otimes t|^2 \qquad (1.18)$$

It is clear that for most practical cases in which P_1 and P_2 are of comparable size, neither of these simple expressions apply. We must now describe the imaging as being *partially coherent*, with coherent and incoherent being extreme limits.

The confocal microscope, however, is far easier to analyse. The only difference in the optical form is that the large-area detector of Figure 1.24 is replaced by a point detector, so that in this case we can write $D(x_2, y_2) = \delta(x_2)\delta(y_2)$. Here it is easy to show by repeated application of equation (1.7) that the image may be written as:

$$I = |(h_1 h_2) \otimes t|^2 \qquad (1.19)$$

Here the imaging is *always* coherent whatever the relative size of the pupils. In essence, the confocal system is simply a coherent optical

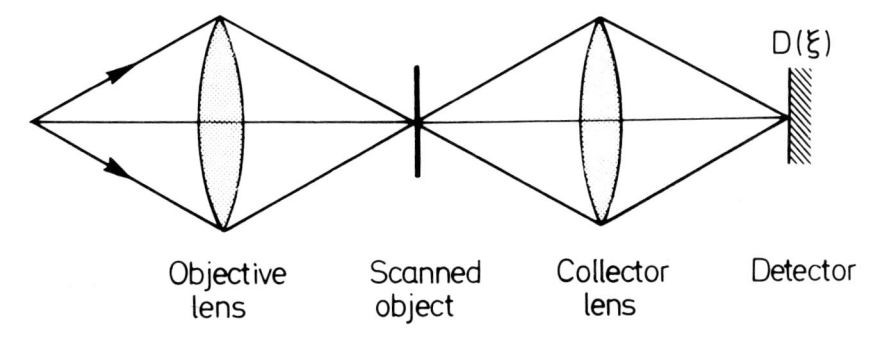

D(ξ)

Objective lens   Scanned object   Collector lens   Detector

Figure 1.24. The form of a conventional scanning microscope.

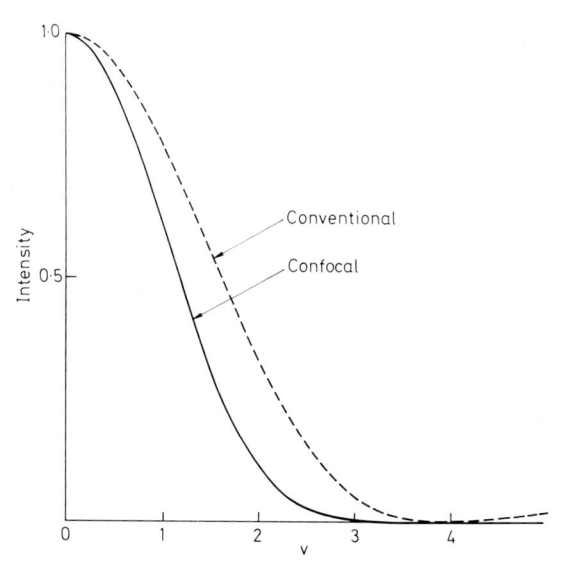

Figure 1.25. The image of a single point object in a conventional and a confocal microscope.

imaging system whose effective point spread function is given by the product of the point spread functions of the two lenses.

If we consider the image of a point object, we find that in the conventional microscope we have:

$$I(v) = \left(\frac{2J_1(v)}{v}\right)^2 \qquad (1.20)$$

whereas, in the confocal case with two equal lenses we find that the image is:

$$I(v) = \left(\frac{2J_1(v)}{v}\right)^4 \qquad (1.21)$$

These functions are plotted in Figure 1.25, and we see that although the functions have the same zeroes, the central peak of the confocal response is sharpened up by a factor of 1.4 relative to the conventional image (measured at half-peak intensity). The sidelobes are also dramatically reduced, and we would thus expect a marked reduction in the presence of artefacts in confocal images. Experimental confirmation of these theoretical predictions has been given by Brakenhoff et al. (1979).

We can further emphasise this improvement in imaging by including the effects of defocus. We do this simply by introducing a quadratic phase factor, $\exp(\frac{1}{2}ju\rho^2)$. The definition we have used for u, equation

Confocal Microscopy

(1.2), is slightly different from the definition in Born and Wolf (1975), and is more applicable at higher numerical apertures. The definitions, of course, agree at low numerical apertures. Under these circumstances we can re-write the amplitude point spread function as:

$$h(u,v) = 2\int_0^1 P(\rho)\exp\frac{1}{2}ju\rho^2 J_0(v\rho)\rho\,d\rho \qquad (1.22)$$

We notice that in reflection systems the defocus of each lens is equal, which means that we can write, for equal pupils, that:

(a)

CONTOUR KEY	
1	0.900E+00
2	0.700E+00
3	0.500E+00
4	0.300E+00
5	0.200E+00
6	0.100E+00
7	0.500E-01
8	0.300E-01
9	0.200E-01
10	0.500E-02

(b)

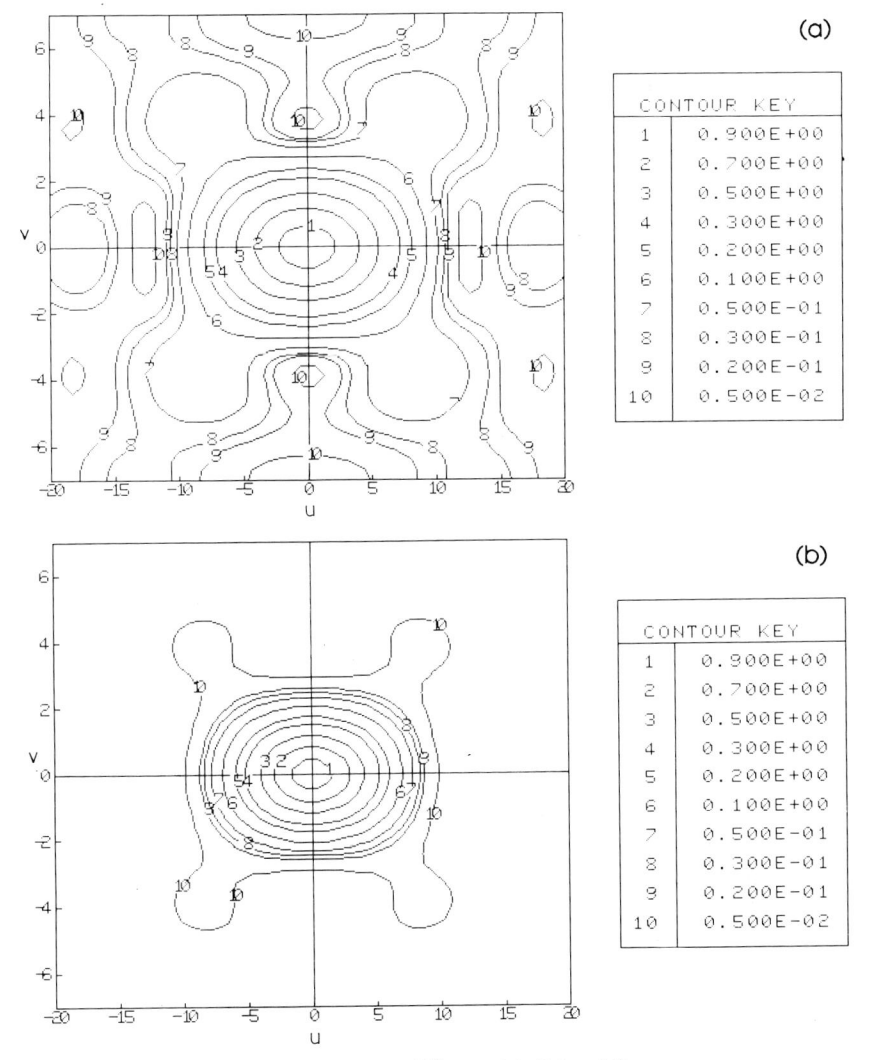

CONTOUR KEY	
1	0.900E+00
2	0.700E+00
3	0.500E+00
4	0.300E+00
5	0.200E+00
6	0.100E+00
7	0.500E-01
8	0.300E-01
9	0.200E-01
10	0.500E-02

Figure 1.26. Contour plots of (a) $|h(u,v)|^2$ and (b) $|h(u,v)|^4$.

$$h_1(u,v) = h_2(u,v) \quad (1.23)$$

whereas, in transmission, the defocus of each lens is equal and opposite, and so we obtain:

$$h_1(u,v) = h_2(-u,v) \quad (1.24)$$

If we now consider again the image of a point object in reflection, we find for the conventional microscope that:

$$I(u,v) = |h(u,v)|^2 \quad (1.25)$$

whereas the confocal microscope gives:

$$I(u,v) = |h(u,v)|^4 \quad (1.26)$$

Figure 1.26 shows these images in the form of a contour plot, and Figure 1.27 as isometric projections. The contour labels in Figures 1.26 (a) and 1.26 (b) are identical, and the improvement in *volume* imaging in the confocal case is apparent.

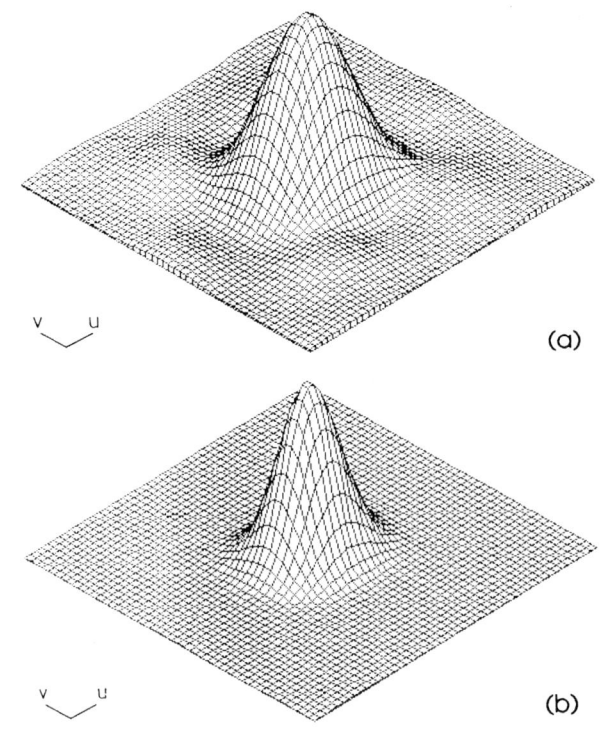

Figure 1.27. Isometric surface projections of (a) $|h(u,v)|^2$ and (b) $|h(u,v)|^4$.

Confocal Microscopy

We have considered the image of a point object in the focal plane in equations (1.20) and (1.21). If we now consider the variation along the optic axis, we find, for the conventional instrument:

$$I(u,0) = \left[\frac{\sin(u/4)}{u/4}\right]^2 \qquad (1.27)$$

and for the confocal:

$$I(u,0) = \left[\frac{\sin(u/4)}{u/4}\right]^4 \qquad (1.28)$$

which is again sharpened up in its central peak and has extremely low side lobe levels.

If we now consider the in-focus confocal image of two closely spaced points, we find that, when the Rayleigh criterion is satisfied, the points are separated by a normalised distance of 0.56 optical units. This is 8% closer than the conventional microscope and 32% closer than in a conventional coherent microscope with equal lenses.

As a final example, we might consider the image of a straight edge. Figure 1.28 shows the images in the conventional and confocal cases. The intensity at the edge in the confocal case is always 0.25, whereas for the conventional microscope it varies between 0.25 and 0.5 depending on the relative sizes of the lenses. When the two lenses are equal, the edge intensity may be shown to be 1/3. In the general case of reflectivity $a_1 \exp(j\phi_1)$ on one side and $a_2 \exp(j\phi_2)$ on the other side, the intensity at the edge is given by [Hewlett *et al.*, 1990]:

$$I(0) = \frac{1}{3}\left[a_1^2 + a_1 a_2 \cos(\Delta\phi) + a_2^2\right] \qquad (1.29)$$

where $\Delta\phi = \phi_1 - \phi_2$. In the coherent imaging case of the confocal microscope, the corresponding result is:

$$I(0) = \frac{1}{4}\left[a_1^2 + 2a_1 a_2 \cos(\Delta\phi) + a_2^2\right] \qquad (1.30)$$

The gradient at the edge is given by:

$$\frac{dI}{dv} = \frac{8(a_2^2 - a_1^2)}{3\pi^2} \qquad (1.31)$$

in the confocal and equal pupil conventional cases, whereas in the coherent conventional case we find:

$$\frac{dI}{dv} = \frac{a_2^2 - a_1^2}{\pi} \qquad (1.32)$$

which is 17.8% sharper. The disadvantage here, however, is that the edge response suffers from considerable overshoot and ringing. A further advantage of the use of a confocal microscope with two equal pupils is that it results in an effective amplitude point spread function which is non-negative, and hence the straight edge response does not exhibit fringing [Sheppard and Wilson, 1978a].

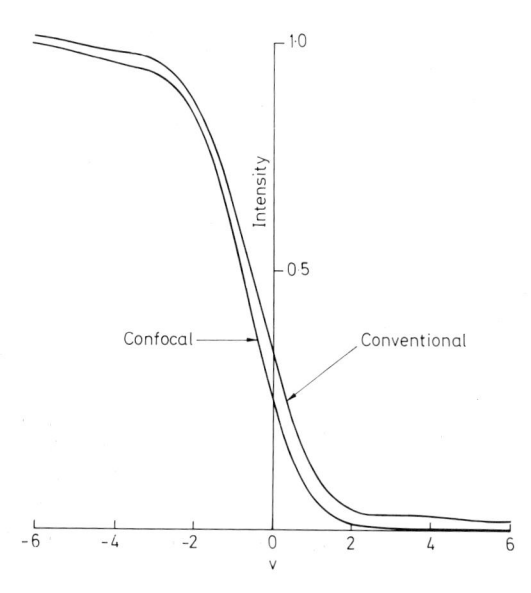

Figure 1.28. The in-focus image of a thin straight edge in the conventional and confocal microscope.

These calculations have been carried out for thin objects. In practice, however, there may be a considerable height, and hence focus, difference between the sides of a step. An example might be a line of metal or photo-resist on an integrated circuit. In this case, images may be calculated by a simple model in the coherent case by merely superimposing the amplitude images from each focal position [Sheppard and Heaton, 1984; Wilson and Carlini, 1986; Wilson et al., 1986]. However, this method proves unsuitable when the step height becomes much more than a wavelength, and when scattering and shading effects become important. In these cases, more complicated theories involving waveguide techniques [Nyyssonen, 1982; Kirk and Nyyssonen, 1985] and/or treating the specimen as a thick surface relief grating [Gaylord et al., 1986] are required. However, these techniques are not without their own difficulties [Sheppard and Sheridan, 1989].

IV.3. Fourier imaging

It is common practice in the analysis of linear electrical filter circuits to describe their behaviour in terms of a transfer function, which indicates how the network transmits signals as a function of frequency. We can, by means of the Fourier transform, express a time-varying signal in term of its frequency content or spectrum, $X(\omega)$. If the network has a frequency response $H(\omega)$, then the output spectrum is given by $X(\omega)H(\omega)$, and the final signal is given by the inverse Fourier transform of $X(\omega)H(\omega)$. The basic idea of Fourier optics is to apply these transfer function ideas to optical *circuits*. Instead of a time-varying input signal, we have our object, $t(x,y)$, and instead of the frequency content of our input signal being measured in Hertz, it is now measured in terms of *spatial* frequencies, i.e., μm^{-1}. In this way we think of the lens as a being a kind of spatial frequency filter. This approach is particularly powerful because it provides a method of comparing optical systems without the need to specialise to the image of a particular object. It also gives a very simple way of predicting the imaging properties of systems merely by considering the symmetry of the transfer function.

We will present a derivation of the transfer function based on our earlier equations. We introduce the frequency content or spectrum of our object as $T(m)$, via:

$$T(m,n) = \int\int t(x,y)\exp\left[2\pi j(mx+ny)\right]dxdy \quad (1.33)$$

such that:

$$t(x,y) = \int\int T(m,n)\exp-\left[2\pi j(mx+ny)\right]dmdn \quad (1.34)$$

where m and n are spatial frequencies. It is straightforward to substitute this expression for $t(x,y)$ into equation (1.19) for the confocal microscope and write the image intensity as:

$$I(x,y) = \left|\int\int c(m,n)T(m,n)\exp\left[2\pi j(mx+ny)\right]dmdn\right|^2 \quad (1.35)$$

where $c(m,n)$ is the coherent transfer function given by:

$$c(m,n) = P_1(\tilde{m},\tilde{n}) \otimes P_2(\tilde{m},\tilde{n}) \quad (1.36)$$

where P_1 and P_2 are the pupil functions of the two objective lenses and $\tilde{m}$ and $\tilde{n}$ are normalised spatial frequencies given by $\tilde{m} = m\lambda/\sin\alpha$ and $\tilde{n} = n\lambda/\sin\alpha$.

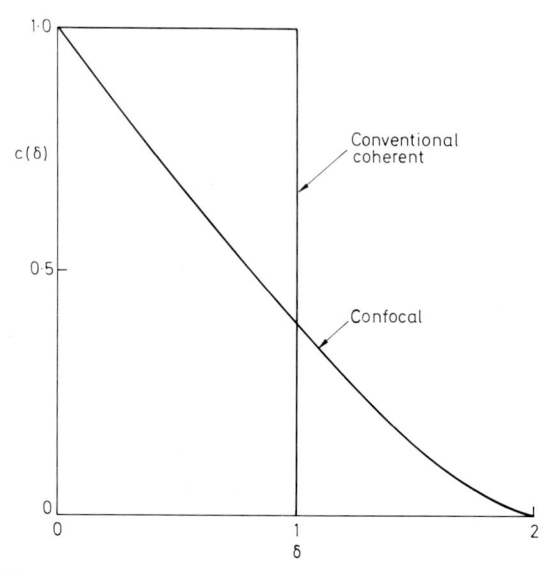

Figure 1.29. The coherent transfer function, $c(\delta)$, of a confocal microscope and a conventional coherent microscope.

This is a particularly important result, because by introducing $T(m,n)$, we have been able to totally separate the effects of the optical system, which determines the form of $c(m,n)$, from the specific object which determines the form of $T(m,n)$. We can, therefore, very easily study the effects of aberrations, defocus, pupil function apodisation, etc. simply by considering their effect on the form of $c(m,n)$.

It is clear that the transfer function has circular symmetry if the pupil functions are circularly symmetric, that is:

$$c(m,n) = c(\delta) \qquad (1.37)$$

where $\delta^2 = \tilde{m}^2 + \tilde{n}^2$. It is also clear that $c(\delta)$ becomes zero when the two pupil functions in the convolution of equation (1.36) no longer overlap. As the pupil functions have a normalised radius of unity, this cut-off condition corresponds to $\delta = 2$, which means that spatial frequencies less than $(2\sin\alpha)/\lambda$ are transmitted, and hence present in the image, whereas those greater than this value are not. In the case of circular, aberration-free pupils it is easy to calculate the convolution [Goodman, 1968] as:

$$c(\delta) = \frac{2}{\pi}\left[\cos^{-1}\left(\frac{\delta}{2}\right) - \left(\frac{\delta}{2}\right)\sqrt{1 - \left(\frac{\delta}{2}\right)^2}\right] \qquad (1.38)$$

which is plotted in Figure 1.29. For the case of the coherent conventional microscopes we find [Wilson and Sheppard, 1984] that:

$$c(\delta) = P_1(\delta) \tag{1.39}$$

which is also plotted in Figure 1.29, where we see that the cut-off spatial frequency is half that of the confocal.

In order to include the effects of defocus, we again simply include a quadratic exponential term in the pupil function as before. If we consider a reflection system, then $u_1 = u_2 = u$ for both pupils and the transfer function is complex. Figure 1.30 shows the form of the defocussed transfer function which has been normalised to unity at $m = 0$ and $u = 0$. These curves also apply to transmission-mode imaging when the defocus is caused by a change in the separation of the lenses. Figure 1.31 presents the data in the form of an isometric plot of $|c(m,u)|^2$. We see that this function is non-zero only for small values of u, that is, close to the focal plane. If, however, the defocus in transmission is caused by a displacement of the object, then $u_1 = -u_2$ and $P_1 = P_2^*$. The transfer function is now wholly real, Figure 1.32. This function is, in fact, identical to the incoherent transfer function of a conventional microscope system [Hopkins, 1953]. We note, however, that in the presence of aberrations the defocussed transfer function becomes complex.

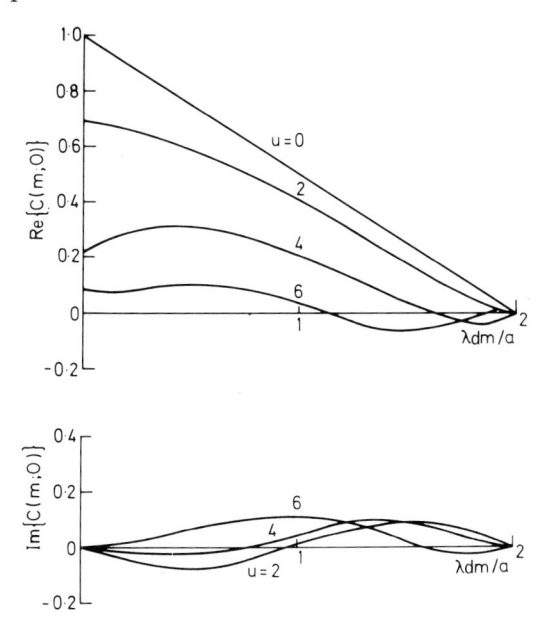

Figure 1.30. The real and imaginary parts of the defocussed transfer function for a reflection mode confocal microscope.

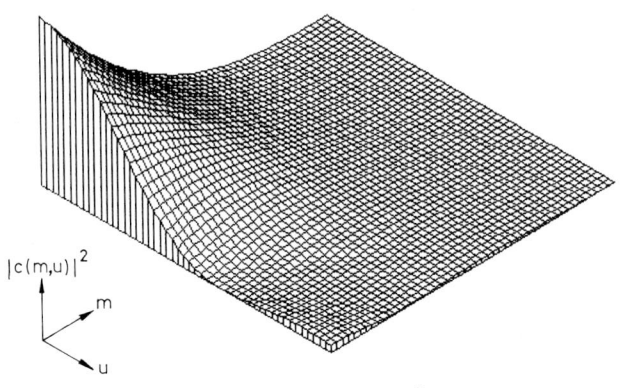

Figure 1.31. An isometric representation of $|c(m, u)|^2$ illustrating that the function only has reasonable value for small values of u.

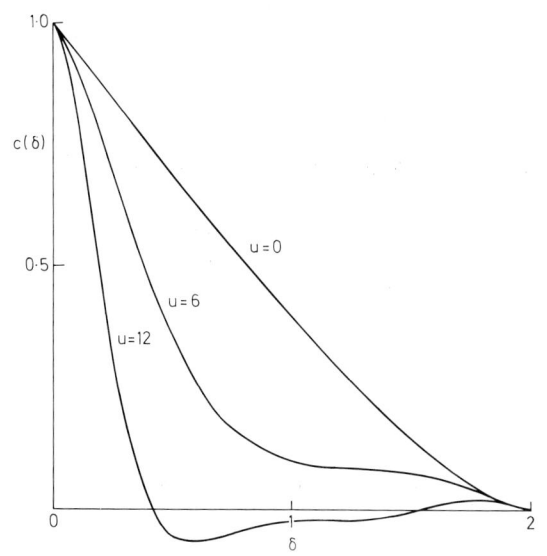

Figure 1.32. The wholly real coherent transfer function for a defocussed transmission confocal system.

We now move on to consider a one-dimensional object whose amplitude transmittance is given by:

$$t(x) = \exp(b \cos 2\pi\nu x) \tag{1.40}$$

and assume that b is sufficiently small that terms in b^2 may be neglected. We also permit b to be complex. The image, equation (1.35), is now given by:

$$I(x) = |c(0)|^2 + 2\text{Re}\{b\, c(\nu)c^*(0)\} \cos 2\pi\nu x \tag{1.41}$$

The strength of the cosinusoidally-varying term is now determined by the function $c(\nu)c^*(0)$, which is called the weak object transfer function. The real part of this function results in the imaging of amplitude variations (related to the real part of b) and the imaginary part in the imaging of phase structure (related to the imaginary part of b). This function is plotted in Figure 1.33, and we see that the function is essentially very small by $u = 6$. This value also roughly corresponds to the first zero in the $V(z)$ response of equation (1.1).

The simple form of equation (1.35) only applies to coherent imaging systems. In general, the expression is more complicated [Hopkins, 1953] and may be written, in one dimension for simplicity, as:

$$I(x) = \int\int C(m;p)T(m)T^*(p)\exp[2\pi j(m-p)x]\,dmdp \qquad (1.42)$$

where p is also a spatial frequency in the x-direction. If we expand equation (1.35), in the one-dimensional case, we obtain:

$$I(x) = \int\int c(m)c^*(p)T(m)T^*(p)\exp[2\pi j(m-p)x]\,dmdp \qquad (1.43)$$

and hence, in this case:

$$C(m;p) = c(m)c^*(p) \qquad (1.44)$$

$C(m;p)$ is a general form of transfer function which is also referred

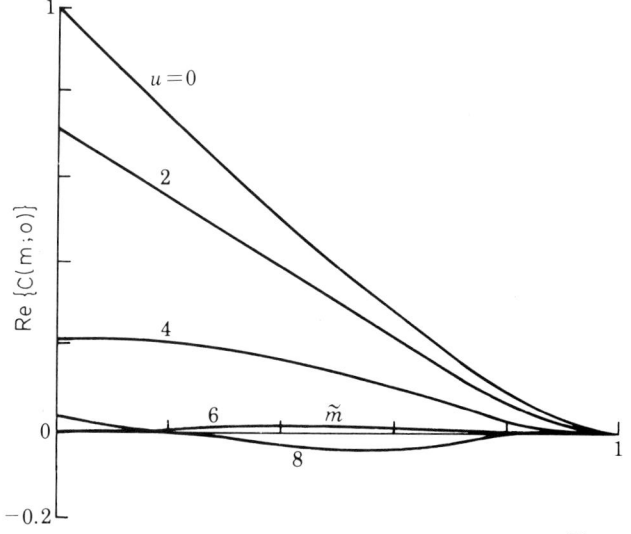

Figure 1.33 cont.

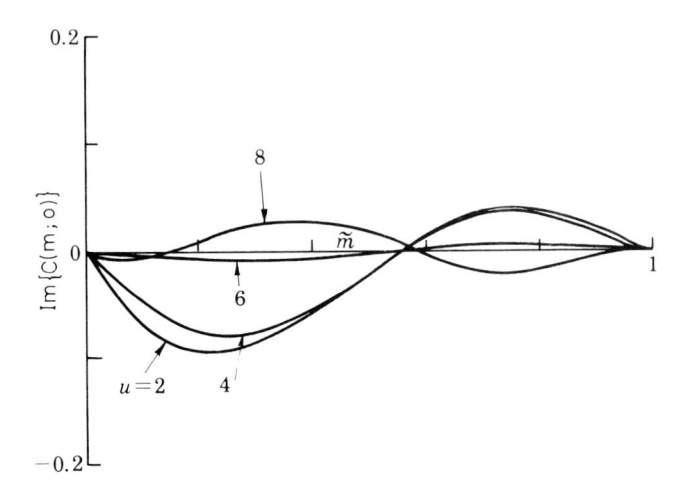

Figure 1.33. Real and imaginary parts of the defocussed weak-object transfer function of a confocal microscope.

to as a transmission cross-coefficient. It gives the magnitude of the spatial frequency component $(m - p)$ in the intensity image [Born and Wolf, 1975]. This latter name arises because it is only in the special case of coherent imaging that the function $C(m;p)$ separates into the product of a function of m and a function of p. In the general, partially coherent case, the function does not separate. In the case of a conventional microscope the function is given by [Wilson and Sheppard, 1984]:

$$C(m;p) = \int\int |P_2(\xi,\eta)|^2 P_1(\xi + \tilde{m},\eta) P_1^*(\xi + \tilde{p},\eta)\, d\xi d\eta \quad (1.45)$$

which is a function of both m and p. This merely underlines the fact that microscope imaging is generally a non-linear process. The form and special limits of this $C(m;p)$ function are discussed in detail in Wilson and Sheppard (1984). We show in Figure 1.34 the form of the $C(m;p)$ surfaces for both confocal and conventional microscopes with two equal circular objective lenses. Figure 1.35 shows contours of constant $C(m;p)$ in both cases. Recall that the spatial frequency in the intensity image is given by $m - p$, so that for a given pair of spatial frequency moduli, the response is higher if they have the same sign (difference frequencies) than if they have opposite sign (sum frequencies). The $m - p$ axis is shown: the greater the distance along the axis, the higher the spatial frequency in the image. For the confocal microscope the response for the sum frequencies is improved, but for difference frequencies is reduced as compared to the conventional microscope. This accounts for the fact that the imaging in confocal microscopy is improved. The different systems may be compared by studying the region of m, p space within which the transfer function is greater than one-half, as illustrated by the shading in the figures. It should be observed that the maximum spatial frequency present in the confocal image is twice that in a conventional image, regardless of the degree of coherence.

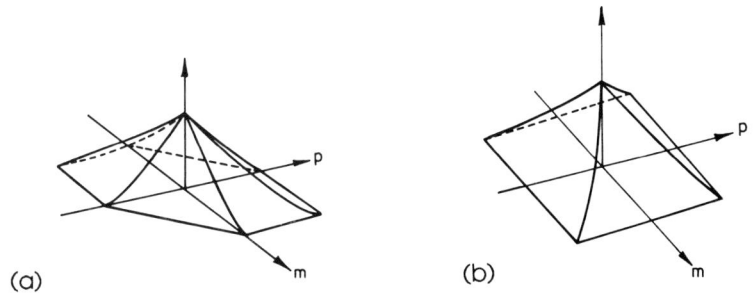

Figure 1.34. The $C(m;p)$ surfaces for (a) a conventional and (b) a confocal microscope.

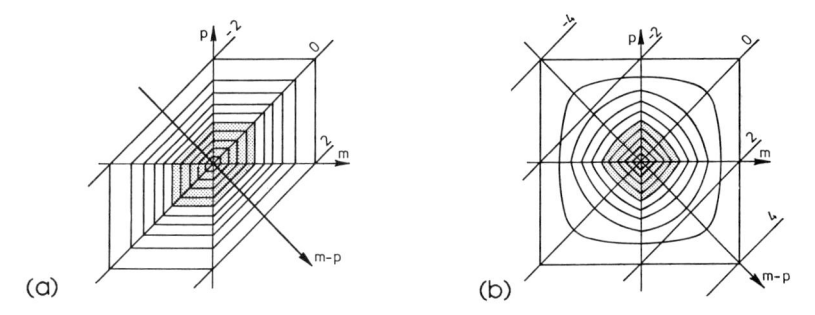

Figure 1.35. Contours of constant $C(m;p)$ for (a) a conventional and (b) a confocal microscope, showing lines of spatial frequency $(m - p)$.

It is a little dangerous, however, to claim that one microscope system is necessarily better than another simply because of the improvement in cut-off frequency. The confocal microscope is a coherent system which operates on amplitude, whereas the incoherent form of the conventional microscope, for example, operates on intensity. The partially coherent case is even more involved, and this is the basic reason why the various systems are difficult to compare in a general way.

Confocal microscopy is often concerned with imaging detail in three dimensions, and in this case it is sometimes instructive to introduce a three-dimensional form of the transfer function [Streibel, 1984; Kimura and Munakata, 1989; Nakamura and Kawata, 1990]. This is discussed in more detail in Chapter 4.

IV.4. Optical sectioning

In Figure 1.9 we illustrated the origin of the optical sectioning property of confocal microscopes by considering a planar reflector, and in Figure 1.10 we showed the result of scanning this reflector through focus. It is now straightforward to use our theory to predict the form that this response should have. As we consider a perfect reflector, $t(x,y) = 1$ and hence $T(m,n) = \delta(m)\delta(n)$. This leads to, equation (1.35):

$$I(u) = |c(0,0)|^2 \qquad (1.46)$$

which, for circular pupils, becomes for a confocal microscope:

$$I(u) = \left| \int_0^{2\pi} \int_0^1 P(\rho,\theta) P(\rho, \pi - \theta) \rho \, d\rho d\theta \right|^2 \qquad (1.47)$$

For unshaded, unaberrated pupils this leads to:

$$I(u) = \left[\frac{\sin(u/2)}{u/2} \right]^2 \qquad (1.48)$$

which is of the general form of Figure 1.10. In the case of the conventional microscope it is clear from the form of $C(0,0)$, equation (1.45), that $I(u) = constant$, and hence, in this case, there is no optical sectioning.

However, as we said earlier, these results apply only to a plane object, and not all objects are planes. If we now consider a point object, we find that the image intensity falls off axially according to equations (1.27) and (1.28) in the conventional and confocal cases respectively. This suggests that for this particular object, both systems exhibit optical sectioning according to this criterion, although the effect is a little stronger in the confocal case.

An alternative approach is to consider the variation of the total power in the image of a point as it is scanned through focus [Sheppard and Wilson, 1978b]. If the image is denoted by $I(u,v)$ then the power, or the integrated intensity, is given by:

$$I_{int}(u) = \int_0^\infty I(u,v) v \, dv \qquad (1.49)$$

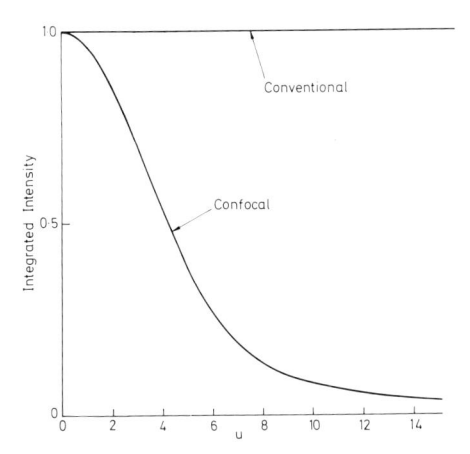

Figure 1.36. The variation of integrated intensity (total power in the image of a point source) with defocus.

In the case of the conventional microscope image, equation (1.25), we can show from Parseval's theorem or by conservation of power that the integrated intensity must be constant. In the confocal case, however, the expression becomes:

$$I_{int}(u) = \int_0^\infty |h(u,v)|^4 \, v \, dv \qquad (1.50)$$

which is plotted in Figure 1.36. For large values of u, this function falls off according to an inverse square law, as is predicted by geometrical optics. These results apply equally to reflection and transmission systems.

IV.5. Flare and scattered light

We have not yet said anything about the role of the pinhole in reducing the amount of flare and scattered light present in the image. This is particularly important and is a major reason for the crisp, clean appearance of many confocal images. We can consider the scattered light to be made up both of light which is scattered by the elements of the optical system irrespective of the object and flare which results from scattering within the object [Cox and Sheppard, 1986]. The former, which is constant in either object- or objective-scanning microscopes, merely serves to reduce contrast, and could, in principle, be measured and subtracted from the detected signal. In the case of beam-scanning microscopes, however, this component is clearly not constant and will vary across the entire image field.

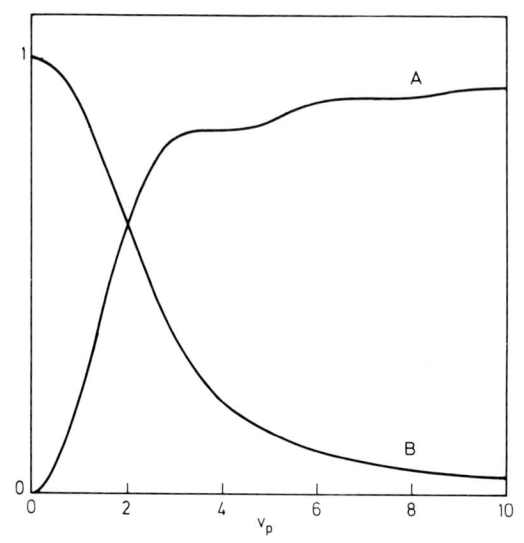

Figure 1.37. Detected signal (A) and the detected signal-to-noise ratio (B) as a function of normalised detector radius, v_p. The normalisation of the detector radius is as in equation (1.13).

In an attempt to quantify these effects [Cox and Sheppard, 1986] we may consider a transmitted light system with no object, but with a finite-sized detector of normalised radius v_p. This permits us to write the detected signal as:

$$I_{det}(v_p) = \int_0^{v_p} \left[\frac{2J_1(v)}{v}\right]^2 v\, dv \qquad (1.51)$$

which, when normalised to unity for a large detector, gives:

$$I_{det}(v_p) = 1 - J_0^2(v_p) - J_1^2(v_p) \qquad (1.52)$$

which is plotted in Figure 1.37. If we now assume that the scattered signal is proportional to the area of the detector, πv_p^2, we can also plot the ratio of signal to flare as a function of the detector size, Figure 1.37. It is clear that the presence of any reasonably small pinhole has a dramatic effect on the rejection of flare and scattered light. We will consider the effect of the finite-sized pinhole on the imaging in general in Chapter 3.

The model we have presented here is a very simple one that is intended to illustrate the main point. Other, more detailed, theories exist [Wells et al., 1989] which model the noise as containing both the shot noise of the signal together with the background. If a signal-to-noise ratio is defined in this way, it is possible to predict an optimally-sized pinhole to maximise the signal-to-noise ratio.

V. Fluorescence microscopy

We turn now to the area of confocal fluorescence microscopy, which has proved to be a very useful and powerful tool in many branches of biology and the life sciences. As it will be discussed in great detail in later chapters, we will do no more here than emphasise the fact that the image formation properties of fluorescence microscopes are *completely different* from those of the brightfield instruments that we have just been discussing. This results from the nature of the fluorescence generation mechanism.

Let us begin by referring back to the conventional microscope of Figure 1.7(c), and imagine that suitable filters are present such that only the fluorescence radiation is detected. We can immediately see that the *resolution* results essentially from the primary, incident radiation and *not* from the *longer wavelength* fluorescence: the collector lens essentially collects fluorescence radiation onto a detector. In a conventional, non-scanning microscope, Figure 1.7(a), this is *not* the case. Here, the primary radiation excites the fluorescence, which is then imaged by the objective lens. Thus, in this case, the resolution results essentially from the *longer wavelength* fluorescence radiation. This leads us to make the very important statement that we expect *scanning* microscopes to be able to produce superior fluorescence images. If we combine these advantages with those concerning dosage and bleaching, together with the optical sectioning of confocal microscopy, we can understand why the technique is so useful [Slomba et al., 1972].

We model the fluorescence generation process as follows. We assume that the fluorescence in the object destroys the coherence of the illuminating radiation and produces an incoherent fluorescent field proportional to the intensity of the incident radiation. If we describe the spatial distribution of the fluorescence generation by f, we can write the field just beyond the object as proportional to $|h_1|^2 f$, where h_1 is the amplitude point response of the first lens imaging at a wavelength λ_1. This field is then imaged at the fluorescence wavelength, λ_2, by the second lens. For the confocal system we can write the image intensity as [Cox et al. 1982; Wilson and Sheppard, 1984]:

$$I = |h_{eff}|^2 \otimes f \qquad (1.53)$$

where h_{eff} is given by:

$$h_{eff} = h_1(u,v) h_2(u/\beta, v/\beta) \qquad (1.54)$$

The optical co-ordinates are defined relative to the primary radiation and $\beta = \lambda_2/\lambda_1$ is the ratio of the wavelengths. We see that the image formation in fluorescence microscopy is incoherent. This is in contradiction to the brightfield case, in which the imaging is always purely coherent.

If we consider the image of a single point object, we can write the image, equations (1.53) and (1.54), for the in-focus case as:

$$I(v) = \left[4\frac{J_1(v)}{v} \cdot \frac{J_1(v/\beta)}{v/\beta}\right]^2 \tag{1.55}$$

which implies that optimal resolution is achieved if $\beta = 1$. This suggests that for the best lateral resolution we should try to operate with the fluorescence wavelength as close as possible to the incident. In fact, over a small range of values of β it is found that the half-width of this response is linearly related to β [Wilson, 1989]. We also note that as $\lambda_2 \to \infty$, the imaging becomes similar to a conventional incoherent microscope with a point spread function $|h_1(u,v)|^2$.

If we now introduce the spatial frequency spectrum of the fluorescence generation function as:

$$F(m,n) = \int\int f(x,y)\exp\left[2\pi j(mx+ny)\right] dx\,dy \tag{1.56}$$

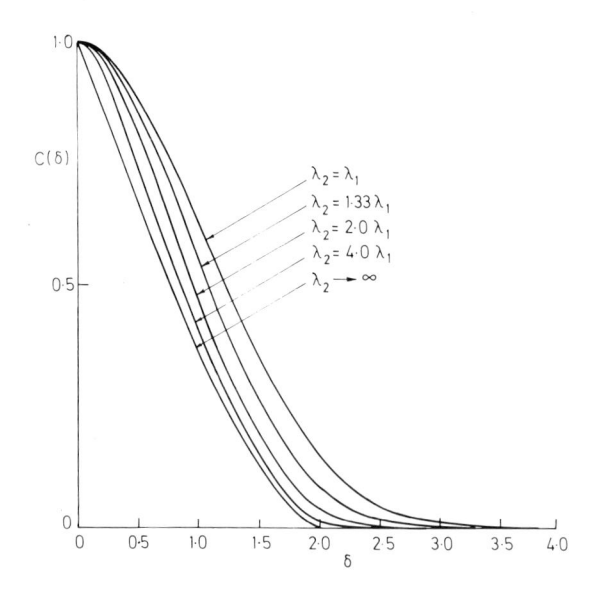

Figure 1.38. The transfer function of the confocal fluorescence microscope for various fluorescence wavelengths in the case of no defocus. The spatial frequency axis is normalised to the incident wavelength.

Confocal Microscopy

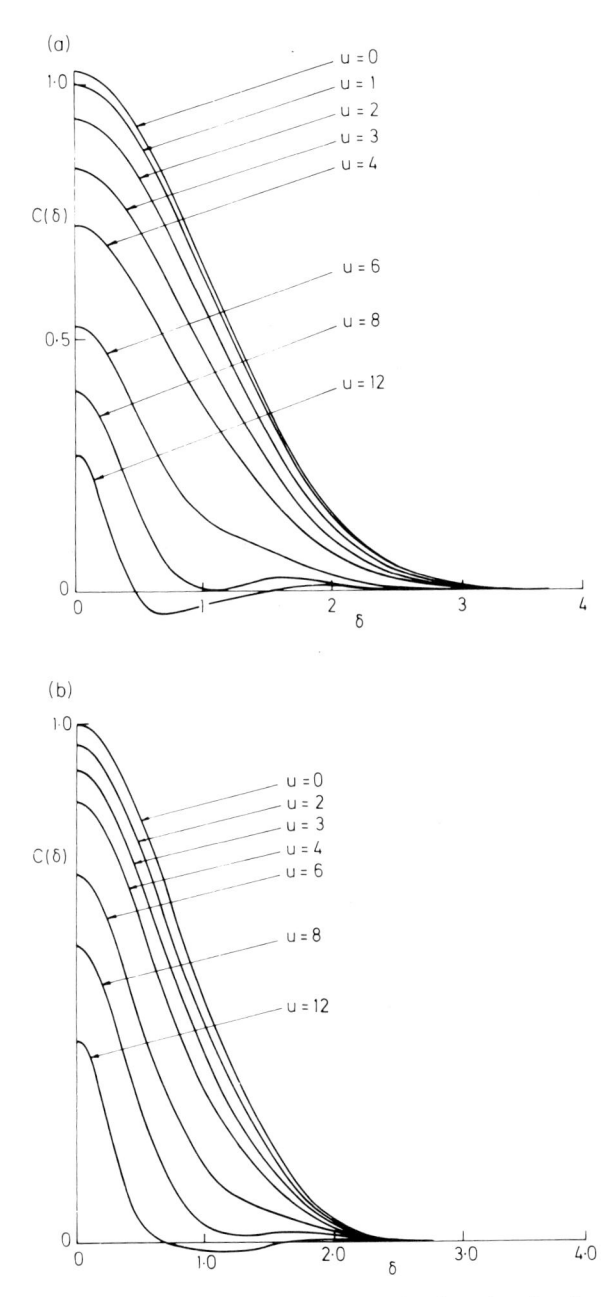

Figure 1.39. (a) The confocal fluorescent transfer function in the presence of defocus for the case of $\lambda_2 = \lambda_1$. The defocus parameter, u, and spatial frequency δ, are normalised to the incident wavelength. (b) As for (a) but for the case of $\lambda_2 = 2\lambda_1$.

we can write the image in the standard form:

$$I(x,y) = \int\int C(m,n)F(m,n)\exp\left[2\pi j(mx+ny)\right]\,dmdn \quad (1.57)$$

Again $C(m,n)$ is a circularly symmetric function which may be written as $C(m,n) = C(\delta)$ as:

$$C(\delta) = [P_1 \otimes P_1^*(\delta)] \otimes [P_2 \otimes P_2^*(\delta/\beta)] \quad (1.58)$$

The spatial frequency cut-off in this case is clearly proportional to $(1/\lambda_1 + 1/\lambda_2)$, which, in the limiting case of $\lambda_1 = \lambda_2$, suggests that the system has twice the bandwidth of a conventional incoherent microscope. However, the situation is not so good, as can be seen from Figure 1.38, where we plot $C(\delta)$ for a range of fluorescence wavelengths and equal circular pupils. The transfer function is very small at the higher spatial frequencies. Again we see that we would expect the best imaging in the $\lambda_1 = \lambda_2$ case.

The effect of defocus is shown in Figure 1.39 for two values of fluorescence wavelength. We notice, *inter alia*, that higher values of defocus result in greater contrast in the image in the $\lambda_2 = 2\lambda_1$ case. This suggests that the rejection of out-of-focus detail, i.e., the strength of the optical sectioning, is not so great as the fluorescence wavelength becomes longer.

We finally confirm these suggestions that the sectioning becomes worse as the fluorescence wavelength becomes larger by considering the form of the signal we would measure as we scan a uniform fluorescent planar object through focus [Kimura and Munakata, 1990], i.e., we set $f(x,y) = 1$ and obtain, equations (1.53) and (1.54):

$$I(u) = \int_0^\infty |h_1(u,v)h_2(u/\beta,v/\beta)|^2\,v\,dv \quad (1.59)$$

which is plotted, for equal circular pupils, in Figure 1.40. There are no zeroes in the response, as there are in the non-fluorescent confocal case, and we see that the sectioning becomes coarse as the fluorescent wavelength increases. Indeed, in the limit as $\beta \to \infty$, the imaging becomes similar to a conventional incoherent microscope, and the sectioning disappears altogether. We can also see that if we detect fluorescent radiation over a range of wavelengths, the final image will contain information from a variety of depths. However, in general, it is clear that to obtain optimal sectioning we should try to operate as close to $\beta = 1$ as possible.

We notice that when $\beta = 1$, the expression in equation (1.59) reduces to that of the integrated intensity in equation (1.50), which we used to describe optical sectioning in brightfield confocal microscopes.

Confocal Microscopy

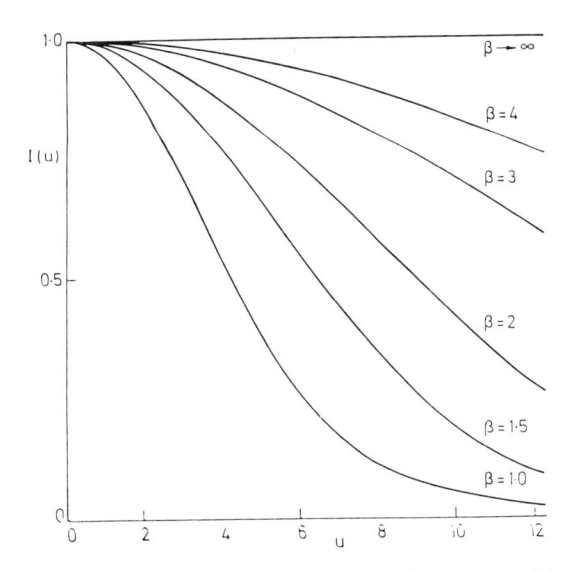

Figure 1.40. The detected signal as a perfect planar fluorescent object is scanned axially through focus for a variety of fluorescent wavelengths, $\beta = \lambda_2/\lambda_1$. The axial distance, u, is again normalised to the primary wavelength, λ_1. We note that if we measure the sectioning by the half-width of these curves, the strength of the sectioning is essentially proportional to β.

The fact that the imaging in fluorescence confocal microscopy is incoherent and described by the simple convolution relationship of equation (1.53) makes the images ripe for enhancement and restoration. In particular, generalised least squares techniques, Wiener filters [Gonzales and Wintz, 1987] and Tichonov regularisation [Bertero et al., 1990; Poggio et al., 1985] may be used. Alternatively, singular value decomposition techniques may be used [Bertero et al., 1987; Young et al., 1989].

VI. The tandem scanning microscope

We now outline the principle of operation of this form of confocal microscope. We will defer a discussion of the theory of the image formation until Chapter 3, where we will also discuss the factors which determine the choice of apertures in the scanning discs and how this affects the strength of the optical sectioning property.

Figure 1.41 illustrates the principle schematically. A transmission system is shown for ease of explanation, although, in practice, the reflection mode is usually employed. The top diagram shows a microscope with one illumination aperture and one detection aperture. An observer viewing the detection aperture would see a confocal image of

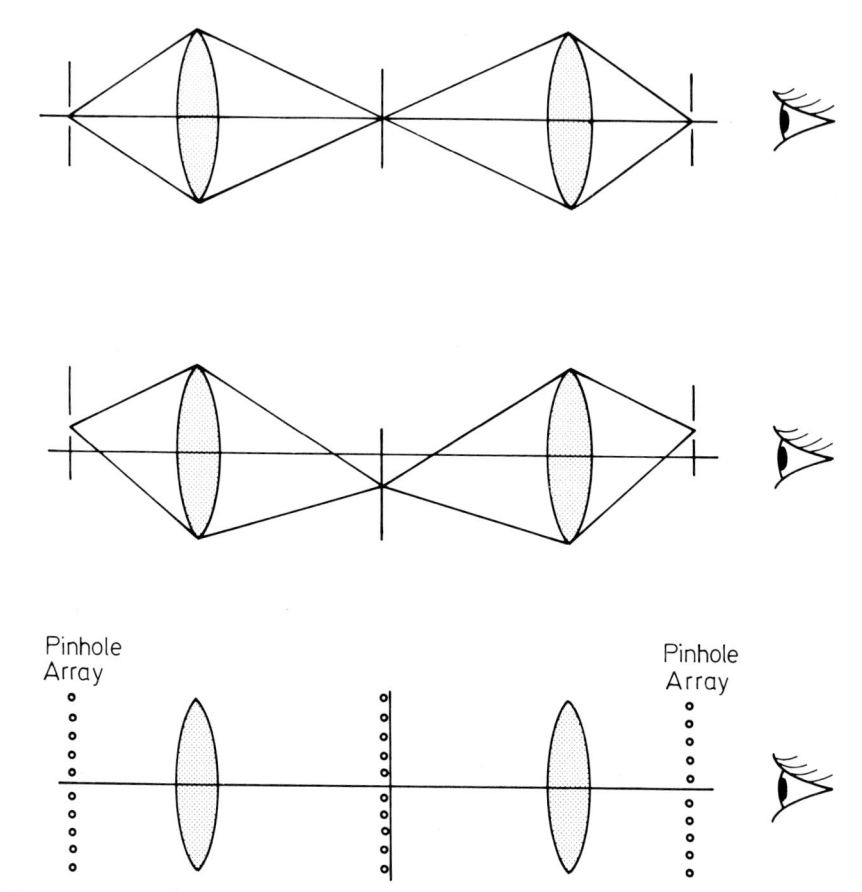

Figure 1.41. The origin of the operation of the tandem scanning direct view confocal microscope.

the illuminated point. If the apertures were now moved to the position shown in the middle diagram, a confocal image would again be seen of the point that is now illuminated. It is clear now that if we want to form an image of the whole object, an array of a vast number of such apertures would be required, as indicated on the lower figure – each imaging its own illuminated spot on the object. In practice, these systems tend to operate in reflection, and the apertures are usually holes in a disc. This disc typically consists of many thousands of holes arranged on interlacing spirals. Finally, the disc is rotated at high speed in order to produce a flicker-free image of the whole field of view. Great care must be taken in the correct choice of size and spacing of these apertures.

The image-formation properties of these instruments are substantially similar to those of laser-based confocal microscopes, but have slight differences which we will discuss later. They do have the advantage of not needing a laser source and giving real-time images at rates substantially faster than television rate. They do, of course, have drawbacks, the main one being light level: the aperture disc dramatically reduces the amount of light available for imaging. Nevertheless, they can produce good images showing good optical sectioning, and can also operate, with the help of image intensifiers, in the fluorescence mode [Boyde et al., 1990].

VII. Non-confocal imaging in scanning microscopes

We will end this introductory chapter by describing briefly some of the other imaging modes which are available with scanning microscopes. These modes may well be more important than confocal in certain applications, and it is thus important to be aware not only that they can be used with scanning microscopes, but also that they can be implemented in very simple ways.

We will begin by making some general remarks about the form the transfer function must take in order for the image intensity to represent some required variation of the object amplitude transmittance. These results prove to be very useful in interpreting the imaging of more complicated systems. We recall that, in the most general case, we describe the image intensity for an object which varies only in the x-direction by equation (1.42). We repeat it here for convenience as:

$$I(x) = \int\int C(m;p)T(m)T^*(p)\exp\left[2\pi j(m-p)x\right]\,dmdp \quad (1.60)$$

from which we see that:

$$C(m;p) = C^*(p;m) \quad (1.61)$$

because the image intensity must be a real quantity.

We now ignore diffraction effects and consider an object whose amplitude transmittance may be written as $t(x) = a(x)\exp[j\phi(x)]$. If

$$C(m;p) = 1 \quad (1.62)$$

then

$$I(x) = |t(x)|^2 \quad (1.63)$$

which is often referred to as perfect amplitude imaging. This is the case of the in-focus imaging we have been discussing. On the other hand, if

$$C(m;p) = mp \quad (1.64)$$

we obtain:

$$I(x) = \left|\frac{dt}{dx}\right|^2 \tag{1.65}$$

which might be called differential contrast. We are also often interested in forming an image which depends on the difference in phase or amplitude, and in these cases if we set:

$$C(m;p) = m + p \tag{1.66}$$

then

$$I(x) = 2a^2(x)\frac{d\phi}{dx} \tag{1.67}$$

which would represent differential phase contrast, whereas when

$$C(m;p) = j(m - p) \tag{1.68}$$

the image becomes:

$$I(x) = 2\frac{d}{dx}\left[a^2(x)\right] \tag{1.69}$$

or differential amplitude contrast. This description is more appropriate in the study of weak objects, for which we may write:

$$a(x) = 1 + a_1(x) \tag{1.70}$$

and on the assumption that $a_1(x)$ is small, equation (1.70) becomes:

$$I(x) = 4\frac{da_1}{dx} \tag{1.71}$$

which is indeed the differential of the amplitude.

We are now in a position to use some of these idealised results to make some general remarks on the form and symmetry of the $C(m;p)$ function. For differential phase contrast, for instance, it must be an odd function, being real for differential phase contrast and imaginary for differential amplitude contrast. These conditions, together with equation (1.61), dictate that the transfer function should possess the symmetry shown in Figure 1.42. These results were, of course, obtained from an idealised system with no diffraction. The main effect of diffraction is to modify the transfer function such that there is a definite spatial frequency cut-off, but not to alter the overall symmetry. It must be mentioned, however, that although many systems may possess the required symmetry, they do not all image equally well. The actual form of the transfer function is still of great importance. We will now discuss examples of how each form of imaging may be obtained in a scanning microscope. We note, however, that more traditional techniques such as Zernike phase contrast and darkfield imaging are

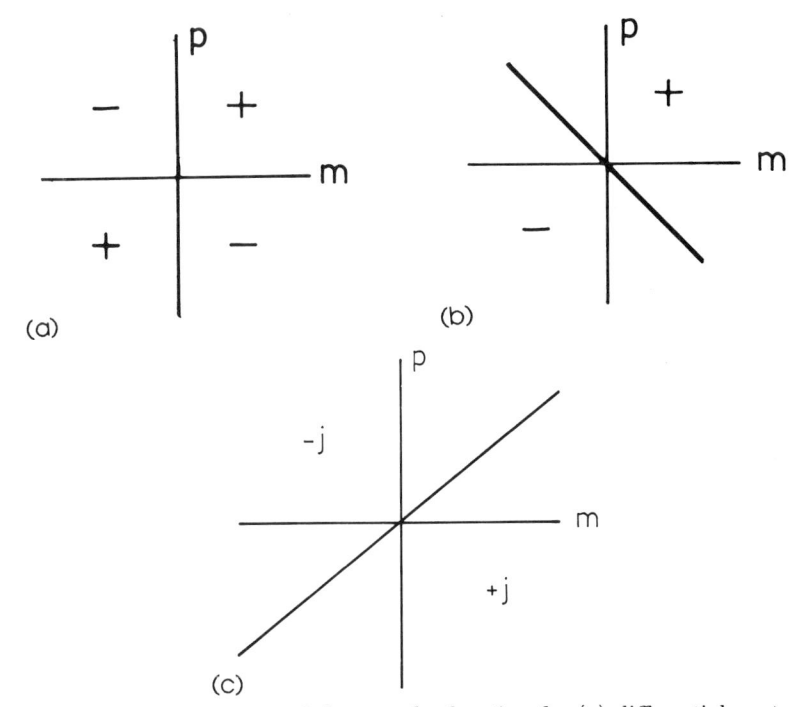

Figure 1.42. The symmetry of the transfer function for (a) differential contrast, (b) differential phase contrast and (c) differential amplitude contrast.

also available in both conventional and confocal scanning microscopes [Sheppard and Wilson, 1980; Wilson and Hamilton, 1985].

VII.1. Differential contrast

This type of contrast may be obtained in a variety of ways. In the confocal microscope, for example, the use of a *split* pupil such that a phase difference of π exists in the transmissivity between the two halves of the lens results in differential contrast imaging [Wilson and Sheppard, 1980]. Alternatively, a filter with linearly varying phase may be placed in the pupil aperture. These techniques essentially involve probing the object with a *differential* probe.

Alternatively [Wilson and Hamilton, 1984], we can obtain the same effect by simply subtracting the image obtained with a confocal microscope from that obtained with a conventional one. The form of the conventional and confocal transfer functions, Figures 1.34 and 1.35, are such that on subtraction they possess the symmetry of Figure 1.42(a). This is confirmed in Figure 1.43.

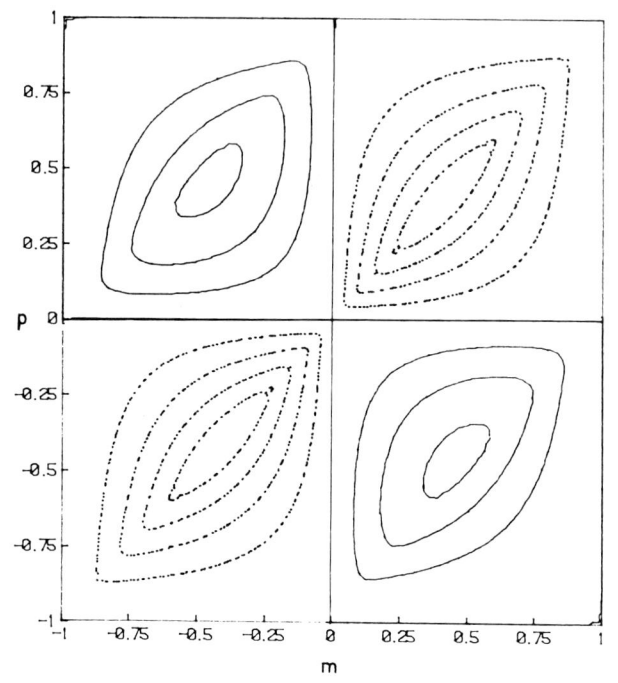

Figure 1.43. Contours of equal values of $C(m;p)$ obtained by subtracting the confocal transfer function from the conventional. The dashed lines represent positive values of transfer function and the full lines negative values. The symmetry is as in Figure 1.42(a), indicating differential contrast.

VII.2. Differential amplitude contrast

The simplest method of obtaining this form of contrast is simply to differentiate the image signal electronically. This has the effect mathematically, equation (1.60), of multiplying the well-behaved transfer function $C(m;p)$ by $2\pi j(m-p)$, which indicates that the transfer function has the appropriate form, equation (1.68), for differential amplitude contrast. An alternative optical method is shown in Figure 1.44. Here the detector is replaced by a large-area detector split into two halves, and an image is formed by subtracting the signals measured by each half. It is clear that a zero signal would be expected for a flat object. An analysis of the geometry shows that the transfer function has the symmetry shown in Figure 1.45 and that differential amplitude contrast does result [Wilson and Sheppard, 1980; Wilson and Hamilton, 1983; Hamilton and Wilson, 1984]. An alternative approach, of course, would be to build a confocal version by replacing the large-area split detector with two closely-positioned point detectors. An advantage of the large-area detector is that it is very simple

Confocal Microscopy 51

to obtain a conventional image merely by adding the signals from the two half-detectors.

A standard method of edge enhancement is to add a differentiated image to a conventional one. This may be done in the present arrangement simply by altering the gains on the two channels. In this way we would obtain a conventional image with an amount of differential added in. Figure 1.46 shows an example of the technique.

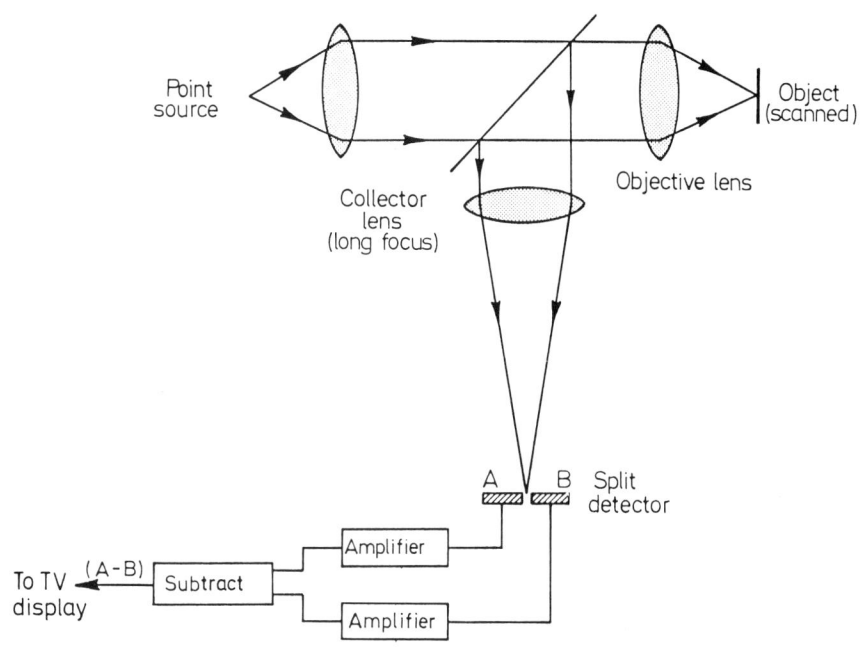

Figure 1.44. An optical arrangement to produce differential amplitude contrast in reflection.

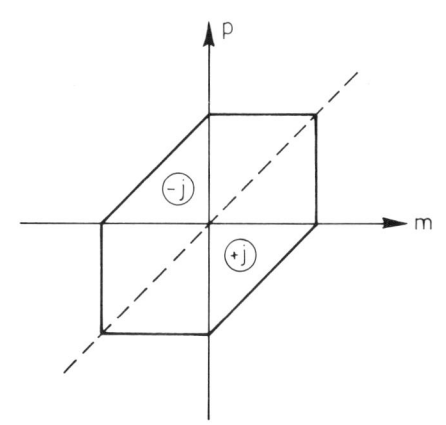

Figure 1.45. The transfer function symmetry for the differential amplitude contrast arrangement of Figure 1.44.

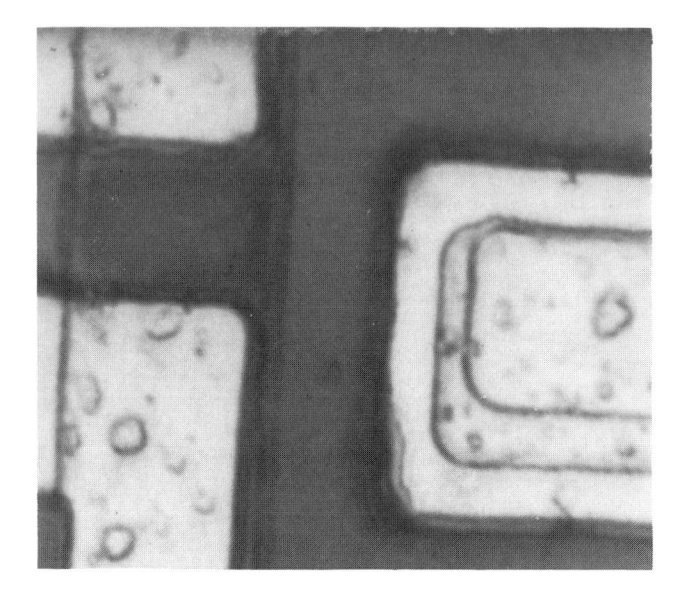

(a)

(b)

Figure 1.46. (a) A conventional image of a portion of a microcircuit and (b) an optically edge-enhanced version.

Confocal Microscopy 53

VII.3. Differential phase contrast

A very simple method of obtaining differential phase contrast in a scanning microscope has been proposed by Dekkers and de Lang (1974), Figure 1.47, which involves replacing the collector lens of a scanning microscope by a large-area photo-detector split into two halves. It is clear that by subtracting the electrical signals from the two halves of the detector, we form a differential phase contrast image which is somewhat similar to that formed by the Nomarksi system of conventional microscopy. Figure 1.48 shows the form of the transfer function. To illustrate the sensitivity of the method, we show in Figure 1.49(a) a differential phase contrast image of a spectroscopic grating. The grooves in this grating are only 2nm deep, yet they are well imaged by this method. Figure 1.49(b) shows another such image of a processed semiconductor device. Here the differential phase contrast has served to reveal the presence of residual photo-resist which has not been adequately removed during the processing.

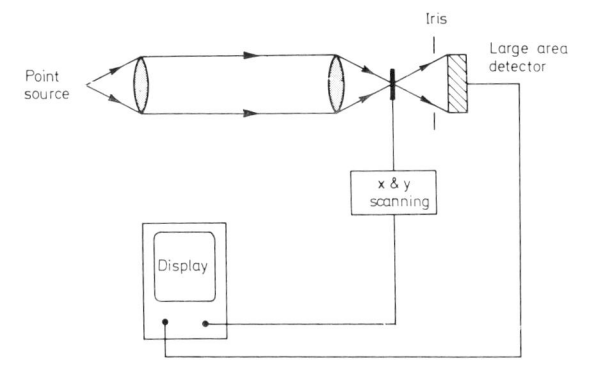

Figure 1.47. A schematic diagram of a differential phase contrast scanning microscope. The large area detector usually consists of two halves. The signals from these are subtracted in order to obtain the differential phase contrast signal. The direction of differentiation may be arbitrarily chosen if a quadrant detector is used.

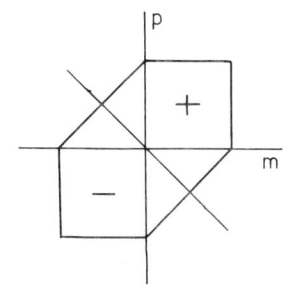

Figure 1.48. The form of the transfer function indicating that differential phase contrast imaging is obtained.

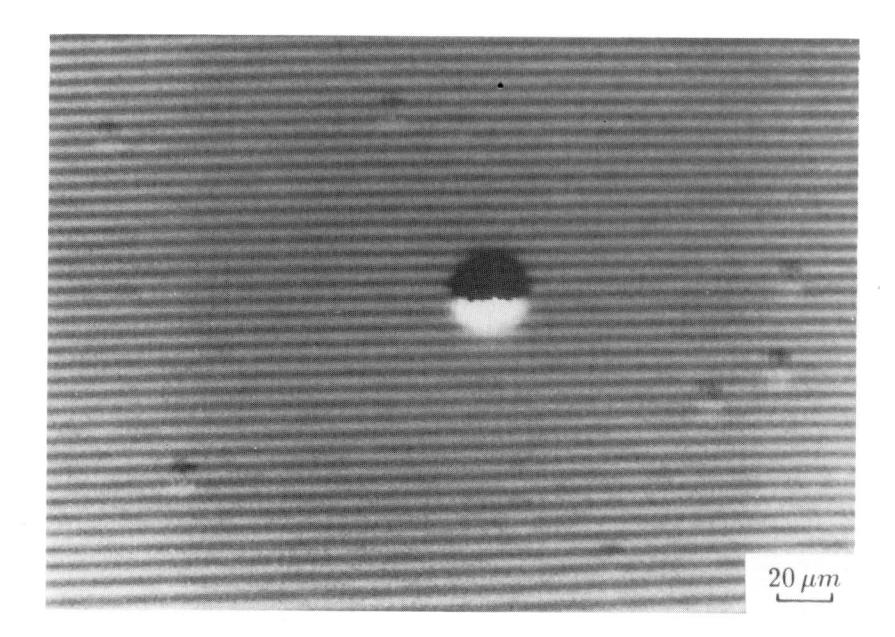

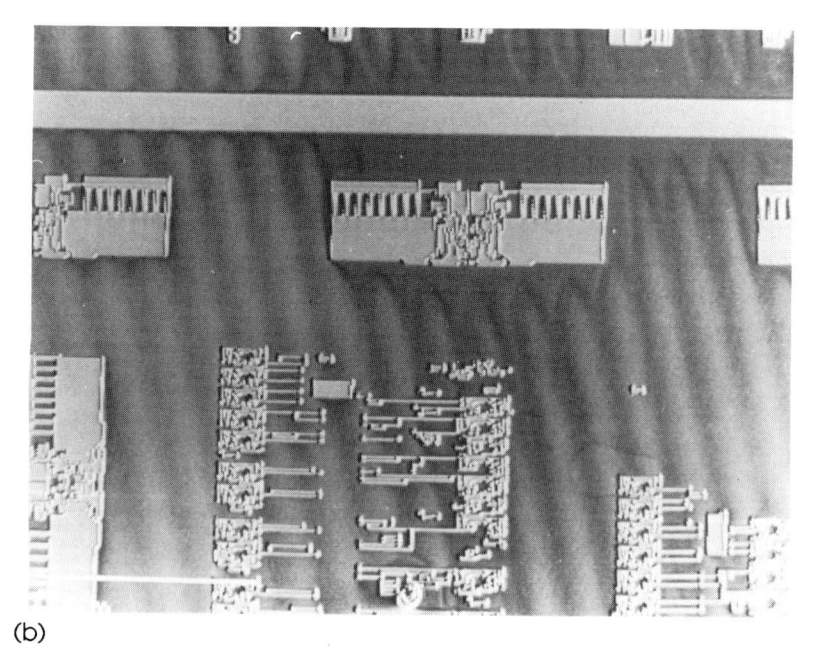

Figure 1.49. Differential phase contrast images (a) of grooves 2nm deep on a spectroscopic grating and (b) revealing the presence of residual photo-resist on a semiconductor wafer.

If we use a detector split into two halves, we are restricted to differentiating in one direction only. The use of a quadrant detector, Figure 1.50, removes this restriction [Wilson, 1987]. Examples are shown in Figure 1.51 of the images of vertical and horizontal lines etched in photo-resist.

The technique is clearly very similar to that of Nomarski (1955) used in conventional microscopy. Figure 1.52 presents comparison images taken of the same specimen. A detailed comparison is given by Hamilton and Sheppard (1984). There is, of course, no reason why the Nomarski method cannot be used in a scanning microscope. The beauty of the Dekkers and de Lang technique lies in its simplicity. We can select the imaging mode we need by choosing whether to add or subtract the signals. Again, edge enhancement is available by varying the gains of the detector channels. The sensitivity of the scheme for small phase gradients may be enhanced by the use of an annular-shaped split, or quadrant, detector [Hamilton and Wilson, 1983; Wilson, 1987].

We note finally that the differential phase contrast schemes we have described do not give rise to optical sectioning or depth discrimination. This can be overcome by suitably combining the technique of Dekkers and de Lang with the confocal microscope. A scheme has been proposed by Benschop (1987). It is shown, in essence, in Figure 1.53. The system consists of two confocal microscopes, but with half their collector lenses obscured [Carlini, 1989; Wilson and Carlini, 1990]. The confocal system ensures that any detail imaged comes from the region of the specimen in the focal plane. The obscuration of the collector lens gives rise to a transfer function which can be represented as a combination of an even function and an odd function. When the images from the two confocal systems are subtracted, it is found that the even functions cancel out and the overall transfer function has the symmetry required to give differential phase contrast.

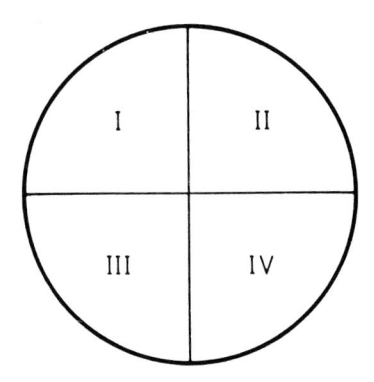

Figure 1.50. The layout of the large-area quadrant detector.

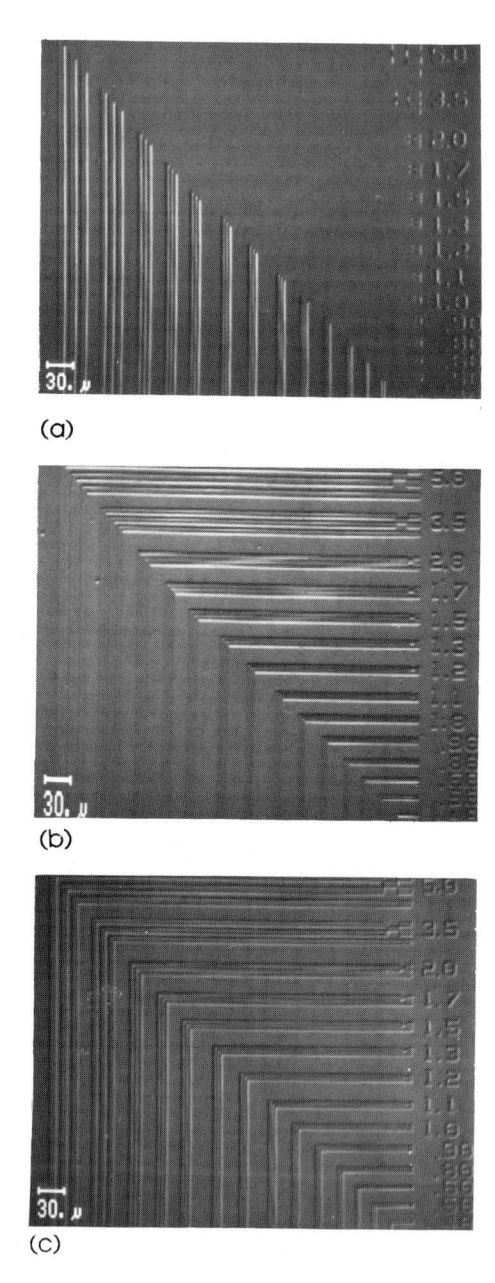

Figure 1.51. (a) Differential phase contrast image formed by displaying quadrants (I + III) - quadrants (II + IV). (b) Differential phase contrast image formed by displaying quadrants (I + II) - quadrants (III + IV) (c) Differential phase contrast image formed by displaying quadrant I - quadrant IV.

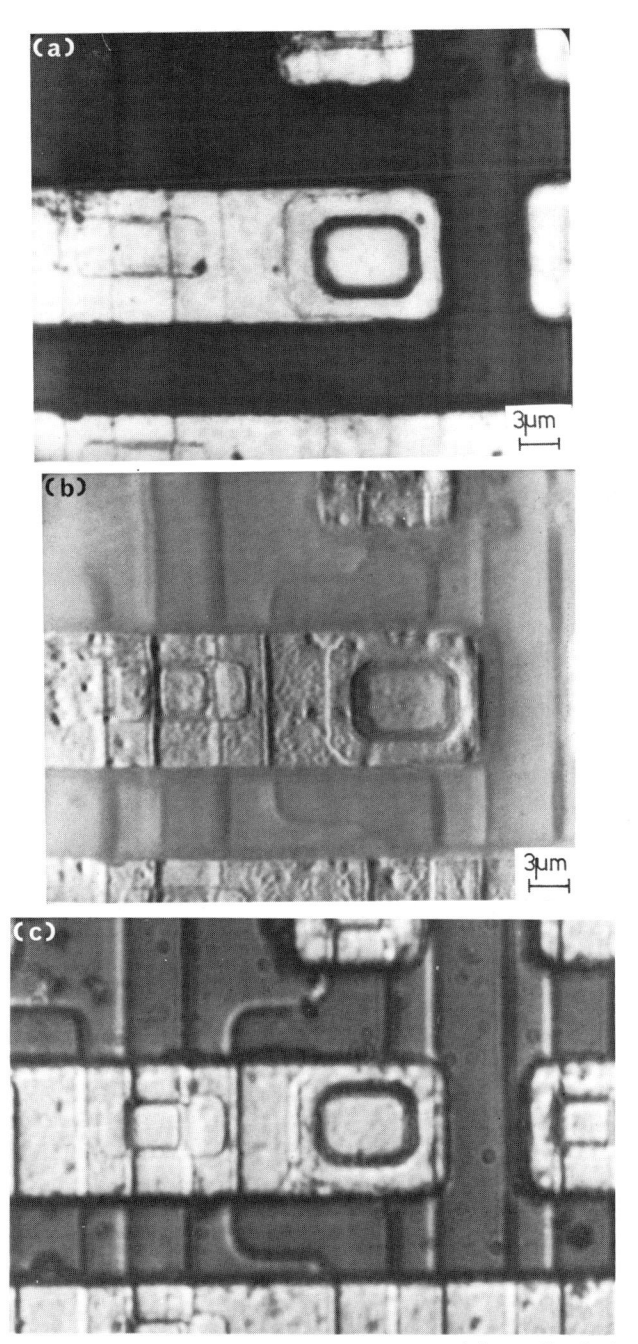

Figure 1.52. An integrated circuit viewed in reflection: (a) amplitude image, (b) differential phase image and (c) a similar region viewed in a conventional Zeiss microscope using Nomarski DIC.

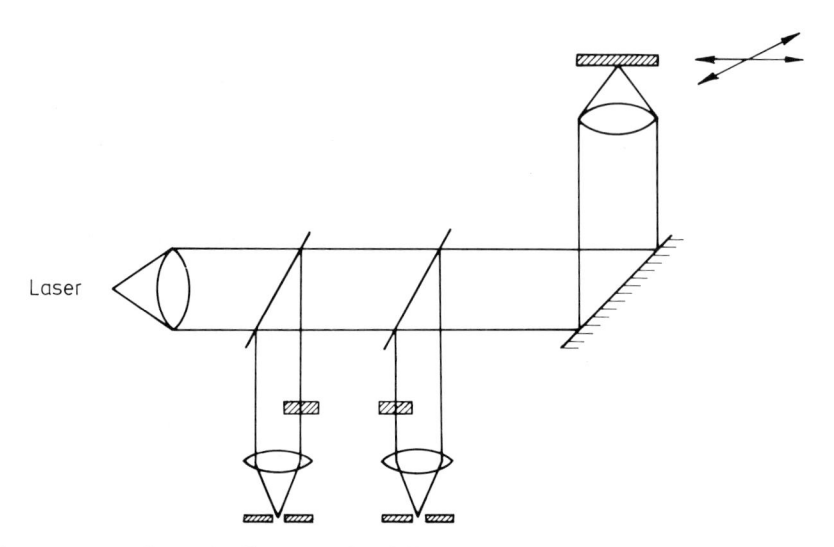

Figure 1.53. Schematic diagram of a differential phase contrast confocal microscope which will exhibit depth discrimination.

VIII. Conclusions

It is clear that the novel feature of the confocal arrangement is its depth discrimination property, which allows us to image volume specimens. The table below gives us an idea of the theoretical three-dimensional resolution which is available. The lateral resolution is measured by the full-width-half-maximum (FWHM) of the image of a single point object, and the depth resolution comes from the FWHM of the $V(z)$ response, Figure 1.11. The figures given correspond to a wavelength of 442nm; the values at other wavelengths may be obtained by simply scaling these values linearly with wavelength.

Numerical Aperture	0.95	1.4 (immersion)
Lateral Resolution (μm)	0.20	0.14
Depth Resolution (μm)	0.34	0.23

The method of achieving confocal operation that we have described is not the only one. In order to build a confocal microscope we need to use a coherent detector. This is straightforward in acoustics [Lemons and Quate, 1974], but more difficult in optics. However, it can also be achieved here in a heterodyne interference system [Sawatari, 1973]. A laser-based scanning optical microscope is also well suited to interference microscopy in general [Brakenhoff, 1979; Sheppard and Wilson, 1980]. The confocal implementation now has the advantage of relaxing the alignment tolerances of the system. It also provides an ideal environment for measuring the combined aberrations of the objective lenses and the optical system. This is discussed more fully in Chapter 15.

The confocal microscope improves the imaging over the conventional instrument by making both lenses contribute equally to the image formation. It is tempting to suggest that the situation would be improved even more if radiation were to be allowed to pass through the object many times [Sheppard and Wilson, 1980]. It turns out, however, that the improvement is not great for the double-pass case, and indeed, for certain classes of object, the imaging is degraded. However, it is quite attractive because it provides an alternative technique to transmission confocal microscopy. Naturally, it does not give exactly the same imaging, but it does make the alignment less difficult. The double-pass system also requires careful alignment, and Johnson *et al.* (1989) have suggested that this may be further alleviated if a phase conjugate mirror were used in place of a conventional mirror. The logical conclusion to ideas involving probing an object many times is to place it within a resonator [Sheppard and Kompfner, 1978; Sheppard and Wilson, 1980].

Although we have discussed many optical imaging arrangements, we have really only considered two types of imaging – brightfield and fluorescence. The choice of which to use will, necessarily, be application-dependent. In the brightfield mode, problems of image interpretation may occur because of interference effects in thick specimens, and also speckle [Mendez, 1982, 1985]. However, there are many other types of imaging available which give totally different, but complementary, information about the object under examination. One example would be a non-linear microscope in which an image would be formed by, say, displaying the second harmonic radiation generated within the specimen by a probe beam [Gannaway and Sheppard, 1978; Gannaway and Wilson, 1979; Wilson and Sheppard, 1979]. Alternatives would be to form an image using the Raman effect [Delhaye and Dhamelincourt, 1975; Sombert *et al.*, 1980], the photothermal effect [Rosencweig and Busse, 1980] or the photoacoustic effect [Quimby, 1984; Zharov and Letkhov, 1986].

In general, then, we can say that there is a great deal to be gained from adopting a scanning approach in microscopy. It opens up the possibility of obtaining a great deal of new information in a relatively straightforward, non-destructive, way.

References

Ash, E. A. and Nicholls, G., (1972). Super-resolution aperture scanning microscopy. Nature, **237**, 510-512.

Baer S. C. (1970). United States Patent 3,547,512. Optical apparatus providing focal plane specific illumination.

Benschop, J. P. H., (1987). Confocal differential phase contrast in scanning optical microscopy. Proc. S.P.I.E., **809**, 90-97.

Bertero, M., Brianzi, P. and Pike, E. R., (1987). Super-resolution in confocal scanning microscopy. Inverse Problems, **3**, 195-212.

Bertero, M., Boccacci, P., Brakenhoff, G. J., Malfanti, F. and van der Voort, H. T. M., (1990). Three-dimensional image restoration and super-resolution in fluorescence confocal microscopy. J. Microsc., **157**, 3-20.

Born, M and Wolf, E., (1975). *Principles of optics*. Pergamon Press, Oxford.

Boyde, A., Jones, S. J., Taylor, M. L., Wolfe, L. A. and Watson, T. F. (1990). Fluorescence in the tandem scanning microscope. J. Microsc. **157**, 39-49.

Brakenhoff, G. J., (1979). Imaging modes of confocal scanning light microscopy. J. Microsc., **117**, 233-242.

Brakenhoff, G. J., Blom, P. and Barends, P., (1979). Confocal scanning light microscopy with high aperture immersion lenses. J. Microsc., **117**, 219-232.

Brakenhoff, G. J., van der Voort, H. T. M., Spronsen, E. A. and Nanninga, N., (1986). Three-dimensional imaging by confocal scanning microscopy. Ann. New York Acad. Sci., **483**, 405-415.

Carlini, A. R., (1989). D. Phil. Thesis, University of Oxford.

Cox, I. J., (1981). D. Phil. Thesis, University of Oxford.

Cox, I. J. and Sheppard, C. J. R., (1986). Information capacity and resolution in an optical system. J. Opt. Soc. Am, **A3**, 1152-1158.

Cox, I. J., Sheppard, C. J. R. and Wilson, T., (1982). Super-resolution by confocal fluorescence microscopy. Optik, **60**, 391-396.

Dekkers, N. H. and de Lang, H., (1974). Differential phase contrast in a STEM. Optik, **41**, 452-456.

Delahaye, M. and Dhamelincourt, P., (1975). Raman microprobe and microscope with laser excitation. J. Raman. Spectr., **3**, 33-43

Egger, M. D. and Petràň, M., (1967). New reflected light microscope for viewing unstained brain and ganglion cells. Science, **157**, 305-307.

Gannaway, J. N. and Sheppard, C. J. R., (1978). Second harmonic imaging in the scanning optical microscope. Opt. and Quant. Elec., **10**, 435-439.

Gannaway, J. N. and Wilson, T., (1979). Imaging properties of the scanning harmonic microscope. Proc. Roy. Microsc. Soc., **14**, 170-174.

Gaylord, T. K., Baird, W. E. and Moharam, M. G., (1986). Zero-reflectivity high spatial-frequency rectangular-groove dielectric surface-relief gratings. Appl. Opt., **25**, 4562-4571.

Goldstein, S., (1989). A no-moving-parts video rate laser beam scanning type 2 confocal reflected/transmission microscope. J. Microsc., **153**, RP1-RP2.

Gonzales, R. C. and Wintz, P., (1987). *Digital image processing.* Addison Wesley, New York.

Goodman, J. W., (1968). *Introduction to Fourier optics.* McGraw Hill, New York.

Hamilton, D. K. and Sheppard, C. J. R., (1984). Differential phase contrast in scanning optical microscopy. J. Microsc., **133**, 27-39.

Hamilton, D. K. and Wilson, T., (1983). Improved imaging of phase gradients in scanning optical microscopy. J. Microsc., **135**, 275-286.

Hamilton, D. K. and Wilson, T., (1984). Edge enhancement in scanning optical microscopy by differential detection. J. Opt. Soc. Am., **A1**, 322-323.

Hamilton, D. K. and Wilson, T., (1986). Scanning optical microscopy by objective lens scannning. J. Phys. E., **19**, 52-54.

Hamilton, D. K. and Wilson, T., (1987). Optical sectioning in infra-red scanning microscopes. Proc. I.E.E., **134**, Pt. I, 85-86.

Hewlett, S. J., Barnett, S. M. and Wilson, T., (1990). Image intensity at the edge of a straight-edge object. J. Mod. Opt.

Hopkins, H. H., (1953). On the diffraction theory of optical images. Proc. Roy. Soc. Lond., **A217**, 408-432.

Johnson, K. M., Cathey, W. T. and Mao, C. C., (1989). Image formation in a super-resolution phase conjugate scanning microscope. Appl. Phys. Lett., **55**, 1707-1709.

Kimura, S. and Munakata, C., (1989). Calculation of three dimensional optical transfer function for a confocal scanning fluorescent microscope. J. Opt. Soc. Am., **A6**, 1015-1019.

Kimura, S. and Munakata, C., (1990). Depth resolution of the fluorescent confocal scanning microscoe. Appl. Opt., **29**, 489-494.

Kirk, C. P. and Nyyssonen, D., (1985). Modelling the optical microscope images of thick layers for the purpose of linewidth measurement. Proc. S.P.I.E., **583**, 179-181.

Lemons, R. A. and Quate, C. F., (1974). A scanning acoustic microscope. Appl. Phys. Lett., **24**, 163-165.

Lichtman, J. W., Sunderland, W. J. and Wilkinson, R. S., (1989). High resolution imaging of synaptic structure with a simple confocal microscope. The New Biologist, 1, 75-82.

Lukosz, W., (1966). Optical systems with resolving powers exceeding the classical limit. J. Opt. Soc. Am., **56**, 1463-1472.

Martin, L. C., (1966). *The theory of the microscope*. Blackie, London.

Matthews, H. J., (1987). D. Phil. Thesis. University of Oxford.

McCutchen, G. W., (1967). Super-resolution microscopy and the Abbé resolution limit. J. Opt. Soc. Am., **57**, 1190-1192.

Mendez, E. R., (1982). Speckle in confocal transmission optical microscopes. Opt. Commun., **43**, 318-322.

Mendez, E. R., (1985). Speckle statistics and depth discrimination in the confocal scanning optical microscope. Opt. Acta, **32**, 209-221.

Minsky, M., (1961). Microscopy Apparatus. United States Patent 3,013,467, Dec. 19, 1961 (Filed Nov. 7, 1957).

Montgomery, P. O'B. (Editor), (1962). Scanning Techniques in Biology and Medicine. Ann. New York Acad. Sci., **97**, 329-526.

Nakamura, O. and Kawata, S., (1990). Three-dimensional transfer-function analysis of the tomographic capability of a confocal fluorescence microscope. J. Opt. Soc. Am., **A7**, 522-526.

Nomarski, G., (1955). Microinterferometric differential à ondes polarisees. J. Rad. Phys., **16**, 9-135.

Nyyssonen, D., (1982). Theory of optical detection and imaging of thick layers. J. Opt. Soc. Am., **72**, 1425-1431.

Poggio, T., Torre, V., and Koch, C., (1985). Computational vision and regularisation theory. Nature, **317**, 314-319.

Quimby, R. S., (1984). Real time photoacoustic microscopy. Appl. Phys. Lett., **45**, 1037-1039.

Rosencweig, A. and Busse, G., (1980). High-resolution photoacoustic thermal-wave microscopy. Appl. Phys. Lett., **36**, 725-727.

Sawatari, T., (1973). Optical heterodyne scanning microscope. Appl. Opt., **12**, 2768-2772.

Sheppard C. J. R., (1987). Scanning optical microscopy. Adv. in Opt. and Electron Microsc., **10**, 1-98.

Sheppard, C. J. R. and Heaton, J. M., (1984). Images of surface steps in coherent illumination. Optik, **68**, 371-380.

Sheppard, C. J. R. and Kompfner, R., (1978). Resonant scanning optical microscope. Appl. Opt., **17**, 513-536.

Sheppard, C. J. R. and Sheridan, J. T., (1989). Micrometrology of thick structures. Proc. S.P.I.E. Optical Scanning and Optical Storage Conference (Paris, France).

Sheppard, C. J. R. and Wilson, T., (1978a). Image formation in scanning microscopes with partially coherent source and detector. Opt. Acta, **25**, 315-325.

Sheppard, C. J. R. and Wilson, T., (1978b). Depth of field in the scanning microscope. Opt. Lett., **3**, 115-117.

Sheppard, C. J. R. and Wilson, T., (1980). Fourier imaging of phase information in conventional and scanning microscopes. Phil. Trans. Roy. Soc., **A529**, 513-536.

Sheppard, C. J. R. and Wilson T., (1981). Effects of high angles of convergence on V(z) in the scanning acoustic microscope. Appl. Phys. Lett., **38**, 858-860.

Slomba, A. F., Wasserman, D. E., Kaufman, G. I. and Nester, J. F., (1972). A laser flying spot scanner for use in automated fluorescence antibody instrumentation. J. Assoc. Adv. Med. Instrum., **6**, 230-234.

Sombert, B., (1980). Etude par spectroscopie et microscopie Raman de phases oxyde de molybdène déposée par alumines. J. Raman. Spectr., **9**, 291-296

Streibl, N., (1984). Depth transfer in an imaging system. Opt. Acta, **31**, 1233-1241.

Wells, K. S., Sandison, D. R., Strickler, J. and Webb, W. W., (1989). Quantitative fluorescence imaging with laser scanning confocal microscopy. In: *Handbook of biological confocal microscopy.* Pawley, J. (Editor), 23-36. IMR Press.

Wilson, T., (1987). Enhanced differential phase contrast imaging in scanning microscopy using a quadrant detector. Optik, **80**, 167-170.

Wilson, T., (1989). Optical sectioning in confocal fluorescent microscopes. J. Microsc., **154**, 143-156.

Wilson, T. and Carlini, A. R., (1986). The image of thick step objects in the confocal scanning microscope. Optik, **72**, 109-114.

Wilson, T. and Carlini, A. R., (1990). Differential phase contrast with optical sectioning.

Wilson, T., Carlini, A. R. and Hamilton, D. K., (1986). Images of thick step objects in confocal scanning microscopes by axial scanning. Optik, **74**, 123-126.

Wilson, T., Gannaway, J. N. and Sheppard, C. J. R., (1980). Optical fibre profiling using a scanning optical microscope. Optical and Quant. Elec., **12**, 341-345.

Wilson, T. and Hamilton, D. K., (1982). Dynamic focussing in the confocal scanning microscope. J. Microsc., **128**, 139-143.

Wilson, T. and Hamilton, D. K., (1983). Differential amplitude contrast imaging in the scanning optical microscope. Appl. Phys., **B32**, 187-191.

Wilson, T. and Hamilton, D. K., (1984). Difference confocal scanning microscopy. Opt. Acta, **31**, 452-465.

Wilson, T. and Hamilton, D. K., (1985). Darkfield scanning optical microscopy. Optik, **71**, 23-26.

Wilson, T. and Pester, P. D., (1987a). The effect of scan velocity on the measurement of semiconductor parameters by optical beam excitation. J. Appl. Phys., **61**, 2307-2313.

Wilson, T. and Pester, P. D., (1987b). Theory of scan speed dependent optical beam induced current images in semiconductors. Optik, **76**, 18-22.

Wilson, T. and Sheppard, C. J. R., (1979). Imaging and super-resolution in the harmonic microscope. Opt. Acta, **26**, 761-770.

Wilson, T. and Sheppard, C. J. R., (1980). Coded apertures and detectors for optical differentiation. Proc. S.P.I.E., **232**, 203-209.

Wilson, T and Sheppard, C. J. R., (1984). *Theory and practice of scanning optical microscopy.* Academic Press, London.

van der Voort, H. T. M., Brakenhoff, G. J. and Boarslag, M. W., (1989). Three-dimensional visualisation methods for confocal microscopy. J. Microsc., **153**, 123-132.

Young, M. R., Davies, R. E., Pike, E. R., Walker, J. G. and Bertero, M., (1989). Super-resolution in confocal microscopy: experimental confirmation in the 1D coherent and incoherent cases. Europhys. Lett., **9**, 773-775.

Zharov, V. P. and Letokhov, V. S., (1986). *Laser optoacoustic spectroscopy.* Springer-Verlag, Berlin.

2. Software and Electronics for a Digital Three-Dimensional Microscope

P.-O. FORSGREN, O. FRANKSSON AND A. LILJEBORG

I. Introduction

This chapter gives a brief description of a confocal scanning light microscope (CSLM), the different ways of controlling a confocal laser microscope, and software methods for analysis of three-dimensional recordings of specimens. The solutions to some of the issues addressed will be shown by the instrument PHOIBOS, designed at our institute [Åslund et al., 1983; Carlsson and Liljeborg, 1988].

The first section shows some examples of the confocal technique and three-dimensional data processing. Section II contains a schematic overview of the building blocks of our system, followed by a description of the actions a user can take when working with the system. The software for a confocal microscope is described in Section III, starting at the top level with a number of display algorithms that have been tested. Some of their merits and drawbacks are discussed. Digital image filtering techniques and storage issues are discussed in Sections IV and V respectively. In Section VI there is a survey of design choices that need to be made in order to interface the hardware of the microscope with a computer. In this context, the computer requirements for different tasks will be discussed, including the post-processing algorithms described in Sections III and IV. Finally, in Section VII, the electronics of some critical parts of a scanning system are described.

The instrument developed at our institute has been used to study a number of different objects. Some of them will be described here as examples of how the technique can be used.

In cooperation with the Department of Physiology III, Karolinska Inst., Sweden, we have recorded the three-dimensional structure of a neuron in the spinal cord of a lamprey [Wallén et al., 1988]. One neuron is selectively stained with a fluorescent dye and then recorded in a series of sections. The data is used to calculate views of the cell from different directions, which facilitates investigation of the three-dimensional structure, Figure 2.1.

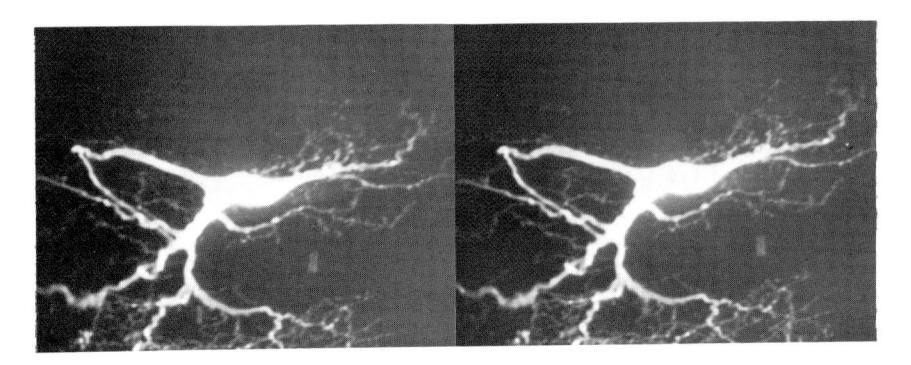

Figure 2.1. Stereo pair of neuron cell from spinal chord of a lamprey. Image size is 100 × 100µm and the resolution 512 × 512 pixels. The stack consists of 285 sections. Difference in viewing angle is 6°. Objective: Zeiss Planapo 100/1.3. Excitation wavelength: 458nm. Specimen provided by P. Wallén, Karolinska Inst., Stockholm.

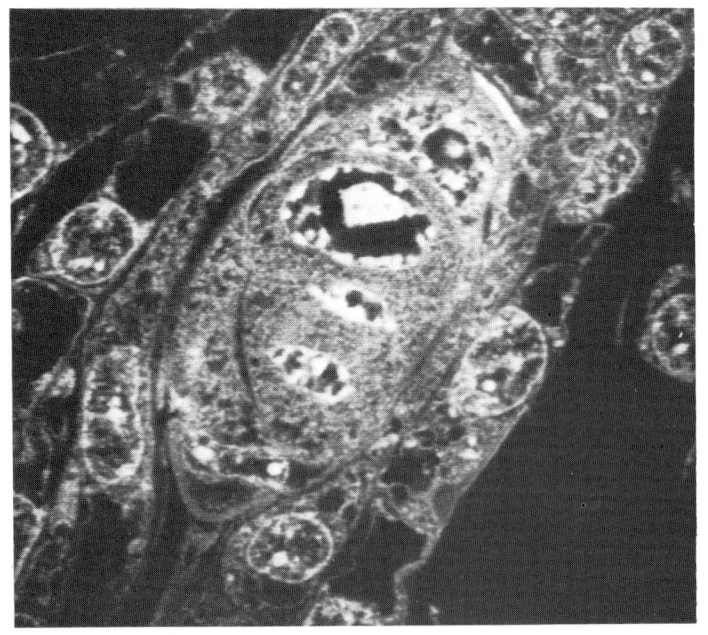

Figure 2.2. Embryo cells from seeds of orchids. Image size is 320 × 320µm and the resolution 256 × 256 pixels. Objective: Zeiss Planapo 40/1.0. Excitation wavelength: 458nm. Specimen provided by M. Fredriksson, University of Gothenburg.

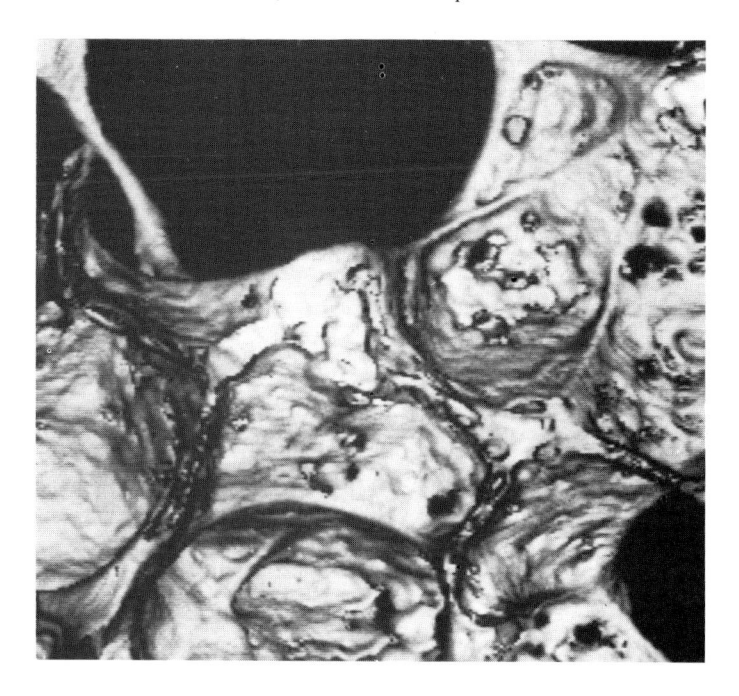

Figure 2.3. Lung tissue from a rat. Specimen size 160 × 160μm, image resolution 256 × 256 pixels, 65 sections. Objective: Zeiss Planapo 63/1.4. Excitation wavelength: 458nm. Specimen provided by E. Oldmixon, Memorial Hospital, Rhode Island.

Ovules from an orchid of the Neuwiedia species (not yet described) were used in a study of embryo cells from seeds of orchids in cooperation with the Department of Systematic Botany, University of Gothenburg, Sweden. The confocal scanning technique offers several advantages over traditional methods in this study. No staining was necessary, since the autofluorescence of the ovules could easily be registered with the laser microscope. The evaluation procedure is much faster, since no embedding with paraffin and subsequent mechanical sectioning of the specimen is necessary, and the sections are perfectly aligned. The sections are also free of distortions caused by the cutting process, and finally the thickness of each section is 1.5μm, compared with 8μm when using a microtome, thus giving pictures that are easier to interpret, Figure 2.2.

The morphology of lung tissue from a rat is studied in cooperation with the Division of Pulmonary Medicine, The Memorial Hospital, Rhode Island, USA. In this study, the aim is to investigate the mechanical structure of the lung, such as lengths and angles of the septa, Figure 2.3.

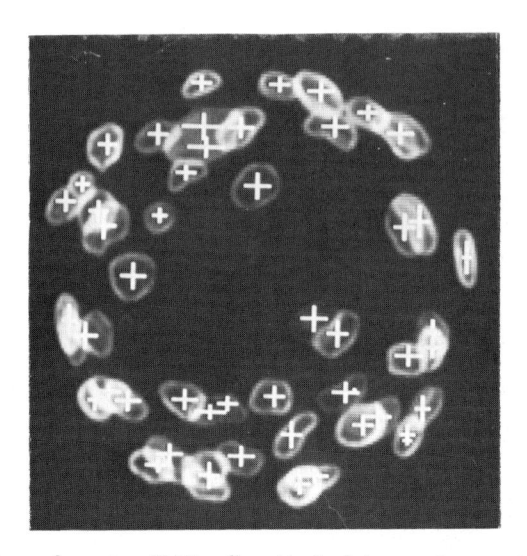

Figure 2.4. Chinese hamster V-79 cells attached to a microcarrier. Specimen area 220 × 220µm, image resolution 256 × 256 pixels, 80 sections. Objective: Zeiss Planapo 40/1.0. Excitation wavelength: 458nm. Specimen provided by E. Szolgay-Daniel, The Gustaf Werner Institute, Uppsala.

The number of cells generated from one single clone cell on the surface of a microcarrier were counted. Since the microcarrier is spherical, and the number of clones is usually larger than 50, it is difficult to count the colonies manually in an ordinary microscope. With PHOIBOS it is possible to section the specimen, visualise it, and count the number of cells with a computer [Brunnström, 1988]. The result of the computer calculation is shown as superimposed crosses on the image, Figure 2.4. This work was carried out in cooperation with the Department of Physical Biology, The Gustaf Werner Institute, Sweden [Szolgay-Daniel *et al.*, 1985].

II. System concepts

A picture showing a commercial scanner system based on the prototype developed at our institute can be seen in Figure 2.5. All scanner systems developed have been built on top of commercially available microscopes (Reichert, Zeiss Universal, Nikon). The microscope need not be modified in any way; the only prerequisite is a photo tube, which is a standard accessory to most high-level research microscopes. The light source is an argon ion laser which can produce wavelengths from 458 to 514nm. A smaller version of the laser can be mounted inside the scanner compartment, resulting in a very compact system.

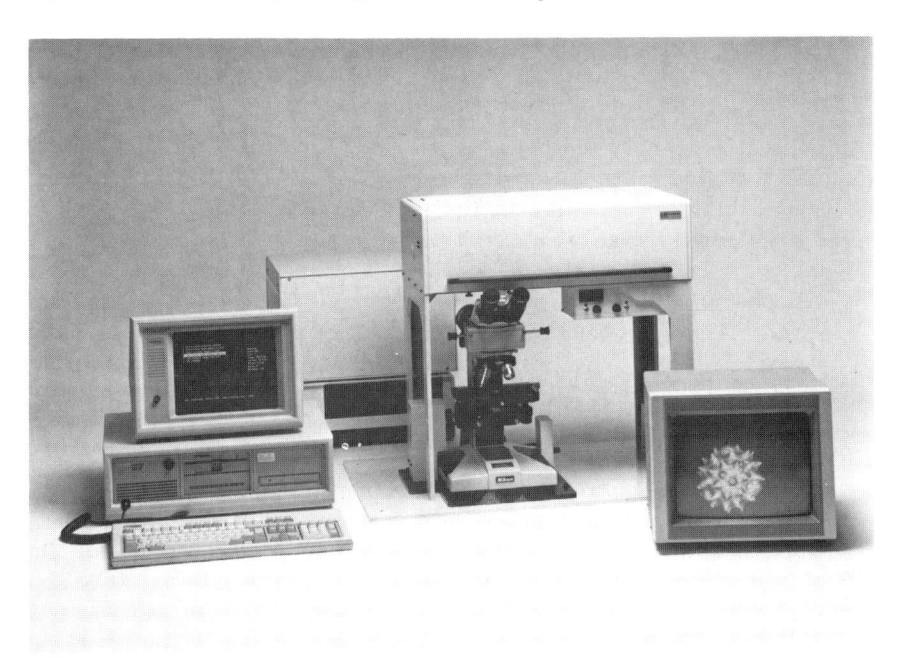

Figure 2.5. Photo of the microscope scanner system PHOIBOS. Host-computer, electronics cabinet, microscope with scanning unit and TV monitor are shown (left to right).

The opto-mechanical principle of the scanner is shown in Figure 2.6. The laser beam is reflected by a beam splitter and two scanning mirrors, one fast vibrating galvanometric mirror and one slow stepping motor-controlled mirror, down through the microscope. The focused laser beam illuminates the specimen and the fluorescent (or reflected) light returns through the microscope along the same raypath as the excitation light. The beam splitter transmits the fluorescent light, which is imaged onto the detector aperture, and the light is converted to an electrical signal by a photo-multiplier tube. A digital image of the specimen can consist of up to 1024×1024 pixels with 256 grey levels. One such image corresponds to an optical slice or section through the specimen in the focal plane of the microscope (confocal microscopy). A three-dimensional representation of the specimen is acquired by scanning consecutive sections with a change of focus setting between each section (optical serial sectioning). This three-dimensional representation can then be presented on a two-dimensional TV screen in several ways, which are presented in Section III. The pixel size is $0.1\mu m$ when using a 100X objective, while the optical lateral resolution is $0.25\mu m$. This means that the digitalisation of the image does not introduce any loss of information. The depth discrimination with the same objective (N.A.=1.3) is 1.0μm.

There are basically three ways in which the operator can interact with the instrument. The most direct way is by means of function buttons directly connected to the instrument. The functions performed include scanning to local image memory (256 × 256, 512 × 512 pixels), setting of some basic display memory parameters (normal, inverted, or pseudo-coloured image) and interactive control of focus and the object stage.

Another way to interact with the instrument is through the use of a computer. The operator then communicates with a program executed on a host-computer (a 32-bit mini- or micro-computer). The program responds to the operator's actions by transmitting commands to the instrument via a communication bus. The instrument then performs the appropriate action and sends any generated data (pictures, etc.) back to the host-computer via the communication bus. This method of interaction gives access to many more functions than via the function buttons. Scanning can be performed at arbitrary resolution, and the focus and object stage positions can be set to absolute values. The user interface is a menu-based program in which the user chooses the appropriate alternative and is prompted for further information to complete the command. Section series can be collected by a single command.

An instrument-independent method of interaction is post-processing of the collected data from the scanner. Data in the form of section series resides in the computer's mass-memory (disk or tape). This data can be processed in several ways, Sections III and IV. It is usually

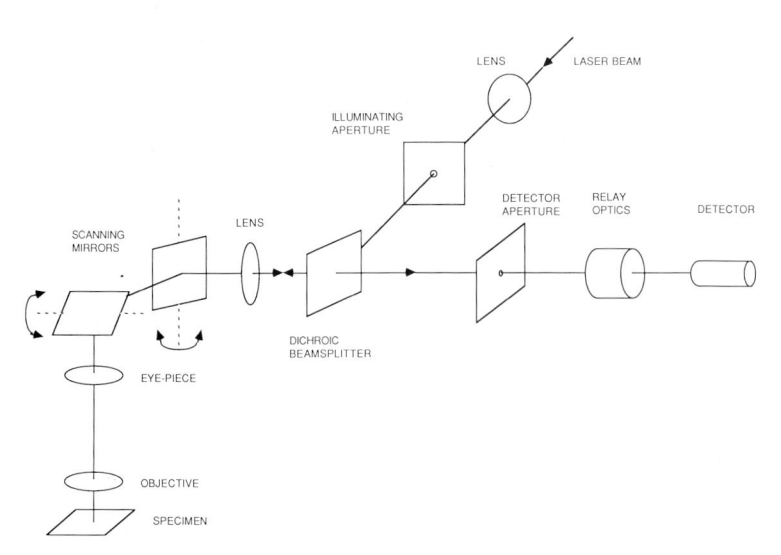

Figure 2.6. A schematic overview of the optical ray-path of PHOIBOS.

filtered and compressed, whereupon a two-dimensional projection of the three-dimensional object is calculated by one of several methods.

III. Display techniques

The primary output of an ordinary microscope are images of small objects. This is also true for other instruments such as electron microscopes and phase-contrast microscopes. A confocal laser microscope is not different in this respect; it also produces images. However, one important difference compared to ordinary light microscopes is its ability to produce images with very little out-of-focus information. This is in itself an improvement compared to ordinary light microscopy, but it can also be used to record a stack of images from different depths, thus forming a three-dimensional recording of the object. This section will focus on problems and solutions in the process of visualising, filtering and storing the information in three-dimensional stacks.

Some terminology of digital image techniques will be explained first. The basic elements that make up the grid of a digital image are called pixels. By analogy, the basic elements of a volume are termed voxels. Voxels and pixels have numerical values which are measures of the intensity in the corresponding points in the specimen.

III.1. Display principles

Confocal microscopes usually record images electronically and store them in a computer. This means that most systems use TV screens or plotters/printers as their output for displaying both individual sections and images revealing three-dimensional structure. There is a lack of good equipment for three-dimensional visualisation: only a few research-systems of various types have been presented. Among these are devices using fluorescence of gas by lasers [Verber, 1977], holograms [Benton, 1982], vari-focal mirrors [Harris and Camp, 1984], and rotating LEDs [Budinger, 1984].

The most common and readily available display equipment is a TV monitor, which can be used in a number of ways to enhance the three-dimensional impression. This can be done by producing stereo images on two TV screens with a stereo adapter in front of them, or merely by anaglyphic colouring. Rapid image-switching on one screen, synchronised with polarising equipment, is another way. A very powerful method is motion, where the three-dimensional impression is created by quickly displaying a succession of projections from different viewing directions of the same object. Motion is best used in the form of rotation, which means that the object is shown from different angles, giving the impression of a rotation of the object.

III.2. Look-through and depth-coded projections

For single images of three-dimensional objects there are a number of different ways to encode and display the information. The problem can be considered in terms of ways of condensing a four-dimensional distribution (x, y, z and intensity) to a three-dimensional distribution on the screen (x', y' and intensity). Two simple ways exist. One is to ignore one of the spatial coordinates and merely add the intensities along a ray through the object, Figure 2.7. This is sometimes referred to as look-through projections. The other way is to neglect the intensity information in the volume and let the intensity on the screen reflect the distance to the closest object point along the ray. This results in what are called depth-coded pictures.

Examples of look-through projections and depth-coded images can be found in Figures 2.1 and 2.8. Look-through projections reveal very little depth information in themselves, but in stereo pairs, as in Figure 2.1, or by a modelled rotation, they can give information about the entire structure in the volume. This is useful for many objects, but not for all. If the object is dense, i.e., the volume is filled with structures, the result can become confusing, as it is difficult to separate many overlaid structures.

Look-through projections can incorporate modelings of absorption of the rays passing through the object, or distance-dependent scaling of the intensities [Lenz, 1983]. When used on the original data, the absorption model can be considered as a way of getting increased depth resolution in the object, viewed from an arbitrary direction. It can also be used on filtered data like intensity gradients, Section IV.

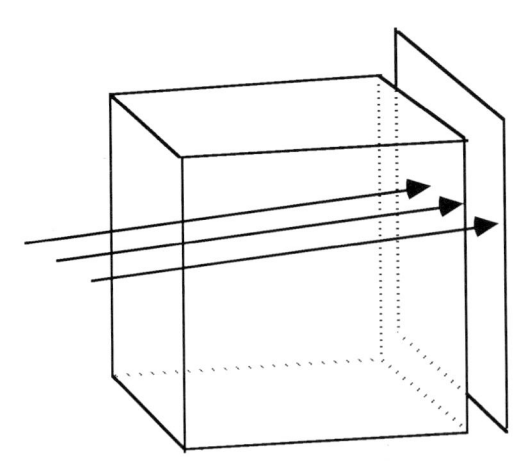

Figure 2.7. The intensities along a ray are summed and form the pixel value in a look-through projection.

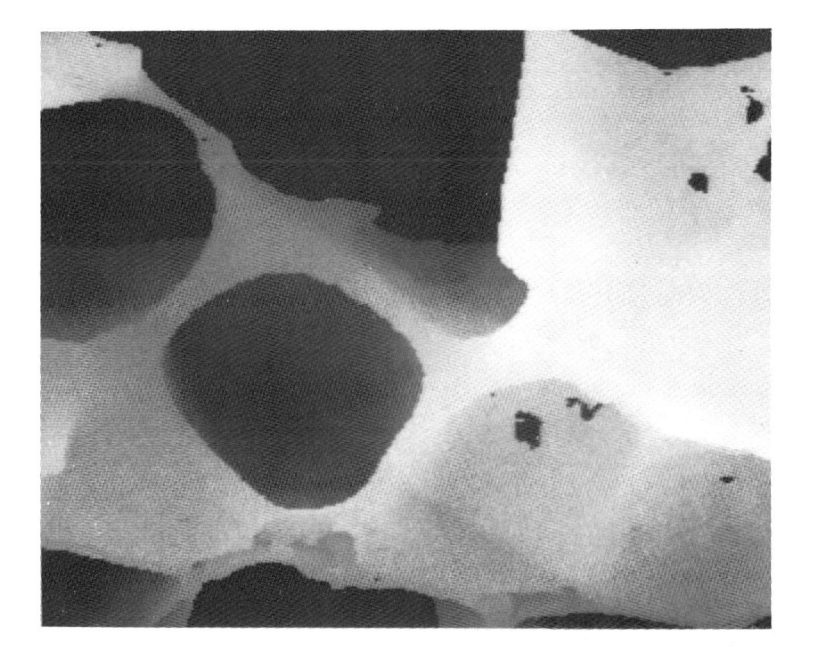

Figure 2.8. Depth-coded projection of volume.

In depth-coded pictures, the intensity of a pixel is normally inversely proportional to the distance from the viewer. This coding is based on the idea that brighter points are interpreted as being closer to the observer. Depth-coded projections can be used to give coarse depth information of different objects or parts of objects. One such example is the lung tissue in Figure 2.8, in which the relation between different planes can be clearly seen. The method requires a z-buffer (i.e., the depth image) in which the depth values are sorted so that the closest is always kept. This hides large parts of the object, with the result that small-scale structures in surfaces cannot be seen.

III.3. Surface shading

Shading by an illumination model is a good method of emphasising small-scale structures in surfaces. In this technique, the aim is to simulate a real-world illumination situation in which there might be several light sources which create specular and diffuse reflections from points on the object. The amount of reflected light in the direction of the viewer (the surface shade) depends on the direction of the surface normal. This can be modeled by diffuse and specular reflections. Diffuse surface models normally reveal more of the surface normal direction.

Figure 2.9. Lambert's law of reflection.

One common way to model the reflectance from a diffuse surface is to use Lambert's law:

$$I = I_0 \frac{\cos \theta}{r^2} \qquad (2.1)$$

where θ is given by Figure 2.9.

To account for intensity due to scattered light, an ambient light term of constant intensity can be added. Without it, shadowed areas become completely dark. The $\cos \theta$ term in equation (2.1) can be formulated with vector algebra instead, which is more illustrative for the computations:

$$I = I_{\text{amb}} + I_0 \frac{(\hat{r}.\hat{n})}{r^2} \qquad (2.2)$$

In this formula ambient light, I_{amb}, has been added to account for the scattered light. If the surface is made up of patches (small surface elements), interpolated shading values over the patches can be calculated with approximations by Gouraud (1971) or Phong (1975). The difference between the two is that the Gouraud method interpolates shading values, whereas the method by Phong interpolates normal directions with better results. It is possible to reduce equation (2.2) further by putting some restrictions on the illumination possibilities. If the distance to the object can be considered large compared to the extent of the object, the denominator becomes a constant. Furthermore, if the position of the light source is the same as the viewing position, the dot product is simplified, since the $\hat{r}$ direction then becomes $(0, 0, 1)$, which reduces the dot product calculation. In some implementations, the exponent of the dot product is assumed to be a variable, γ, which gives the surface different characteristics. Normally it varies between 1 and 2.

$$I = I_{\text{amb}} + I_1 (\hat{r}.\hat{n})^\gamma \qquad (2.3)$$

Examples of variations of the parameter γ can be seen in Figure 2.10.

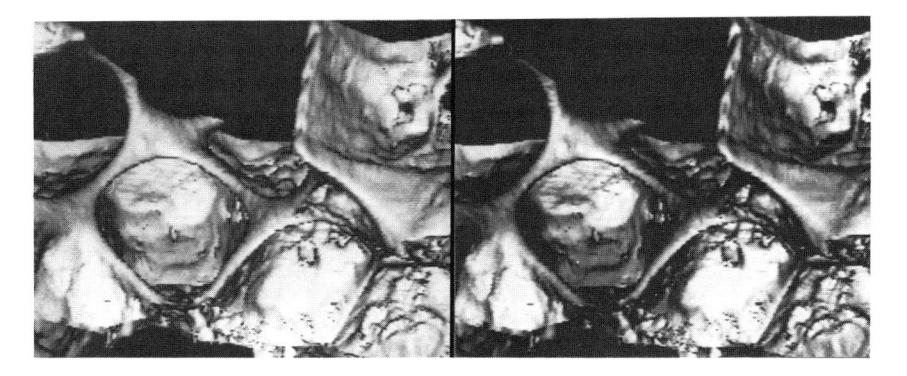

Figure 2.10. Surface shaded projections with different γ in shading formula. Left: $\gamma = 1$, right: $\gamma = 2$.

A fundamental problem when calculating surface shading is to find good ways of defining surfaces and their normal directions. Systems can generally be divided into two classes, graphical and voxel-based. Graphical systems are based on vectorial descriptions of the objects, whereas voxel-based systems keep the full three-dimensional description. Graphical systems normally work with surface patches which have three or more corners (polygon representation). These patches need not be planar, but can be splines or other curved surfaces. For data from three-dimensional imaging devices, it is possible to adapt a graphical method, but this requires the surface patches to be extracted from the original data in some automatic way. It is also possible to work directly with the data in voxel form.

There exists an algorithm by Gordon and Reynolds (1985) that calculates the surface normal from the depth map of an object. As can be seen in Figure 2.11, the principle is simple: the difference in depth values (z) between successive pixels is a measure of the slope. The surface normal is:

$$\bar{n} = \left(\frac{\partial z}{\partial x}, \frac{\partial z}{\partial y}, 1\right) \qquad (2.4)$$

The method is elegant because no three-dimensional filtering is necessary: only a simple depth-coded image is used as input.

This approach is simple but has drawbacks: it is difficult to get good estimates of small slopes, as the discrete depth values introduce large relative errors in the partial derivatives. Moreover, it is difficult to distinguish between steep slopes and discontinuities, a problem which is normally overcome by not shading steep slopes correctly or leaving black borders where one object occludes another.

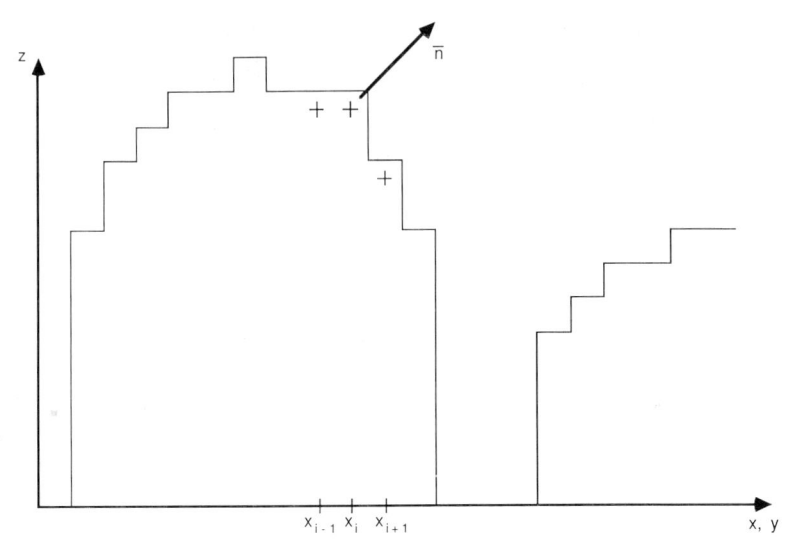

Figure 2.11. Surface normal calculation from depth map. The three + marked depth points are used to compute $\partial z(x_i)/\partial x = (x_{i+1} - x_{i-1})/2$.

A computationally more intensive approach is to calculate the true surface gradient of all surface elements in the volume. As this calculation can be made on the basis of the intensity in the volume, much better numerical behaviour can be anticipated. The surface normal is then:

$$\bar{n} = \left(\frac{\partial i}{\partial x}, \frac{\partial i}{\partial y}, \frac{\partial i}{\partial z} \right) \quad (2.5)$$

The different results of the two approaches can be seen in Figures 2.10 and 2.12.

Ray-tracing is a method that produces very realistic images of objects in the macroscopic world. In ray-tracing, fictive light rays are traced in the three-dimensional volume, where they are affected by the reflectance properties and refractive indices of different materials. The rays which reach the image plane form a realistic image of the three-dimensional volume. It is not clear how to use this method on a microscopic level, because different objects have different optical properties.

III.4. Rotations

There are a number of computational approaches toward performing modeled rotation. Many medical systems for Computer Aided Tomography are voxel-based, i.e., they store all of the volume information.

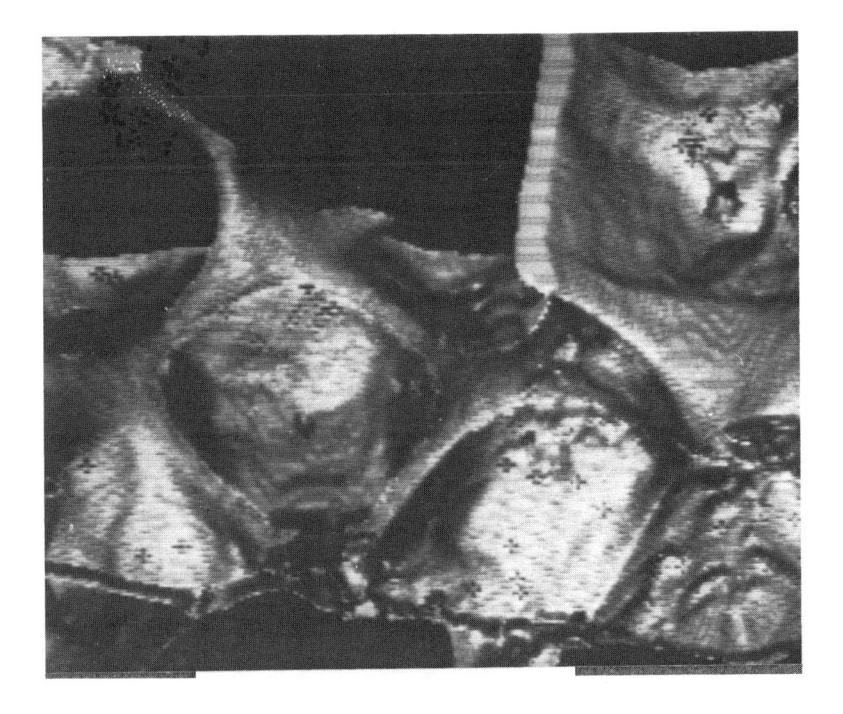

Figure 2.12. Surface shaded projection calculated from depth-coded image.

This is imperative for the analysis of computer tomography volumes, as it is often necessary to be able to threshold the volume at different levels of intensity, corresponding to different types of tissue. Systems that are able to perform real-time manipulation of volumes on this basis are mostly dealing with 256^3 resolution cubes. They are often of the parallel-processor type, dividing the total volume into smaller parts that can be treated with standard memories and processing elements. These systems are generally big, both physically and in terms of memory size: a 256^3 cube needs 16 MBytes of memory if 8 bits are sufficient for intensity information [Goldwasser and Reynolds, 1987; Lindskog, 1988].

Another approach is to use graphical displays, which are based on polygon models of the objects. In graphical systems, data is usually entered in coordinate form. If a graphical system is to be used, one of the problems is to extract a polygon representation from the three-dimensional voxel data produced by the CSLM instrument. Furthermore, very few of the commercial systems support modes other than surface-based, which means that the intensity distribution cannot be observed. Many research results and methods are implemented on standard computers, which of course cannot give real-time responses.

We have developed a method that is similar to the one used in graphical systems, but which avoids the polygon formation. By filtering the volume with certain criteria, we select interesting points and store these together with their coordinates in a vector description. The criteria for extracting the interesting points are various: simple thresholding, surface extraction and gradient filtering have been used. The vector description requires larger storage room for an individual point than does the raster method. On a standard computer, each voxel in the raster representation requires one byte if 8 bits are sufficient for intensity information (giving the range 0-255), whereas the vector description requires four or eight bytes depending on the size of the volume (256-coordinates can be stored in one byte each, larger volumes require two bytes for each coordinate). In most of the volumes we have worked with, large data reductions are achieved in any case. This is because of the lack of fluorescence in large parts of the specimens, which enables the system to discard these voxels in the vector storage. The advantages of this reduction go beyond storage purposes. The amount of data that has to go into the computations is also reduced. Since the vectors now contain the geometrical information, transformation to a new coordinate system can easily be achieved by matrix calculations. Hence, the position of a voxel in the new coordinate system is given by:

$$r' = A(r - r_0) + R_0 \qquad (2.6)$$

where A is the rotation matrix, r_0 is the rotation centre, and R_0 is a translation vector to fit the projection coordinates onto the screen. As r_0 and R_0 are constants, the calculations can be reduced to nine multiplications and nine additions per voxel. It is possible to make the vector description more compact [Forsgren, 1988], and at the same time reduce the computations, by taking into account the fact that the vectors are normally ordered: i.e., all voxels in the same plane are sequential, hence the z-value of a plane needs to be stored only once.

IV. Filtering of volumes

As the amount of data in a three-dimensional recording is very large, the filtering methods have to be chosen carefully, and are restricted in complexity if reasonable response times are required.

A common problem with recordings of biological specimens is low signal-to-noise ratio, owing to photon quantum noise [Carlsson, 1988; see Section VII.3]. The noise can of course be reduced by increasing the integration time, but there are practical limits on the length of time that can be used, Sections VI.4 and VII.3. Detection of surface elements by simple thresholding of the intensity produces poor results, especially if surface shading methods are to be used for display. Thus it is often advantageous to do some sort of smoothing operation in order to remove some of the high spatial-frequency noise. Smoothing is normally performed by a filtering or convolution operation on the image: the sum of the intensities of the neighbourhood voxels weighted by kernel values forms a new voxel value. Averaging of 3 × 3 × 3 voxels is an example of such filtering with uniform kernel weights. Many arguments in image analysis speak in favour of gaussian-like smoothing kernels. Gaussian kernels use the normal gauss-distribution to define the components of the filter mask, equation 2.7.

$$f(x,y,z) = \frac{1}{\sigma^3 2\pi\sqrt{2\pi}} \exp\left(-\frac{x^2+y^2+z^2}{\sigma^2}\right) \quad (2.7)$$

Different kernels are characterised by their σ, which is a measure of how wide they are and how much smoothing they perform. A most valuable feature of gaussian kernels, from a computational view, is that they are separable. This means that a three-dimensional kernel can be separated into three one-dimensional kernels. For three-dimensional processing, this can reduce the amount of computation considerably. A gaussian kernel of width N in all directions consists of N^3 coefficients, whereas a separable version of the same filter only holds $3N$ coefficients. The number of operations needed to perform a filtering is proportional to the number of coefficients. We have mostly used 3 × 3 × 3 or 5 × 5 × 5 gaussian kernels, as these give a good reduction of the high-frequency variations and are not too computing-intensive.

Amplification of rapid changes in intensity is sometimes wanted. This can be performed with a gradient filter that detects areas of high contrast, which are often borders between different types of materials or different objects. The gradient magnitude m is:

$$m = \sqrt{\left(\frac{\partial i}{\partial x}\right)^2 + \left(\frac{\partial i}{\partial y}\right)^2 + \left(\frac{\partial i}{\partial z}\right)^2} \quad (2.8)$$

The gradient filter decreases the already low signal-to-noise ratio mentioned above, since difference operators amplify high spatial frequencies. The gradient filter we have mostly used calculates the differences in a 3 × 3 × 3 cross of six pixels. As this is very sensitive to noise, a smoothing operation is often performed prior to gradient filtering. Alternatively, a gradient with a larger width for the difference calculations can be used.

V. Data storage

The method of storing the information from a confocal laser microscope is dependent on the computer systems used and on the particular implementation. Therefore, this section will only point out some very general issues of the topic and then describe our particular implementation.

The amount of data that a three-dimensional microscope produces can be very large. 256 × 256 resolution of individual sections, and a number of sections that varies between 30 and 300, adds up to between 1.9 and 18.8 Mbytes. By doubling the resolution, the amounts grow by a factor of four, to between 7.5 and 75 Mbytes. It is often not necessary to scan sections at the same distance as the spacing between consecutive pixels, as the resolution in depth for a CSLM is lower than the lateral.

In our implementation, which is made for the UNIX operating system, data is kept in a number of files. Each section, vector description, and projection image is an individual file. To combine these into stacks of sections or a set that simulates a rotation, and to store recording parameters, a number of information files are stored as well. All of these taken together constitute a volume, Figure 2.13, which may consist of several adjacent stacks, but from the same object within a specimen.

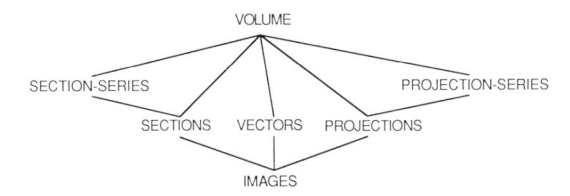

Figure 2.13. Volume data structure.

```
Volume name:       /img/po/example/
    Section number:              1
    Creation time:   Fri Jul 15 11:00:40 1988
    xsize:                       512
    ysize:                       512
    unsigned char picture
    type:                        original
    table position in x/y/z:     10 10 1
    start position in x/y:       0 0
    increment in x/y:            2 2
    laser wave length:           458
    beam splitter wave length:   550
    barrier filter wave length:  530
    aperture size:               50
    integration time:            25
    number of accumulations:     0
    photomultiplier voltage:     800
    magnification of objective:  40
    internal magnification:      1.25
    photometric offset:          0
```

Figure 2.14. Example of the information available for an original section from the microscope.

The information files also keep parameters of the recording situation or the calculations performed. An example of information attached to an original section scanned with the microscope is shown in Figure 2.14.

VI. Instrument control and computer requirements

The PHOIBOS confocal microscope has the following functions and requirements which should be handled by electronic hardware and software:
(1) Movable parts, such as deflecting mirrors.
(2) Image data acquisition.
(3) Image memory for storage and display.
(4) Post-processing of the three-dimensional data.

In the sections **Control processor selection** and **Communication bus selection** the demands on hardware and software for different configurations are described. **Computer considerations for data-processing** shows how processor and communication bus selection affect the choice of computer and its use in a confocal microscope system. The last section, **Instrument control**, shows which functions are needed to control the instrument, and where in a system different tasks should be performed.

VI.1. Control processor selection

By control processor, we mean a processor that is tightly coupled to the microscope scanner. The control processor must handle all time-dependent tasks without any delays. Collection of image data by direct memory access is a task that must run uninterrupted, or else data might be lost.

If the microscope is intended to be used only for viewing digitised sections (confocal or non-confocal) of the specimens, a simple control processor is sufficient, such as custom 8-bit microprocessor systems or IBM PCs, Figure 2.15, configuration A.

The digital representation of the microscope specimen and the unique properties of confocal microscopes (thin sections, fitting of consecutive sections, fitting of adjacent volumes, and three-dimensional

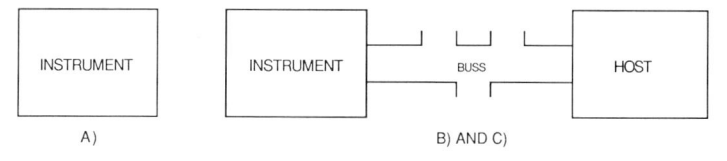

Figure 2.15. System configurations A, B, and C.

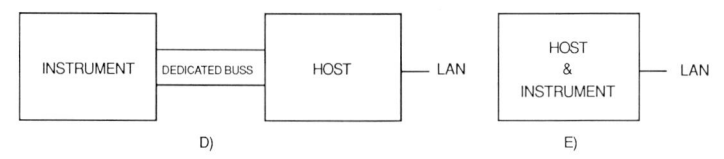

Figure 2.16. System configurations D and E.

photometry) [Franksson et al., 1987] offers new possibilities of storing images, visualising and computing numeric values of properties (length, volume etc.) of the specimen. To utilise these new possibilities, it is necessary to have a reasonable amount of computer power for computations, Section III, as well as user interaction.

One way to perform the computations and achieve user interaction is to have a simple (8-bit) control processor in the instrument and transfer the image data to a powerful main computer during scanning, Figure 2.15, configuration B. The first version of PHOIBOS is an example of this approach [Åslund et al., 1983]. A drawback is that some of the computing power of the main computer must be spent on low-level instrument control. Further on, more complicated functions, such as administering the image memory, burden the main computer which it is better suited to perform post-processing of image data.

If a 16-bit microprocessor system, or an IBM PC/AT, is used as control processor, the requirements on main computer interaction for instrument control decrease, Figure 2.15, configuration C. This concept is used for the second and third versions of PHOIBOS [Carlsson and Liljeborg, 1988].

The introduction of cheap 32-bit microprocessors, such as the Motorola 68020 and Intel 80386, has made it possible to acquire powerful 32-bit systems at a reasonable cost. A 32-bit system can handle basic control, as well as image-display memories, advanced user interaction, and post-processing of three-dimensional data.

A development of configuration C is to connect an Intel 80386 32-bit system closely to a custom 16-bit microprocessor system, Figure 2.16, configuration D. The current version of PHOIBOS is an example of this configuration, which allows fast communication with the instrument and good computational power.

A development of the 32-bit microcomputer systems is the ability to do both data-processing and instrument control with the same computer, Figure 2.16, configuration E. This opportunity is currently being investigated for PHOIBOS and may appear in the future.

VI.2. Communication bus selection

To communicate between different units in the system, the communication buses GPIB (IEEE 488) and Ethernet (IEEE 802.3, UDP/IP) are used.

For stand-alone systems using concept A, no bus is needed. If a main computer is involved, as in concepts B and C, a fast communication bus capable of handling several units is necessary. In concept D, a dedicated fast bus for two units is needed between the microcomputer and the instrument. Concept E requires no communication bus, as all functions are integrated in one system.

Since all instrument control and post-processing can be handled by the powerful 32-bit microprocessor, it is not necessary to attach a bus to other main computers in concepts D and E. However, the rapid development of network hardware and software offers interesting possibilities (shared file systems, etc.). If there are other computers in the neighbourhood of the instrument, it is easy to connect it to the computers with some type of network (LAN). This network is usually not sufficient for direct instrument control (collecting scanning data, etc.) since the instrument and the computers are loosely coupled, and there is no control of real-time events. In this way, however, users who wish to process data collected from PHOIBOS can easily access it from other computers.

GPIB offers transfer rates up to 1 Mbyte/s and Ethernet up to 1.25 Mbyte/s. As everyone who has worked with networking knows, the effective transfer rates are much lower: for GPIB about 200 kbyte/s and for Ethernet 10-50 kbyte/s (UDP/IP protocol).

Ethernet is well adapted for use in configuration C and E, to communicate with "main" computers. In configuration B, the demands on transfer of control information, owing to the restricted capabilities of the control processor, makes Ethernet less attractive. For configuration D, the transfer rates offered by Ethernet are too low, and a dedicated Ethernet channel is quite expensive.

GPIB is well suited for configuration B and D. Configuration D needs a high-speed cheap dedicated channel, and B needs a high-speed channel (even if multiple masters are a bit complicated on GPIB).

VI.3. Computer considerations for data-processing

An 8-bit microprocessor would be sufficient to control the instrument except for handling of images and image memory. A $256 \times 512 \times 8$ bits image consist of 65536 bytes (64 kbytes), which fill the whole address space of an 8-bit microprocessor. For both 8-bit and 16-bit microprocessors, data volumes larger than 64 kbyte present problems (16 bits can form an address within 64 kbyte maximum), which force hardware and/or software developers to use different types of memory segmentation. This limitation becomes very severe when post-processing of three-dimensional data is considered. Programming becomes complex and somewhat slow because of this memory segmentation. Often the maximum memory size including the segmentation is limited to one megabyte, which is not sufficient for efficient calculations on three-dimensional data. Control and simple data management could be handled by a 16-bit microprocessor. Advanced data management, extensive user interaction, and data processing demand the power of at least a 32-bit microprocessor.

A typical recording and analysis session with the PHOIBOS instrument may produce some 60 optical sections, each consisting of 512 × 512 pixels. A smoothing and interpolating operation creates more sections in order to match lateral resolution with depth increment between successive sections. In this way the digital volume elements (voxels) become cubic, and correct geometry of the specimen is maintained. Vector descriptions are also generated from the voxel data and used to generate projections. All in all, the total data may amount to about 40 Mbytes. For convenient work conditions, it is necessary to be able to contain several volumes at the same time in the computer. With these requirements, a minimum of 100 Mbytes free disk space is needed, and it is very advantageous to have more. It is important to have a backup station and media that can handle the same amounts of data, with a backup time that is at least as fast as the scanning.

Frame buffers (image-display memories), which store and display image data, are necessary to visualise the acquired and processed digital image data. 8-bit pixels can contain values between 0 and 255, which an image memory translates to one red, one green, and one blue intensity on a TV monitor. In this way 16.8 million colour combinations can be assigned to each pixel value (i.e. pseudo-colours). An image memory handled by the control processor is very useful, since the user can see the result of the scanning immediately.

In a computer system the user/users and the hardware resources, such as hard disks, are administered by software: the operating system (OS). A common and primitive single-user and single-process OS is MS-DOS. On an IBM-PC-type system running MS-DOS, there is no problem in controlling real time processes such as moving stepping motors and capturing image data from the laser beam scanning the microscope specimen. Multi-user and multi-process systems, such as UNIX, have several processes which share the same hardware and software. This means that the programmer does not have complete control over the system timing, and real time control is hard to achieve. There are hardware and software extensions to UNIX systems which improve real time performance, but to control custom hardware in realtime is still rather complex in a multi-process environment.

In the development of system concepts D and E, these problems have to be solved. In our opinion a powerful 32-bit system should serve several users and several processes, thus allowing users to share resources such as printers and disks. Different computer systems should run the same OS, since the users would then have the same environment on all computer systems. Also, a program written on one computer could easily be moved to another type of computer. The only system today that can accomplish this is UNIX, which runs on desk-top computers as well as supercomputers. The problems of real-time control can be solved in hardware, or with intelligent co-processor boards which have their own microprocessors.

The time-critical parts of the PHOIBOS instrument control are stepping motor control and data collection. Stepping motor control can be solved with careful device driver programming (the interface between the user's processes and the hardware/software of the system) and some extra hardware, such as counters. Data collection demands a buffer memory into which the collected data is transferred immediately, or a frame buffer (image memory) with an image data input which accepts analogue or digital image data without any delays caused by software overheads.

VI.4. Instrument control

In order to have a well-structured user interface, the instrument control should be partitioned into simple basic functions. Typically, some low-level functions are to move the object table a number of stepping motor steps in the x-direction or setting up the scanning mirrors in their start positions. This partitioning enables the programmer to change the user interface of the instrument without changing the low-level instrument software and hardware. An instrument with a less powerful control processor forces the main computer to use some amount of computer power for instrument control and user interaction.

The basic functions in the PHOIBOS instrument are:
(1) Movable parts demanding real-time control.
    (a) Object table movements in two perpendicular directions of the specimen.
    (b) Focus motor.
    (c) Slow scanning mirror.
    (d) Filters.
    (e) Dichroic beam splitters.
(2) Image data acquisition, demanding sustained high data rate and uninterrupted data flow.
    (a) Collect pixel data.
    (b) Collect pixel data and add to previous pixel data.
(3) Image memory for storage and display demanding large address space.

(a) Read and write rectangular blocks of pixels.
(b) Edit pixels.
(c) Text generation.
(d) Graphics.
(e) Cursor.

Some more complex functions are also performed by the microprocessor directly controlling the instrument. A section can be scanned in the x and z directions, obtaining an image of a vertical slice of the specimen. In the x-z image the user can find the interesting area, mark it with a cursor, and an ordinary x-y image can be registered at the marked z (i.e., focal) depth.

If the specimen is stained with chemicals which give fluorescence in several different wavelengths at the same time, multiple detectors can be used to detect light in different wavelengths simultaneously. In this case, image data is coming from several sources and has to be sorted according to which detector generated the data. The images from the different wavelengths could be sorted by software after all the image data has been collected in the (image) memory. The image would not appear correctly until after the sorting is finished. Special hardware could perform the sorting in real time during the scanning, and in this case the images would be correctly displayed immediately.

If a large object, such as a nerve cell, Section I, is studied, a stepping motor-controlled object table can be used to fit adjacent stacks of serial sections together to one large data volume. It is important that the mechanical and electronic precision of the object table is higher than the smallest pixel size of the instrument. This is also necessary for those microscope scanners that perform the scanning with the object stage. Much work is eliminated if the stacks are aligned pixel by pixel when the registration is taking place. It is difficult and time-consuming to fit the digitised stacks together by computer methods. The optical image distortion must be taken into account if a good fit of the adjacent stacks is to be achieved.

High scanning speed is important in specimens which change their properties over time, or if many objects are to be processed in a limited time. The sensitivity of the PHOIBOS instrument is limited by the number of photons reaching the detector for the duration of one pixel. As a consequence, the noise in the photometric signal results from the quantum nature of the illuminating light. This means that increased scanning speed results in proportionally decreased image data values and increased noise, Section VII.3. If the same image quality is required for increased scanning speed, several images must be recorded and accumulated, or the illuminating power must be increased. Because of fading and saturation of the fluorescent stain, it is often not possible to get more fluorescence when the illuminating power is increased.

The maximum scanning speed is given by the required minimum image data quality. In the PHOIBOS instrument it takes about 6s to collect a 256 × 256 image, which is a compromise between scanning speed, illuminating power, and image quality. This compromise means that a useful image of a fluorescent specimen can be obtained with one registration and reasonable illuminating power.

VII. Electronics

VII.1. Microprocessor system

The PHOIBOS scanner is controlled by an Intel 80186 16-bit microprocessor. The microprocessor system consists of several boards conforming to the Multibus I standard. There are five boards used in the system, a microprocessor board with an Ethernet interface, an image memory and display board, a parallel I/O board, a stepping motor control board, and an IEEE-488 bus interface. These boards reside in a Multibus I rack which is integrated with the rest of PHOIBOS electronics (analogue-to-digital (A/D) conversion boards, stepping motor drive boards and power supplies).

VII.2. Position encoding of the scanning mirror

A vibrating mirror scans the laser beam over the specimen. The mirror is mounted on a moving iron galvanometer which is driven by an analogue signal. The galvanometer has an integrated position sensor which produces an analogue position signal. This signal is proportional to the angular excursion of the mirror.

In order to obtain correct geometric information of the specimen, the photomultiplier signal, Sections II and VII.3, must be A/D converted at specific intervals during the scan-line. There are several ways to generate these intervals: one of the simplest is to generate a burst of pulses of equal length starting at some predetermined threshold of the position-signal. This method assumes that the motion of the mirror during the scan-line is linear in time within acceptable error limits. This method is sensitive to drift in the servo-amplifier driving the galvanometer, and it is difficult to achieve a linear scan in time due to ringing and/or damping in the servo-loop controlling the galvanometer, Figure 2.17. This method has not been used in the instrument.

Another method is to divert part of the laser-light reflected by the galvanometer mirror, and let it pass through a transmission grating. The light is then focused onto a detector which generates pulses corresponding to each line in the grating. A phase-locked loop is used

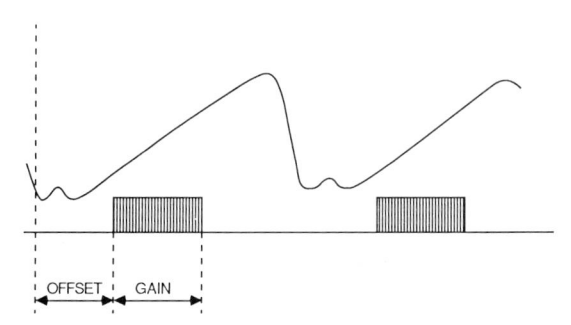

Figure 2.17. Analogue position signal and pixel clock burst. The position signal from the galvanometer is shown at the top. Underneath is shown the burst of digital pulses generated by the digital position encoder. These pulses are used to start the A/D conversions of pixels along the scan-lines. The specimen is scanned unidirectionally and is controlled by a sawtooth-shaped signal.

to increase the frequency of the pulses to achieve the desired number of samples per scan-line. This method compensates both for irregularities in scanner motion and tangent error in flat-field scanning applications. It has not been used in this instrument because of the loss of laser excitation light and fluorescence light resulting from introducing an extra beam-splitter in the ray-path, Figure 2.18 [Tweed, 1985].

The encoding method used in this instrument is based on the analogue position signal coming from the transducer in the galvanometer. The intervals between consecutive samples on one scan-line are given by the analogue position signal via a digital position encoder developed at the institute [Liljeborg, 1988]. This method does not give as high accuracy as the grating method mentioned above, e.g., the tangential error due to flat-field scanning is still present. The encoding is done by comparing the analogue position signal with a digitally-generated signal, Figure 2.19. When the analogue position signal exceeds the digitally-generated signal, a counter is incremented and its accumulated contents are converted to analogue form, which is compared to the analogue position signal. The digitally-generated signal then becomes larger than the analogue position signal and the comparator is reset. This process is repeated until the counter reaches its maximum value. Then the counter is reset and awaits the next line-scan.

The galvanometer is driven by a sawtooth-shaped signal and the scanning is done unidirectionally. A first attempt was made to scan bidirectionally, but the hysteresis in the position transducer was too large to obtain the desired resolution. The sawtooth-shaped movement of the mirror minimises the loss of duty cycle due to unidirectional scanning.

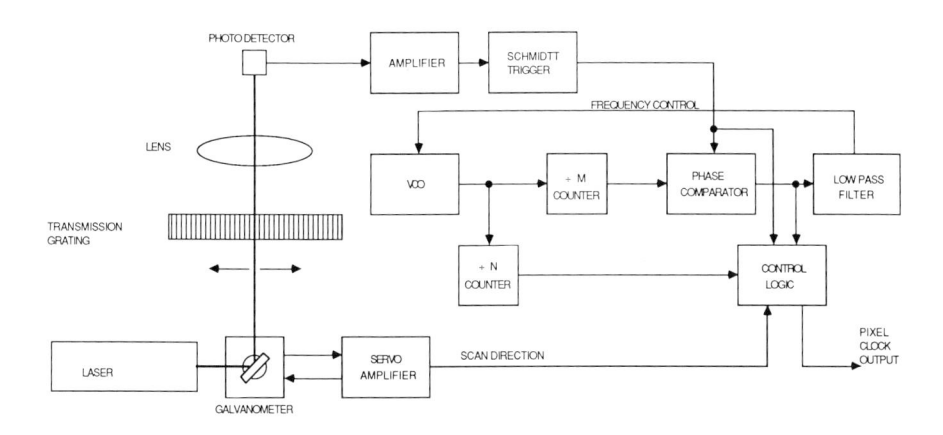

Software and Electronics for a Digital 3-D Microscope   89

Figure 2.18. Transmission grating used for position encoding. A phase-locked loop is used to increase the frequency of the pixel clock derived from the grating.

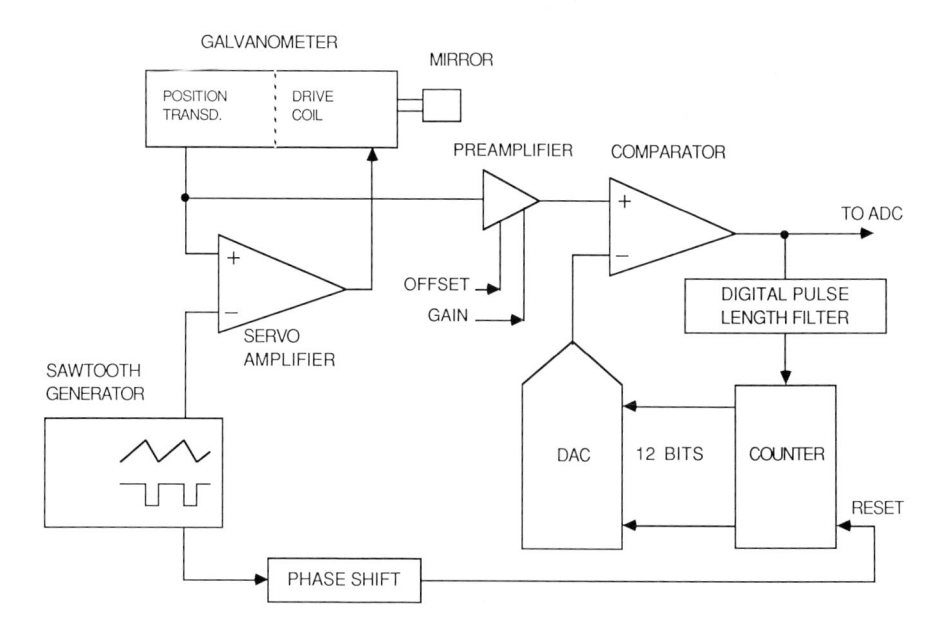

Figure 2.19. Schematic of digital position encoder. See text for a description of the principle of operation.

VII.3. Detector signal enhancement

The reflected or fluorescent light from the specimen is projected via the scanning mirrors and the detector aperture onto a photomultiplier tube, Figure 2.6. The output signal of the PM-tube is A/D converted to 8-bit data yielding 256 grey-levels. Prior to the A/D conversion the signal is integrated by a hardware integrator, Figure 2.20. In this way the sensitivity is increased and the noise is suppressed. The integration time can be controlled from the host computer, but the maximum time is limited by the time between two consecutive pixels on a scan-line. The maximum time is thus dependent on which digital resolution is used. E.g. a 256×256 pixel image scanned at full resolution would give a maximum integration time of $\sim 5\mu s$ while a 256×256 pixel image scanned over the entire field of view would give $\sim 20\mu s$ maximum integration time.

The instrument signal-to-noise ratio is limited by the number of photons reaching the detector [Carlsson, 1988]. By assuming a reasonable signal-to-noise ratio (100), the necessary power of excitation light to achieve this ratio was roughly calculated. This power proved to be larger than what is actually used, which implies that it is hard to achieve the assumed signal-to-noise ratio because of the difficulty of getting a sufficient amount of photons through the system to the detector.

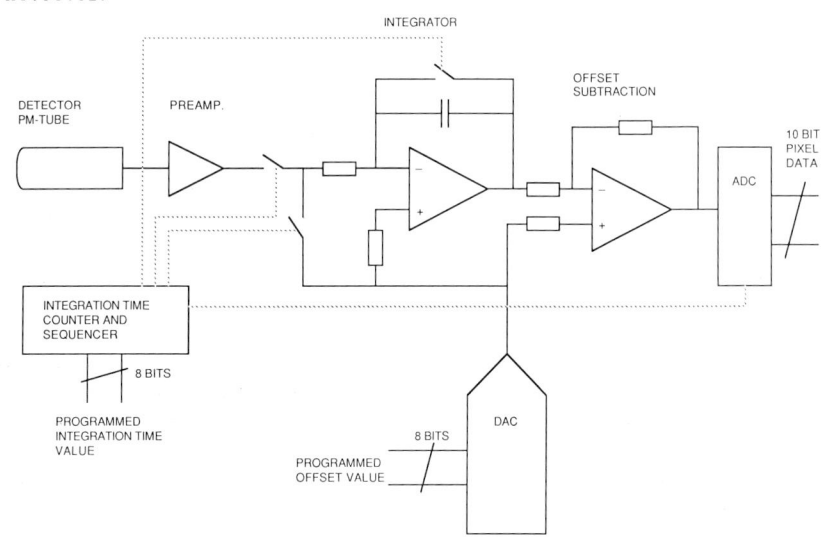

Figure 2.20. The hardware integrator is controlled by analogue switches opened and closed in a predefined sequence. The control processor programs the integration time, which alters the sequence timing of the switches. There is also a programmable offset that can be used to remove unwanted background fluorescence in the images.

Another independent way to find out which part of the detection system is limiting the signal-to-noise ratio is to measure the noise in the digitised signal. Noise arising from photons is proportional to the square-root of the signal intensity. Several measurements both of laser-excited fluorescent light and transmitted light from a tungsten lamp were made at different signal levels. All measurements indicate the photon-limited relationship between noise and signal. This shows that the signal-to-noise ratio is limited by the number of photons reaching the detector and not by the following signal processing circuitry.

References

Åslund, N., Carlsson, K., Liljeborg, A. and Majlöf, L., (1983). PHOIBOS, a microscope scanner designed for micro-fluorometric applications, using laser induced fluorescence. Proc. 3rd Scandinavian Conference on Image Analysis, Studentlitteratur, Lund, Sweden. 338-343.

Benton, S. A., (1982). Survey of holographic stereograms. Proc. S.P.I.E., **367**, 15-19.

Brunnström, K., (1988). Counting cells on spherical micro-carriers using three-dimensional image-processing techniques. TRITA-FYS-4014, The Royal Institute of Technology, Sweden.

Budinger, T. F., (1984). An analysis of three-dimensional display strategies. Proc. S.P.I.E., **507**, 2-8.

Carlsson, K., (1988). Ph.D. dissertation. Physics IV, The Royal Inst. of Technology.

Carlsson, K. and Liljeborg, A., (1988). A confocal laser microscope for digital recording of optical serial sections. J. Microsc. To be published.

Forsgren, P-O., (1988). Architecture of a simple real-time display device for three-dimensional data. TRITA-FYS-4012, KTH, Stockholm.

Franksson, O., Liljeborg, A., Carlsson, K. and Forsgren, P-O., (1987). Confocal laser microscope scanning applied to three-dimensional studies of biological specimens. Proc. S.P.I.E., **809**, 124-129.

Goldwasser, S. M. and Reynolds, R. A., (1987). Real-time display and manipulation of three-dimensional medical objects: the voxel processor architecture. Comp. Vis. Graph. and Image Proc., **39**, 1-27.

Gouraud, H., (1971). Continuous shading of curved surfaces. I.E.E.E. Trans. Comput., **C20**, 623-628.

Harris, L. D. and Camp, J. J., (1984). Display and analysis of tomographic volumetric images utilizing a vari-focal mirror. Proc. S.P.I.E., **507**, 38-45.

Lenz, R., (1986). Reconstruction, processing and display of 3D-images. Dissert. no. 151, Linköping University, Sweden.

Liljeborg, A., (1988). Digital position encoding of galvanometer scanner in a laser microscope. Opt. Eng., **27** (9).

Lindskog, B., (1988). PICAP3, An SIMD architecture for multi-dimensional signal processing. Dissert. no. 176, Linköping University, Sweden.

Phong, B-T., (1975). Illumination for computer generated pictures. Comm. ACM 18, No. 16, 311-317.

Szolgay, D., Larsson, B., Brunnström, K. and Forsgren, P-O., (1985). A new method for colony formation of Chinese hamster cells on microcarriers. Proc. 2nd Int. Symp. on Neutron Capture Therapy, Tokyo University.

Tweed, D. G., (1985). Resonant scanner linearization techniques. Opt. Eng., **24**, 1018-1022.

Verber, C. M., (1977). Present and potential capabilities of three-dimensional displays using sequential excitation of fluorescence. Proc. S.P.I.E., **120**, 62-67.

Wallén, P., Carlsson, K., Liljeborg A. and Grillner, S., (1988). Three-dimensional reconstruction of neurons in the lamprey spinal cord in whole-mount, using a confocal laser scanning microscope. J. Neuroscience Methods, **24**, 91-100.

3. Optical Aspects of Confocal Microscopy

T. WILSON

I. Introduction

The key difference between the confocal scanning microscope and the conventional instrument is that the confocal uses a point detector rather than a large-area detector. In all other respects the optical systems are identical. As all the considerable advantages of confocal microscopy follow directly from the use of a point detector, it is important to know how small a practical detector must be in order to approximate, as closely as possible, a true point detector. It is usual to form the "point" detector by placing a circular pinhole in front of a suitable photodetector. It is reasonable, in general, to try to use as small a detector as possible, although in many applications we may not have this freedom because of signal-to-noise problems. This is particularly true, for example, in the imaging of weakly fluorescing biological tissue. It is therefore appropriate to discuss how the strength of the optical sectioning property deteriorates as the detector becomes larger. We shall consider both brightfield imaging and fluorescence imaging. It is important to realise that the image formation properties of these two modes are *completely* different, and that it is thus potentially very dangerous to apply design criteria intended for one mode to the other.

As we have seen in Chapter 1, the ideal confocal microscope is a coherent imaging system with an effective point spread function, h_{eff}, given by the product of the point spread functions of the two imaging lenses. This means that both lenses contribute equally to the imaging, and also, sadly, that the aberrations of both lenses are equally important. This is the price we pay for the improved three-dimensional imaging of confocal microscopy – both lenses must be very well corrected if the instrument is to operate at its optimum.

The form of the effective point spread function gives us the freedom to modify the imaging of a confocal microscope by introducing pupil plane filters into the objective lenses. This permits us to alter the form of h_{eff} in particular ways. Pupil plane filters will be discussed in detail in Chapter 5, but we will discuss two simple examples later in this chapter. We will conclude the chapter with a discussion of image formation in tandem scanning, or direct view, confocal microscopes, and we will compare their imaging with that of laser-based systems before finally reviewing the various methods of scanning which can be employed.

II. Brightfield image formation with finite-sized detectors

II.1. The point object and the planar reflector

In order to simplify our discussion we will consider only point objects and planar objects. In doing so, we will be able to highlight the role of the finite size of the detector in the imaging. However, we must always keep in mind the fact that our remarks will apply strictly only to these specific objects.

We begin by considering brightfield imaging in a scanning microscope with a finite-sized detector of intensity sensitivity D. We can easily use the methods of Fourier optics outlined in Chapter 1 to permit us to write the image of a point object in such a microscope system as [Sheppard and Wilson, 1978]:

$$I = |h_1|^2 \left\{ |h_2|^2 \otimes D \right\} \qquad (3.1)$$

where h_1 and h_2 are the point spread functions of the two imaging lenses, and the symbol $\otimes$ again denotes the convolution operation. The true confocal case corresponds to D being represented mathematically by a Dirac delta function, and hence $I = |h_1 h_2|^2$. The conventional case requires D to be constant, which leads to $I = |h_1|^2$. If we simply consider the in-focus image, we have the well-known point images of:

$$I_{CONV}(v) = \left[\frac{2J_1(v)}{v} \right]^2 \qquad (3.2)$$

in the conventional case and:

$$I_{CONF}(v) = \left[\frac{2J_1(v)}{v} \right]^4 \qquad (3.3)$$

in the confocal case where v is the normalised optical co-ordinate given by:

$$v = \frac{2\pi}{\lambda} r \sin \alpha \qquad (3.4)$$

It is intuitively clear that as the detector size increases, the image intensity will broaden gradually from the confocal to the conventional

Optical Aspects of Confocal Microscopy

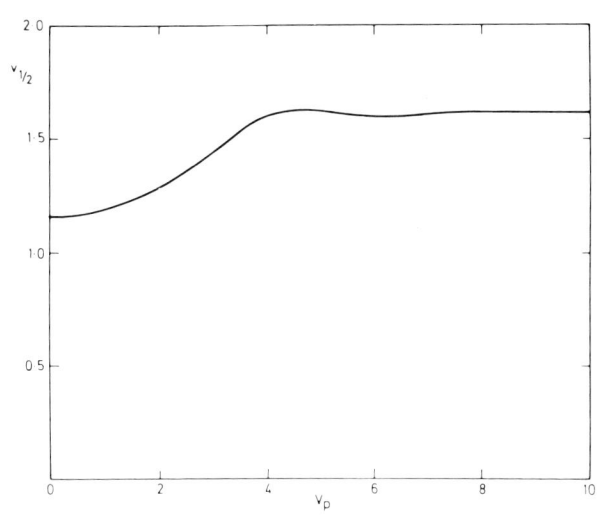

Figure 3.1. The half-width of the image of a single point object as a function of detector pinhole size.

response. If we adopt as a measure of the lateral resolution the value of v at which the image intensity falls to one-half of its value at $v = 0$, we can plot this value, $v_{\frac{1}{2}}$, as a function of normalised detector radius, v_p [Wilson and Carlini, 1987]. Figure 3.1 shows that the $v_{\frac{1}{2}}$ resolution is relatively insensitive to v_p for v_p less than about 0.5, which means that $v_p \sim 0.5$ would be a reasonable choice of pinhole size for true confocal operation that also maximises the detected signal. As an example, for a 40X, 0.6-numerical-aperture lens, this criterion suggests that a 6.7μm-diameter pinhole is needed when red, 0.6328μm wavelength, helium-neon-laser light is used. It is clear that the magnification of the lens is an important factor in determining the degree to which a system is confocal. This has important implications in beam-scanning systems, in which the objective lens must be changed in order to achieve a full range of magnification. Changing objectives can, therefore, lead to change in the extent to which the imaging is truly confocal. Object-scanning systems do not suffer from this problem, as they can be made to cover the full magnification range without the need to change lenses.

We now move on to consider the effect of detector pinhole size on the axial resolution, or depth-discrimination property, by considering the variation in detected signal as a perfect reflector is moved axially through focus. We recall that for a true confocal microscope:

$$I(u) = \left[\frac{\sin(u/2)}{u/2}\right]^2 \qquad (3.5)$$

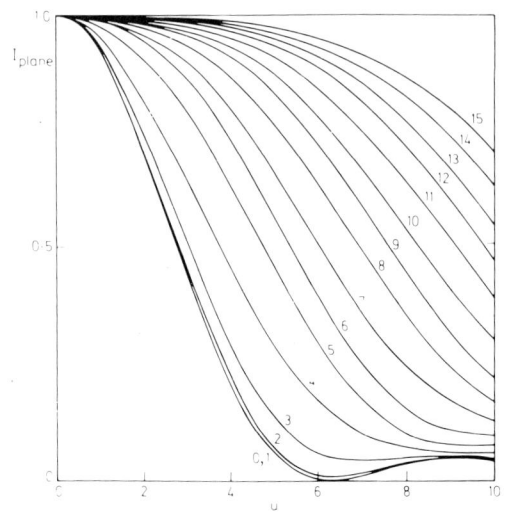

Figure 3.2. The variation of $I(u)$ against u for a variety of values of detector pinhole size, v_p.

where the normalised axial coordinate, u, is related to z by:

$$u = \frac{8\pi}{\lambda} z \sin^2\left(\frac{\alpha}{2}\right) \tag{3.6}$$

whereas in the conventional case, $I(u) = constant$. In the case of a finite-sized circular detector $D(v)$ we can write [Wilson and Carlini, 1987]:

$$I(u) = \int |h(2u,v)|^2 D(v)\, v\, dv \tag{3.7}$$

This is plotted in Figure 3.2 for a variety of values of v_p, the normalised radius of the detector, showing the increasing reduction in the strength of the sectioning as the detector becomes larger in size. It is clear from this figure and Figure 3.1 that the depth-discrimination property is considerably less sensitive to pinhole size than is the lateral resolution. The curves corresponding to $v_p = 0$ and $v_p = 1$ are indistinguishable on the scale of Figure 3.2 This is further emphasised in Figure 3.3, where we plot the half-width, $u_{\frac{1}{2}}$, of the curves of Figure 3.2 as a function of v_p. We see that for v_p less than about 2.5, the half-width, and hence the depth discrimination, remains constant.

We emphasise that we are measuring v_p relative to the object plane, and that to obtain the actual value of pinhole size we must multiply v_p by the magnification of the lens system between the object and the detector. This means that if we require $v_p \leq 2.5$ for true confocal sectioning, we must have:

$$\frac{M}{\sin\alpha} \geq \left(\frac{\pi}{2.5}\right)\frac{d}{\lambda} \tag{3.8}$$

where M is the magnification and d is the actual diameter of the pinhole. This has implications for beam-scanning systems where it is necessary to change objective lenses in order to cover the entire range of magnification. It is clear from equation (3.8) that by changing the magnification we also change the degree of "confocality".

It is worth pointing out that the geometrical-optics approximation, which requires the detector to collect all the signal until $2u = v_p$, and therefore to follow an inverse law $(v_p/2u)^2$, predicts that $u_{\frac{1}{2}} \sim v_p/\sqrt{2}$. It is clear from Figure 3.3 that, although the values of v_p and u that we have considered are large, they are not large enough for geometrical optics to apply reliably.

Figure 3.4 confirms these theoretical predictions [Wilson and Carlini, 1987] by showing the responses obtained by scanning a mirror through focus in a scanning microscope using red light (0.6328μm wavelength) and an objective of nominal numerical aperature 0.5 and 32X magnification. In particular, we see that the curves for 5 and 10μm-diameter pinholes are essentially indistinguishable, whereas the curves become distinctly broader by the time a 150μm-diameter aperture is used. However, we can expect that a degree of depth discrimination will still be present in instruments using such large pinholes.

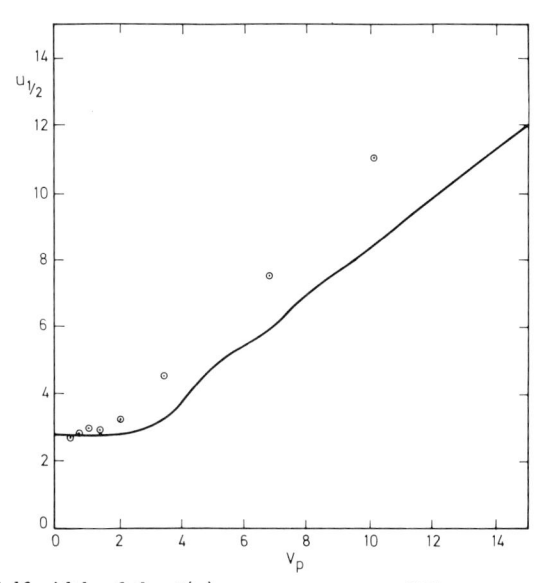

Figure 3.3. Half-width of the $I(u)$ versus u curves of Figure 3.2 as a function of normalised detector radius, v_p. The theoretical curve is shown as the full line and the experimental results as the dots for a numerical aperture of 0.44.

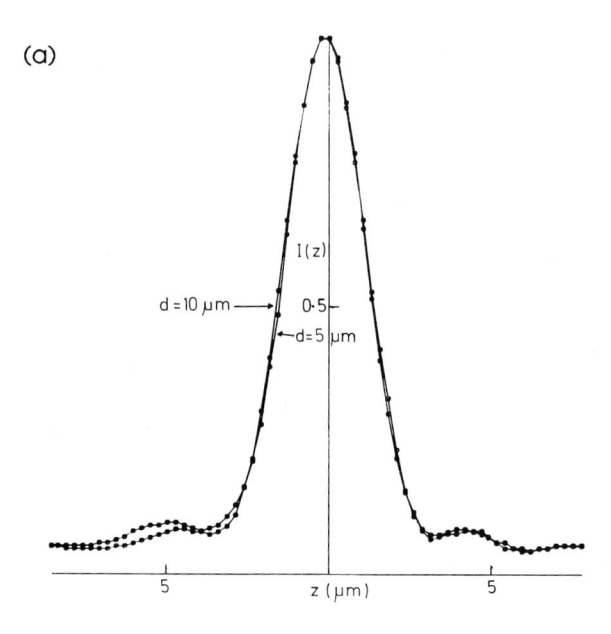

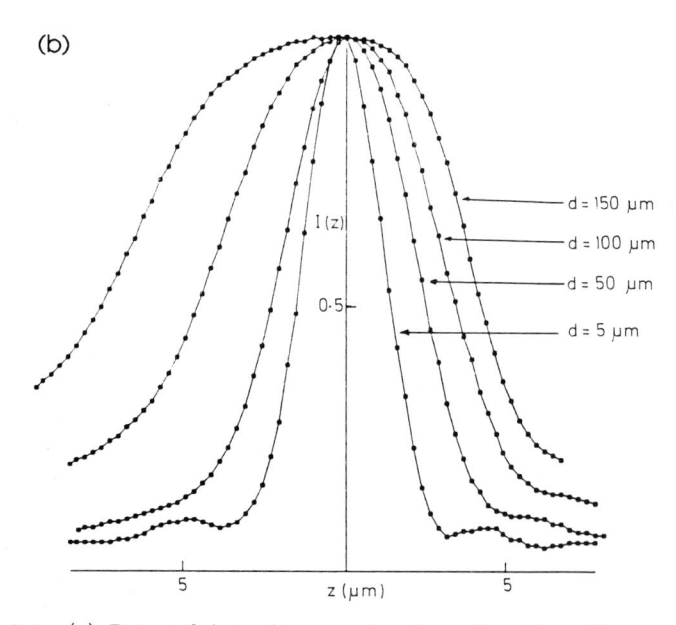

Figure 3.4. (a) Detected intensity as a function of axial position, as a plane mirror is scanned axially through focus in a scanning microscope with 0.6328μm radiation and a 32X, 0.5 numerical aperture objective lens. The curves are for detector aperture diameters of 5μm and 10μm. (b) As (a), but for a much larger range of aperture sizes.

We can also see that although the curves of Figure 3.4 show the same trends as the theoretical curves, they do not agree perfectly. Figure 3.3 compares the half-widths of these curves with theory. In particular, the experimental results are somewhat asymmetric, and the side lobes do not go down to zero. There are a variety of reasons for this. The theory we have used to obtain our theoretical predictions is essentially scalar and paraxial. A fuller theory, taking into account high angle effects [Sheppard and Wilson, 1981a], shows that the side lobes should not go down to zero for high values of numerical aperture. However, they are predicted to be a lot smaller than we have measured here. The most likely explanation for the shape of the curves is that the pupil function is not perfectly uniform across the aperture. The simple theory assumes that the pupil function is given by a circ function. It has recently been shown by an interference technique [Matthews, 1987] and by curve fitting [Corle et al., 1986] that this is not usually the case.

The asymmetry in the curves most likely results from the presence of aberration and shading in the objectives. As we will see later, this method of quantifying the depth discrimination is *very* sensitive to the presence of any pupil function aberration or other deficiency. The shading, or falling-off in transmissivity toward the edge, of the objectives may have a variety of causes, including Fresnel loss at each of the curved surfaces of the glass elements comprising the lens. This permits us to *crudely* quantify this effect by a numerical aperture somewhat lower than the nominal value. We will return to these points later, and again in Chapter 15, where an interference technique will be described which, under some circumstances, permits the pupil function of the lens to be measured directly.

II.2. Images with a finite-sized detector

We now move on to consider the images of a specific object, in order to see how the finite size of the detector affects the resolution and depth-discrimination property. We choose the resistor pattern in the control circuitry of an EPROM as a suitable specimen, because it has sufficient surface relief to make any change in optical sectioning strength quite clear. We would expect from Figure 3.3 that the depth-discrimination property should not significantly degrade until we use pinhole sizes greater than, say, 30μm. This is borne out in Figure 3.5, where we see that the depth discrimination is essentially unchanged over the range of detector pinholes. The microscope was focussed, for all these images, on the top layer of metallisation, and in order to produce a meaningful comparison, each image was normalised in the computer so that a chosen region of metal had the same average brightness in all images.

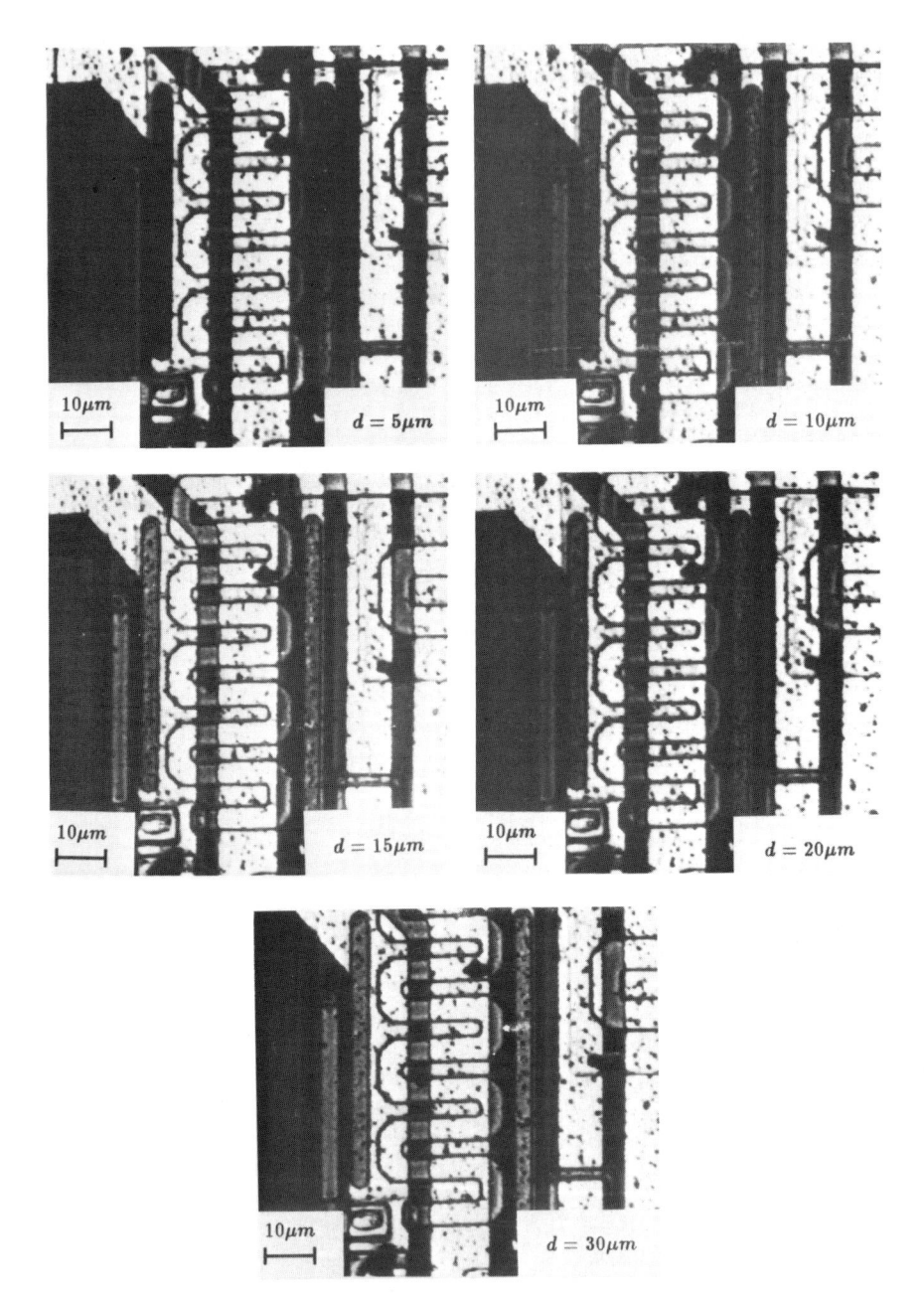

Figure 3.5. A series of images of an EPROM taken with 5, 10, 15, 20 and 30μm-sized detector pinholes.

Optical Aspects of Confocal Microscopy 101

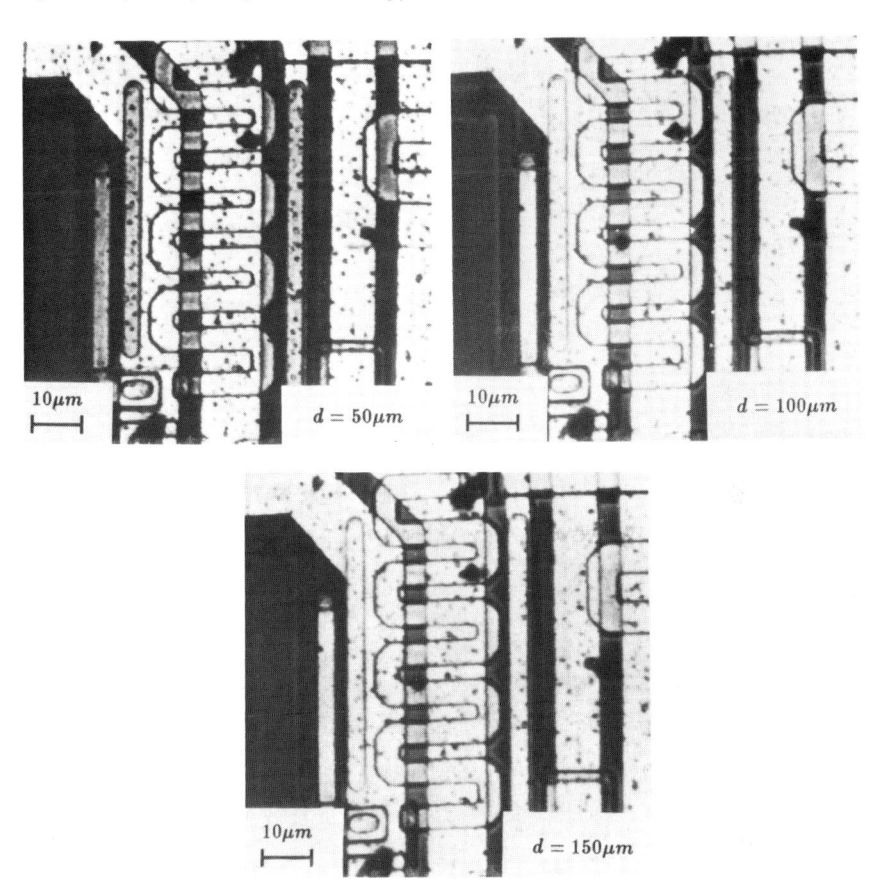

Figure 3.6. A further series of images illustrating the deterioration of the imaging as the detector pinhole size is increased to 50, 100 and 150μm in diameter.

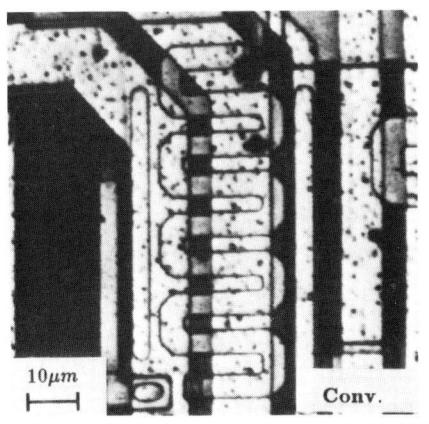

Figure 3.7. Conventional scanning microscope image of the EPROM.

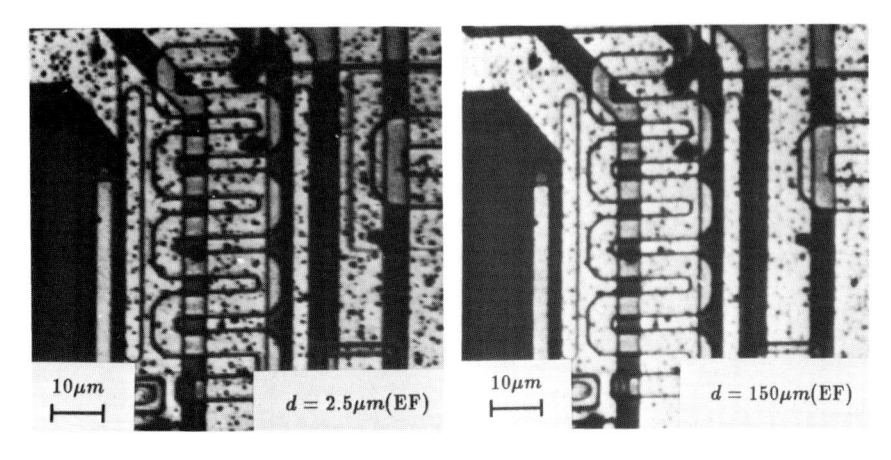

Figure 3.8. Extended-focus image created from data obtained with a 2.5μm and 150μm diameter detector pinholes.

As the pinhole size is increased further (Figure 3.6), we see that the depth discrimination becomes less strong, and that at 150μm, for this specimen, the image becomes very similar to the conventional image (Figure 3.7) obtained by removing the pinhole altogether. The effect of noise on the crispness of the images may also be seen from this set. In general, the smaller the pinhole, the more superior the image. This is particularly true of situations in which the pinhole is too large to give an improvement in the single point resolution.

As much of the processing of data from confocal microscopes involves the manipulation of a set of through-focus images, it is important to see how the finite size of the detector aperture might affect these images. We choose, as a specific example, the extended-focus technique, where we merely sum (integrate) a series of images obtained at different focal settings. Figure 3.8 shows the images created from data obtained using different-sized pinholes, and illustrates, as might be expected, that acceptable, although inferior, images can be obtained even when very large detector pinholes are used. Similar conclusions apply to auto-focus and height images [Wilson and Carlini, 1988].

II.3. The role of aberrations

As we have said before, the fact that both lenses contribute equally to the imaging of a confocal microscope means that the aberrations of each lens also "contribute" to the imaging. In Chapter 15, an interference technique will be discussed which, in certain cases, enables us to measure directly the effective pupil functions of the objective lens, and

Optical Aspects of Confocal Microscopy

hence to measure the wavefront aberration [Matthews, 1987]. We will, therefore, present an alternative method here, in which we measure instead the modulus square of the point spread function, $|h(u,v)|^2$. We do this because it is directly related to the function which actually probes the specimen, and also because it is obtainable, very easily in practice, from the $V(z)$ responses.

The basis of the method is as follows. A sufficiently small detector, correctly placed on the optic axis of the confocal microscope, gives a $V(z)$ response, equation (3.7), as:

$$I(u) = |h(2u, 0)|^2 \qquad (3.9)$$

whereas if the detector were offset laterally by a normalised distance v_D, the response would be given by:

$$I(u) = |h(2u, v_D)|^2 \qquad (3.10)$$

If we now take many traces at various settings of v_D, we can easily construct contours of equal intensity from the data. The results, for a particular objective lens, are shown in Figure 3.9. Theoretically

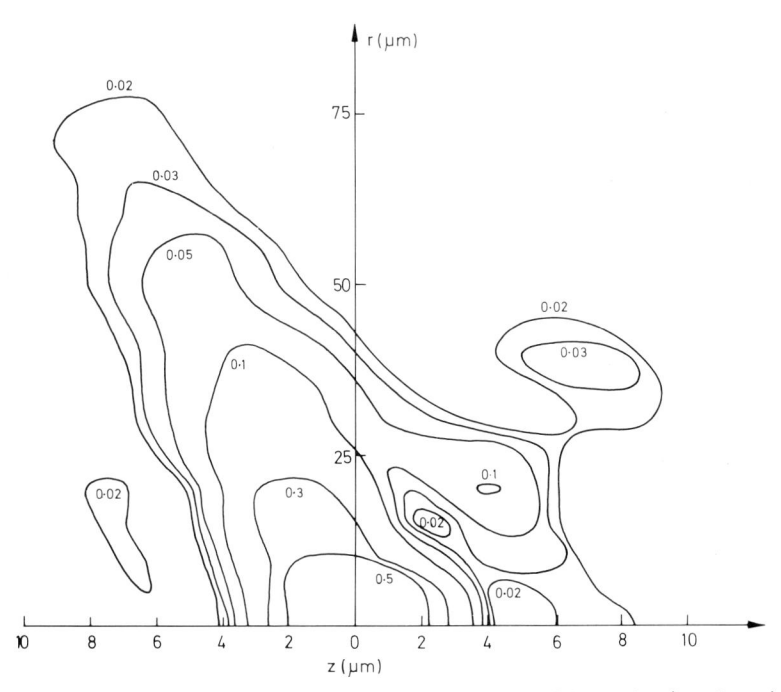

Figure 3.9. Experimentally determined contours of equal intensity (isophotes) of the function $|h(z,r)|^2$. Note that the objective lens magnification was 32, and that this factor affects the radial scale, r.

these contours, or isophotes, should have the same form as those we plotted in Figure 1.26 in Chapter 1. We repeat these contours in Figure 3.10(a), but this time we plot the function as $|h(kz, kr)|^2$ for the particular numerical aperture, 0.5, of our test lens where $k = 2\pi/\lambda$. This is important, because although the optical coordinates u and v are tremendously useful, as one curve applies for all values of numerical aperture, they do have a drawback when they appear together as axes, as in this case. This is because they tend to distort the perspective, as the constant factors relating optical "distance" to real distance are different for u and v, equations (3.6) and (3.4).

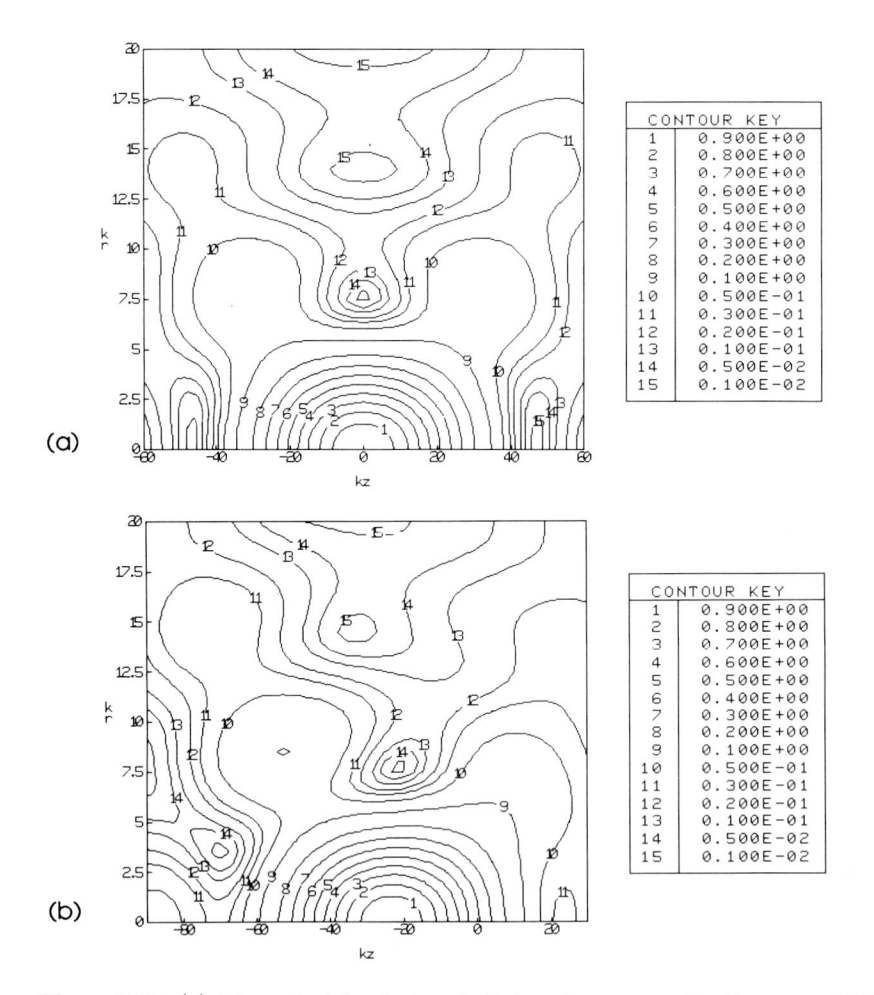

Figure 3.10. (a) Theoretical isophotes plotted on linear axes for the case of 0.5 numerical aperture. (b) As (a), but with spherical aberration corresponding to $A = 0.5$.

Optical Aspects of Confocal Microscopy 105

It is clear, however, that even with this perspective correction, Figures 3.9 and 3.10(a) are very different, and that the difference results from aberration. We model this by introducing a pupil function, in polar coordinates, as:

$$P(\rho,\theta) = \exp\left(\frac{1}{2}ju\rho^2\right) \exp[j\Phi(\rho,\theta)] \qquad (3.11)$$

where we have separated the effects of defocus from those of wavefront aberration, Φ. A suitable form of Φ which describes the main primary aberrations is:

$$\Phi = 2\pi\left(A\rho^4 + B\rho^3\cos\theta + C\rho^2\cos^2\theta\right) \qquad (3.12)$$

where A, B and C represent the coefficients of spherical aberration, primary coma, and primary astigmatism respectively [Zernike and Nijboer, 1949; Stamnes, 1986]. Figure 3.10(b) shows the effect on the contours of the point spread function of introducing an amount of spherical aberration corresponding to $A = 0.5$, which confirms that this particular lens suffers from a degree of spherical aberration.

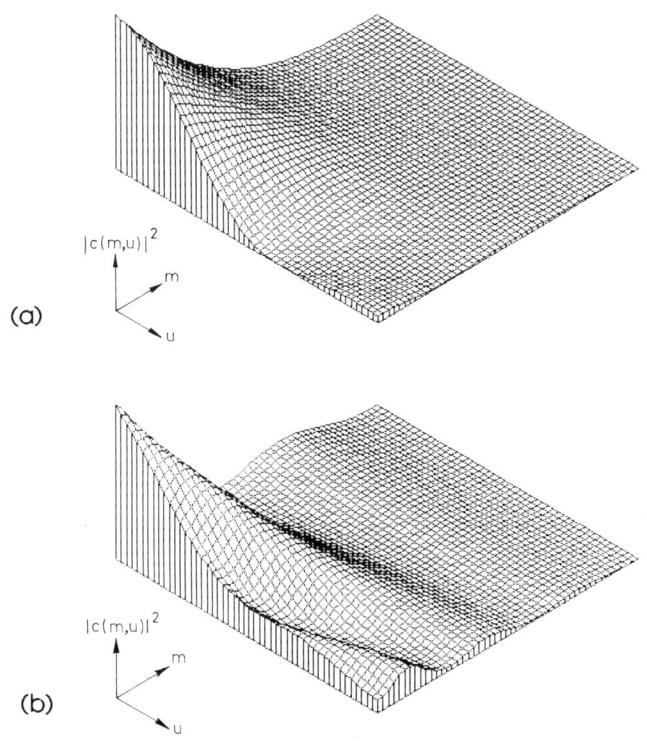

Figure 3.11. Isometric surface plots of the transfer function for the case of (a) zero aberration and (b) aberration corresponding to $A = 0.5$, $B = 0.3$ and $C = 0.3$.

These results were obtained from a series of displaced $V(z)$ responses, and it is important to understand how these responses themselves are affected by aberration before interpreting contour plots such as Figure 3.9. In order to do this, it is useful to use the alternative expression for $V(z)$ of equation (1.47), which is reproduced here as:

$$I(u) = \left| \int_0^{2\pi} \int_0^1 P(\rho, \theta) P(\rho, \pi - \theta) \rho d\rho d\theta \right|^2 \qquad (3.13)$$

from which it is clear, equation (3.12), that any amount of primary coma will have *no effect* at all on the $V(z)$ response, and therefore will not be present in the contours of Figure 3.9. We also see that the $|h|^2$ that we have mapped out is one whose effective wavefront aberration is 2Φ. In other words, the actual aberration of the objective lens is only one-half the value we have measured here.

Although we have shown that coma does not affect the form of the $V(z)$ response, it does, of course, play a part in the imaging of general objects. We show in Figure 3.11 the form of the defocussed transfer function $|c(m, u)|^2$ in the presence of no aberration and in the presence of spherical aberration, primary coma and primary astigmatism. It is clear that when aberrations are present, more information will be present from image detail at greater values of u — that is, the degree of optical sectioning will be reduced [Wilson and Carlini, 1989].

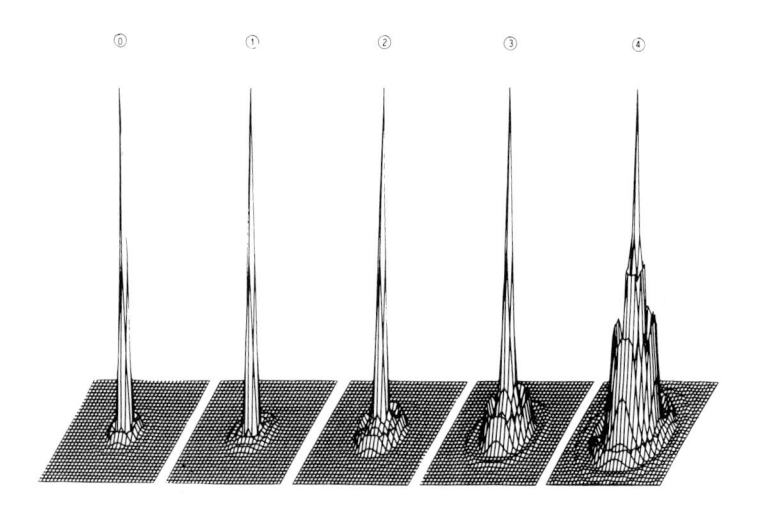

Figure 3.12. Change in intensity distribution with increasing penetration into a watery medium with a Planapochromat. (Courtesy of Dr. H. E. Keller, Carl Zeiss, Inc.)

It is particularly important to get a feel for the effect of aberrations in confocal microscopy, because the three-dimensional imaging capabilities of confocal microscopes almost inevitably mean that we must focus deep into a specimen, and hence run the risk of unavoidably introducing spherical aberration. Figure 3.12 emphasises this point further by showing the change in the intensity distribution of the focussed spot with increasing penetration into an object in a watery medium, with a Planapochromat 1.4 numerical aperture oil immersion objective.

A further instance where depth resolution is impaired is the case in which spherical aberration is introduced by variations in coverslip thickness. Figure 3.13 illustrates the change in the $V(z)$ depth responses for coverglass thicknesses of $120\mu m$, $170\mu m$ (nominal) and $220\mu m$. Figure 3.14 shows the changes in the half-width of these curves as a function of coverglass thickness. Good quality coverglasses have tolerances of $\pm 10\mu m$, and even this can lead to a severe reduction in the optical sectioning at the highest numerical apertures. As the numerical aperture increases beyond, say, 0.5, it is very important to choose the correct coverglass thickness for dry and water immersion lenses. Even oil immersion lenses such as Planapochromats perform optimally only with a 0.17mm coverglass thickness [Hellmuth, et al., 1988; Keller, 1989].

We should also mention that tube length is a very important parameter which must be properly set to optimise the depth response. When all the standard adjustments have been made, it is sometimes possible to optimise the $V(z)$ response further by placing a weak correction lens into the optical system. This is particularly straightforward for object-scanning systems, and is discussed further in Chapter 8.

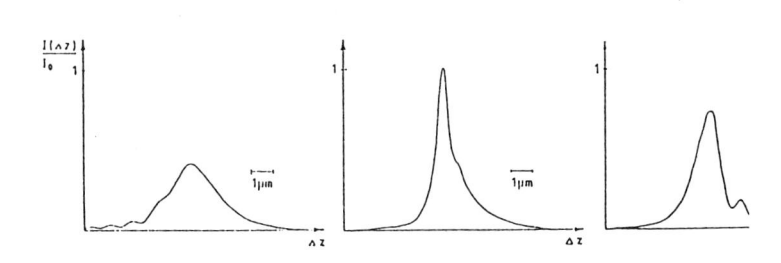

Figure 3.13. Change in resolution and depth detection through windows of 120, 170 (nominal), and $220\mu m$ thickness with a Plan-Neofluar, 1.2 numerical aperture oil immersion objective. (Courtesy of Dr. H. E. Keller, Carl Zeiss, Inc.)

FWHM [μm]

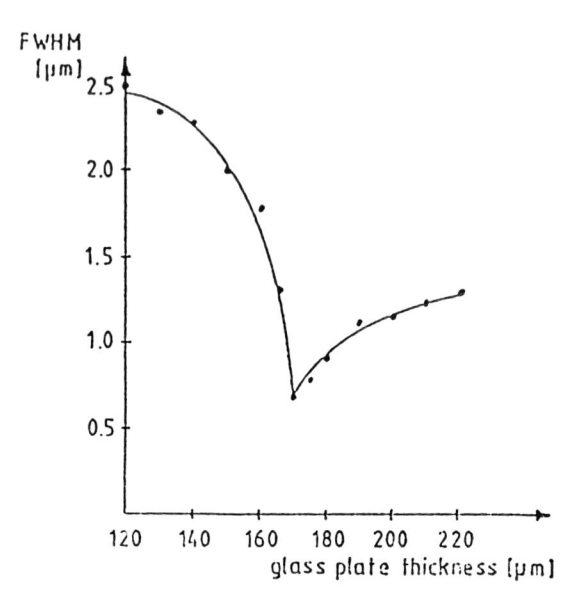

[Graph showing FWHM vs glass plate thickness, with values on y-axis from 0.5 to 2.5 μm and x-axis from 120 to 220 μm. Curve starts near 2.5 at 120, decreases sharply to minimum around 180, then rises slightly to ~1.2 at 220.]

glass plate thickness [μm]

Figure 3.14. Changes in the half-width of the $V(z)$ responses of Figure 3.13 with changing window thickness. (Courtesy of Dr. H. E. Keller, Carl Zeiss, Inc.)

The other aberration of great importance in fluorescence microscopy is chromatic aberration. The factors affecting the choice of objective here are discussed in detail by Keller (1989), but one should try to choose an objective with the best — or, at least, identical — correction at both excitation and emission wavelengths.

II.4. Alternative detector geometries

We have concentrated so far on circular detectors, but there are many other forms of detector which can be used. These include squares [Awamura et. al, 1987] and slits [Koester, 1980; Sheppard and Mao, 1988; Wilson, 1989a]. These are all compromises to some degree, but the use of a slit may have certain advantages. A slit, for example, permits a larger signal to be detected than a pinhole with a diameter equal to the slit width. A drawback, however, is that by using a non-circular detector, the image of a circular object, such as a point, is no longer circularly symmetric, equation (3.1). The degree of asymmetry is shown in Figure 3.15, and may not be significant in applications where line structures are being imaged, such as integrated circuit metrology. We note that the "image" perpendicular to the slit is actually sharper than in the true confocal case.

Optical Aspects of Confocal Microscopy 109

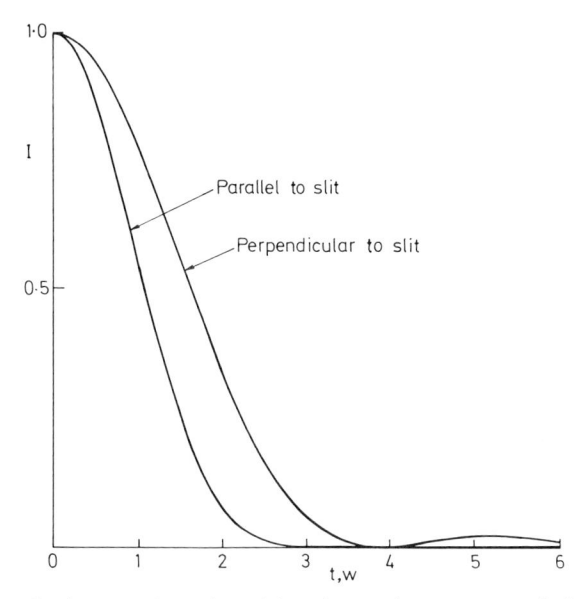

Figure 3.15. The image of a point object in a microscope consisting of a point source and an infinitely narrow slit detector. The image is shown in directions parallel and perpendicular to the slit. The parameters t and w are optical coordinates parallel and perpendicular to the slit, and are defined in the same way as in equation (3.4).

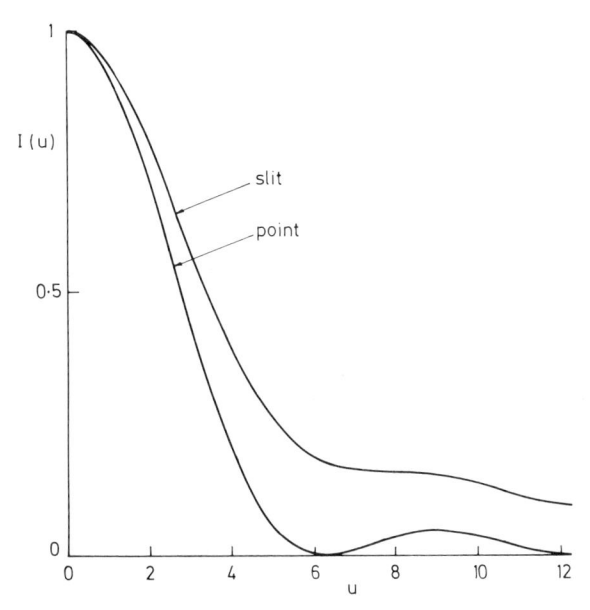

Figure 3.16. The variation of $I(u)$ against u for the case of an infinitely thin slit detector and an ideal point detector.

If we now move on to consider the sectioning strength of a slit detector, we find that for a point source and an infinitely narrow long slit, the response is given by Figure 3.16 [Wilson, 1989a]. There is no simple mathematical expression for this curve. The main observations are that the sectioning is less strong, and that the response has no zeroes and takes much longer to die away, Figure 3.17. The question of the correct choice of slit width can also be addressed, and a whole family of curves similar in trend to Figure 3.2 may be derived. Again, if the half-width is taken as the metric of sectioning, the curve of Figure 3.18 results, from which we see that we really need to try to use a slit of a width less than about one optical unit to achieve maximum sectioning. This means that we must use a slit considerably smaller than the circular pinhole in order to obtain optimal sectioning, and that there may also be directional effects.

In Figure 3.19 we show some comparison images taken with pinhole and slit detector systems. In these images the slit was oriented horizontally. It is clear that in the focal region, Figure 3.19(a), the images are very similar, whereas in the presence of defocus, the slit image tends to become washed out in the direction parallel to the slit, Figure 3.19(b). This asymmetry is discussed in detail in Wilson and Hewlett (1990a). It results from the fact that the imaging of object features which are parallel to the slit is similar to conventional imaging, whereas those perpendicular to the slit are imaged somewhat as in a confocal microscope. This is an oversimplification, but it is reasonable to assume that the behaviour of the system with defocus will be considerably different in the two cases.

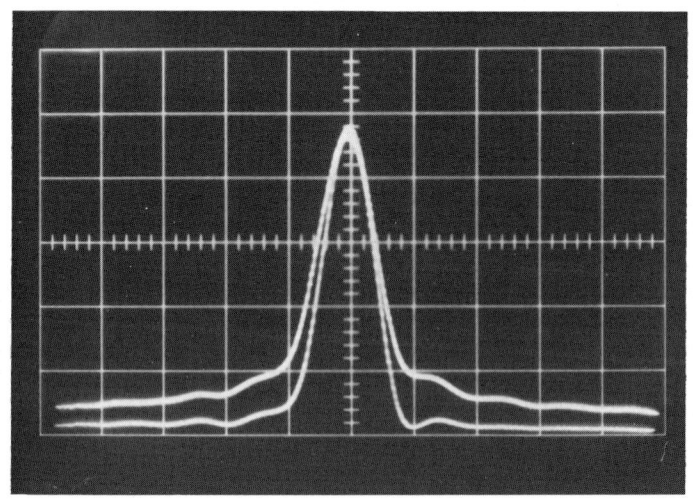

Figure 3.17. Experimental confirmation of the predictions of Figure 3.16. In this case, the full horizontal scale represents an axial scan of the mirror of 25.0μm.

Optical Aspects of Confocal Microscopy　　　　　　　　　　　　　　111

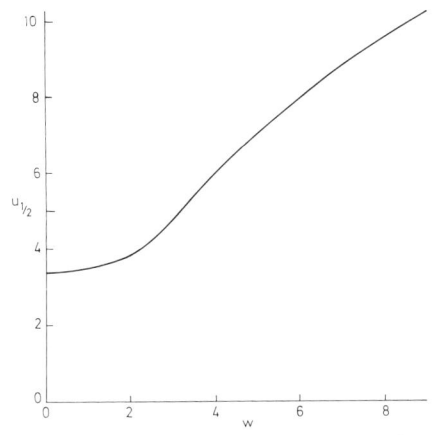

Figure 3.18. The half-width of the $I(u)$ curves for a slit-shaped detector of half-width w optical units.

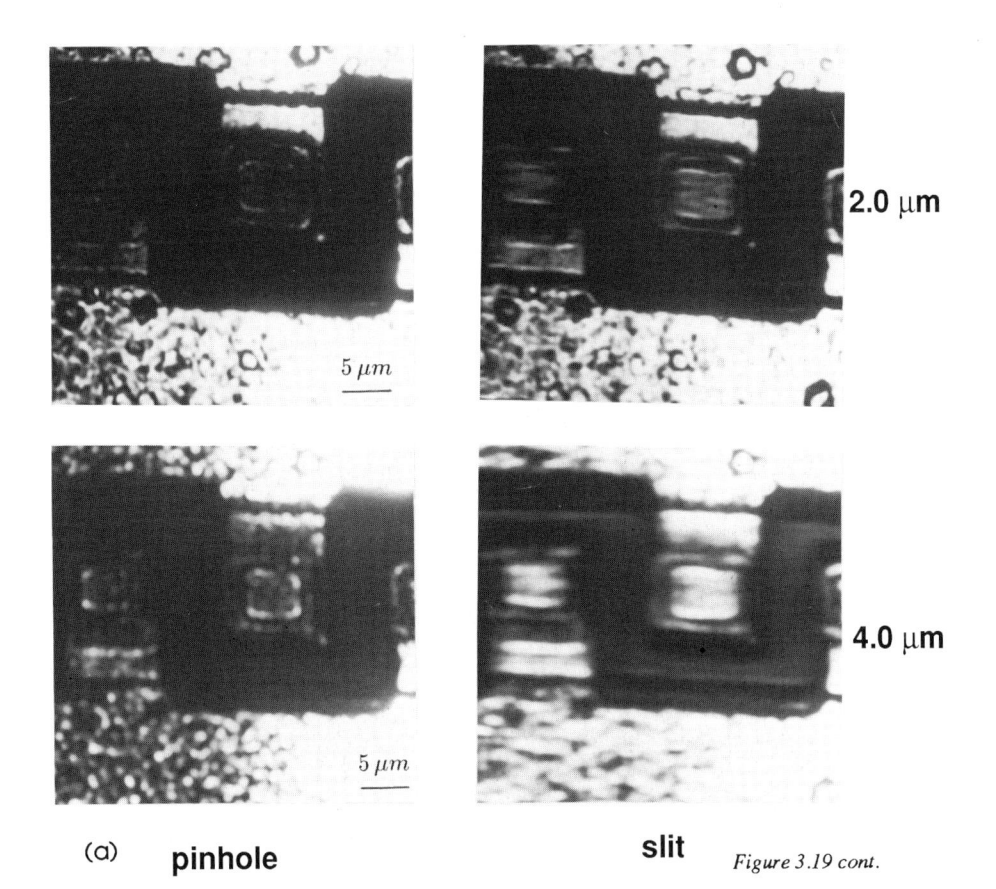

2.0 μm

4.0 μm

(a)　　pinhole　　　　　　　　　　　slit　　　　Figure 3.19 cont.

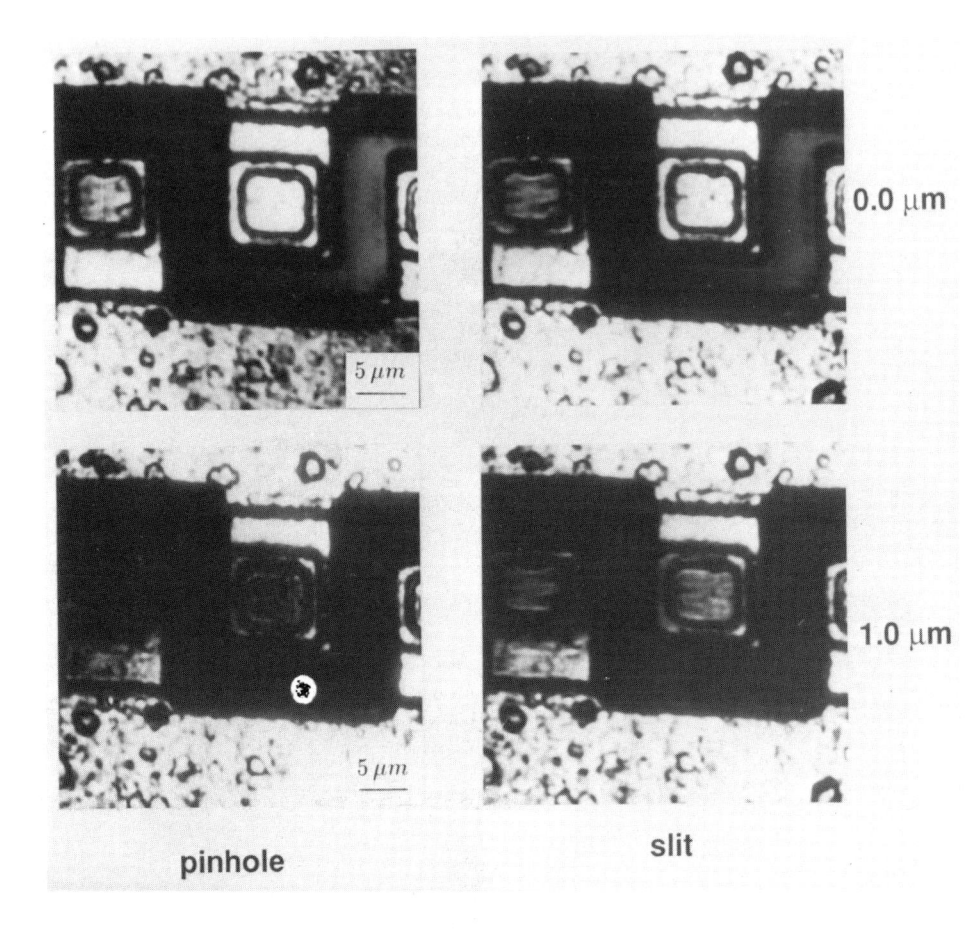

(b)

Figure 3.19. (a) Images taken with point detector and slit detector microscopes at a nominal focal plane and focussing 1.0μm farther into the specimen. (b) As for (a), but at focal settings of 2.0μm and 4.0μm. The slit was oriented horizontally.

In Figure 3.20, we show images which depend on the three-dimensional aspects of the imaging. The images shown here were obtained with a slit detector, and despite the asymmetry of the component images, the extended-focus image, height image and isometric representations are all quite respectable.

Another form of detector which is becoming more important as semiconductor technology advances is the detector array. This is useful because, in principle, it permits us to choose whatever detector

Optical Aspects of Confocal Microscopy

distribution we want. In particular, the detector sensitivity can be made negative at some points. This freedom permits us to realise confocal microscopy in a completely different way. As an example, if we were to build a detector with intensity sensitivity $2J_1(av)/(av)$, we would find that [Wilson 1989a, 1990] for $a > 2$ the axial image of a planar object would be exactly as given by equation (3.5). Although this may not be the best way of achieving confocal operation, it does point to the new possibilities which open up.

The fact that, by using a detector array, we can sample the field at all points in the detector plane gives us more information, and should enable us to produce better images which correspond more closely to specific object features. An example of this is the "type III" microscope [Schutt et al., 1988]. The idea here is that instead of displaying, at each picture point, the intensity we detect on the optic axis, we display instead the maximum signal we measure in the detector plane. This scheme is thought to be particularly sensitive to phase edges in biological tissue.

We shall return to the question of detector arrays later, when we discuss methods of tuning the strength of the optical sectioning property, as well as imaging in direct-view or tandem scanning microscopes.

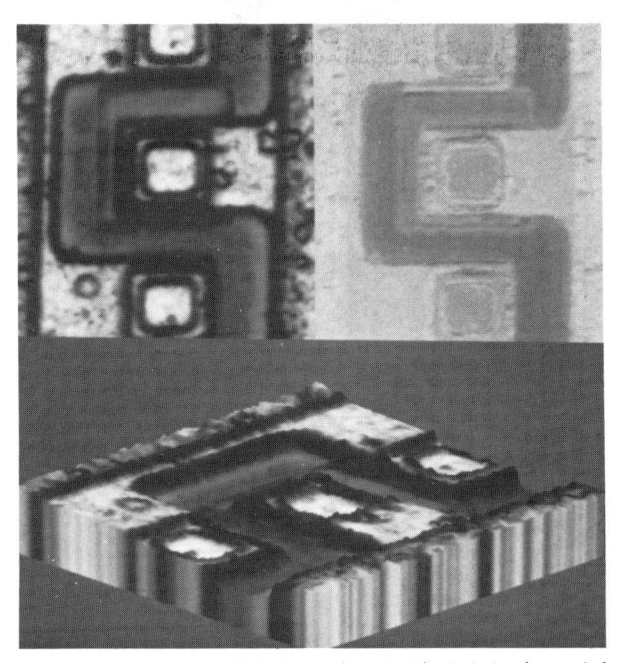

Figure 3.20. A montage of extended-focus (top left), height (top right) and isometric projections of the specimen of Figure 3.19. These images were generated with a slit detector microscope.

III. Fluorescence imaging

We can make very similar arguments and calculations about the size of the detector in the fluorescence case. In this mode, the image formation is always incoherent whatever the size of the detector. The basic imaging equation is [Wilson, 1989b]:

$$I = |h_1 h_{2eff}|^2 \otimes f \qquad (3.14)$$

where h_1 is the amplitude point spread function of the exciting lens and h_{2eff} is the effective point spread function of the fluorescence imaging lens given by:

$$|h_{2eff}|^2 = |h_2|^2 \otimes D \qquad (3.15)$$

where D is the sensitivity function of the detector. In the case of a large detector we find $I = |h_1|^2 \otimes f$ as in a conventional fluorescence microscope, whereas for a vanishingly small detector we find $I = |h_1 h_2|^2 \otimes f$, as we discussed in Chapter 1. We also remarked at that time that in order to obtain the optimum sectioning, we should try to operate with as small a difference in wavelength as possible between the exciting and the fluorescence radiation, Figure 1.40.

If we again measure the strength of sectioning by scanning our uniform fluorescence object axially through focus, we obtain the curves of Figure 3.21. The sectioning becomes progressively worse as the detector becomes larger, but, as in the brightfield case, there is a range of detector apertures over which the sectioning remains essentially constant. However, this range becomes considerably smaller as the wavelength ratio, $\beta = (\lambda_2/\lambda_1)$, becomes larger, Figure 3.22.

(a)

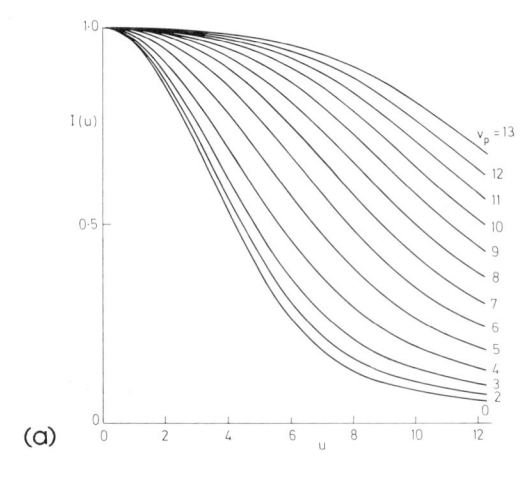

Figure 3.21 cont.

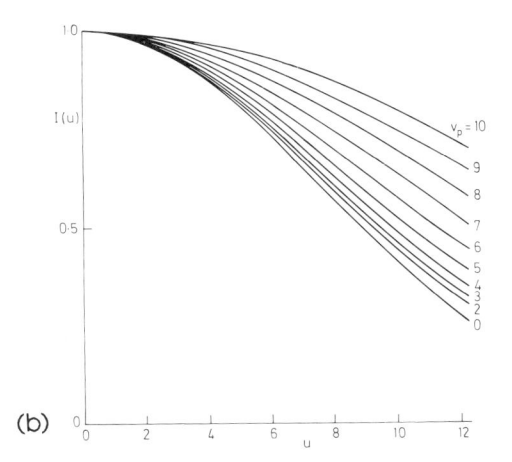

Figure 3.21. (a) The detected signal as a perfect planar fluorescent object is scanned axially through focus for a variety of normalised detector radii, v_p. The parameters, u and v_p, are normalised to the incident wavelength. The curves are drawn for the limiting case of $\beta = 1$. (b) As for (a), but for the case of $\beta = 2$.

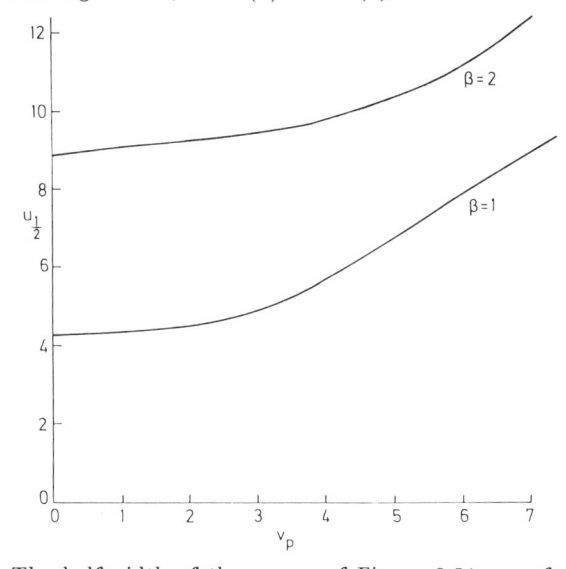

Figure 3.22. The half-width of the curves of Figure 3.21 as a function of the normalised detector radius, v_p.

III.1. Fluorescence imaging with a slit-shaped detector

Here is a case in which the increase in signal level afforded by the use of a slit-shaped detector may well be useful. The strength of the sectioning is a little worse than for the case of a point detector, and again deteriorates with slit width. Figure 3.23 presents the results for the limiting case of $\beta = 1$.

The question of the asymmetry of the images of specific objects such as points and lines is discussed by Wilson and Hewlett (1990a), where it is found that in the $\beta = 1$ case, the image of a point object is identical to the brightfield image. The axial image of a line object is found to be almost identical whether it be oriented perpendicular or parallel to the slit, but the asymmetry increases as the wavelength difference becomes larger. However, the asymmetry is not great, and it is to be expected that good fluorescence images will be obtained from slit detector systems, Figure 3.24.

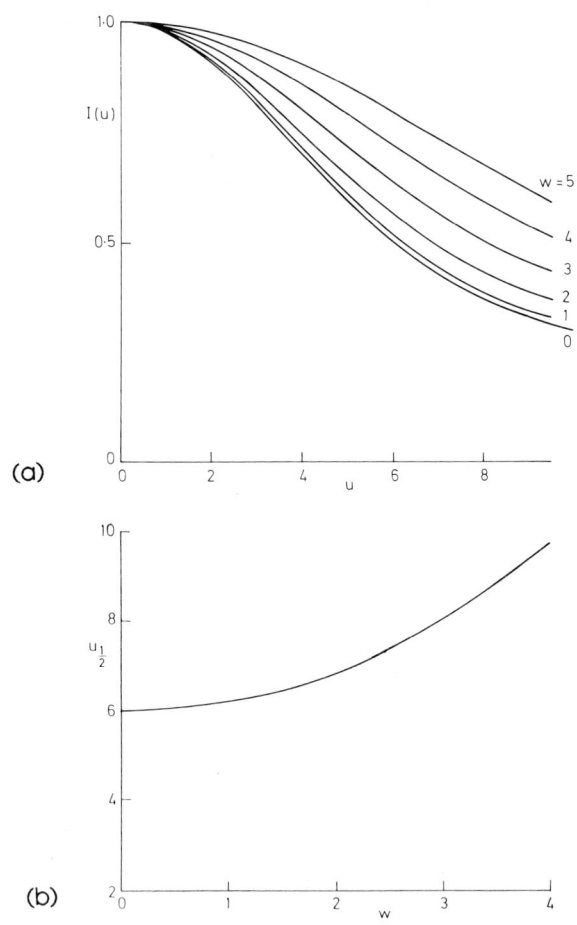

Figure 3.23. (a) The detected signal as a perfect planar fluorescent object is scanned axially through focus for a variety of normalised slit widths, w. The normalised coordinates are all normalised to the incident wavelength. The curves are drawn for the limiting case of $\beta = 1$. (b) The half-width of the curves of (a) as a function of the normalised slit half-width, w.

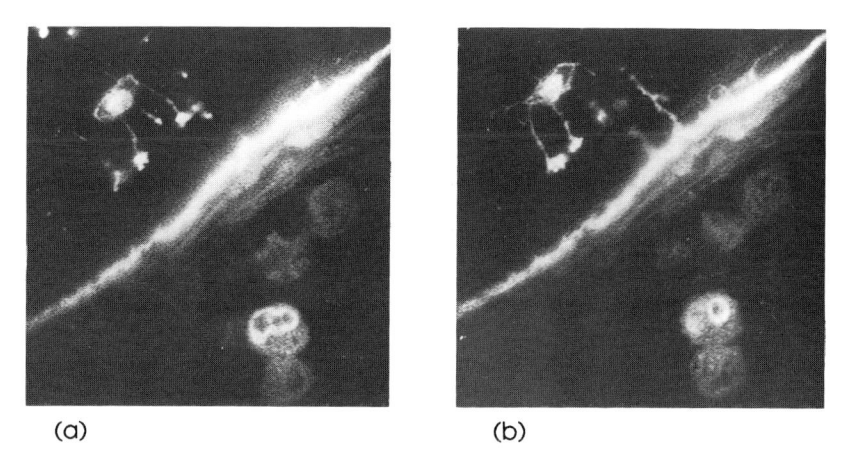

Figure 3.24. Two confocal sections of a rhodamine-stained bone-marrow sample, with (a) recorded 2μm lower than (b). The horizontal field of view is 60μm, and a 1.25 numerical aperture objective was used. (Courtesy of Dr. A. Draaijer, T. N. O., Delft, whose work was financially supported by Tracor Northern.)

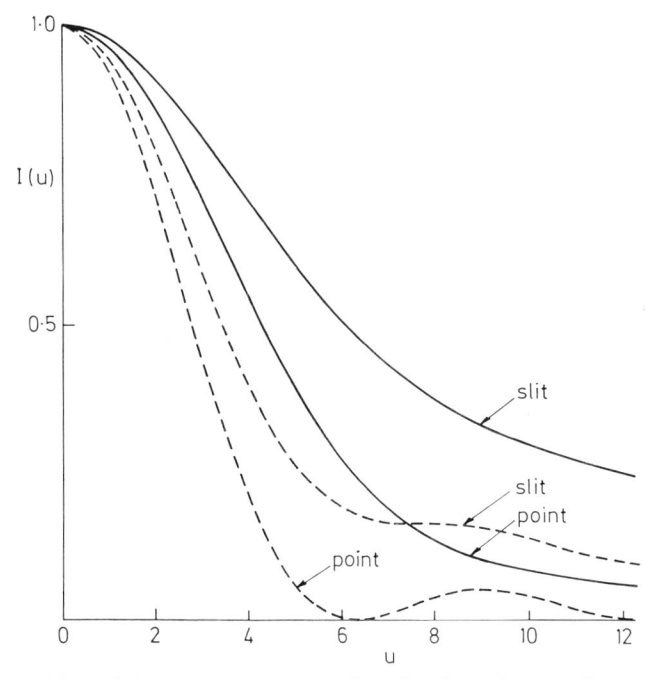

Figure 3.25. The axial response to appropriate "perfect planar reflectors" fluorescent (solid line) and non-fluorescent (broken line) microscopes using both slit and central point detectors. The fluorescent curves correspond to the limiting case of $\beta = 1$.

Although we may not, in practice, have a choice between brightfield and fluorescence imaging, and it may be more straightforward and convenient to interpret the results from one imaging mode rather than another in a specific case, we should bear in mind that the strength of the optical sectioning in fluorescence microscopy is always inferior to that in brightfield. We compare the various systems in Figure 3.25.

Our discussion has been based on a scalar, paraxial theory. In the case of fluorescence imaging numerical results are available based on a vector theory of imaging [van der Voort and Brakenhoff, 1990]. These are more applicable to high numerical aperture imaging.

IV. Coherent detectors in scanning microscopy

In our discussion of finite-sized detectors so far we have inevitably considered only incoherent detectors, that is, those which respond to the intensity of the light which falls upon them. If the amplitude of the field at the detector plane were U, then these detectors give a signal, I:

$$I = \int |U|^2 dS \tag{3.16}$$

where the integral is taken over the area of the detector. In optics, unlike acoustics, it is not possible to have a coherent detector directly, that is, one which responds to amplitude. In this case the detected signal would be given by:

$$I = \left| \int U \, dS \right|^2 \tag{3.17}$$

It is interesting to ask if it is possible to modify a confocal brightfield system to behave like a system employing a large-area coherent, amplitude-sensitive detector. The motivation, of course, is that the imaging in this case is purely coherent, and may be written as, equation (1.19):

$$I = |h_1 h_2 \otimes t|^2 \tag{3.18}$$

If we now consider a scanning microscope with a point source and a finite-sized detector of amplitude sensitivity $D(x_2, y_2)$, we can write the image intensity (using the methods of Fourier optics of Chapter 1, section IV.1) as:

$$I(x,y) = \left| \int \int \int \int h_1(x_0, y_0) t(x_0 - x, y_0 - y) h_2(x_0 + x_2, y_0 + y_2) \right.$$
$$\left. \times D(x_2, y_2) \, dx_0 dy_0 dx_2 dy_2 \right|^2 \tag{3.19}$$

If we now introduce [Wilson and Hewlett, 1990b]:

Optical Aspects of Confocal Microscopy

$$h_{2eff}(x_0, y_0) = \int\int h_2(x_0 + x_2, y_0 + y_2) D(x_2, y_2)\, dx_2 dy_2 \quad (3.20)$$

we can write equation (3.19) in exactly the same form as equation (3.18), i.e., as:

$$I = |h_1 h_{2eff} \otimes t|^2 \quad (3.21)$$

Thus, if we can build a lens having the appropriate amplitude point spread function h_{2eff}, we can use the confocal arrangement to mimic a system using a large-area coherent detector. It is clear from the form of equation (3.20) that if we simply take the Fourier transform of both sides, we obtain an effective pupil function, P_{2eff}, as:

$$P_{2eff} = F_D . P_2 \quad (3.22)$$

where F_D represents the Fourier transform of D. It is now apparent that all we need to do is to introduce a pupil plane filter whose amplitude transmittance is F_D into the pupil function of one of the lenses of a confocal microscope.

As a specific example, we will briefly consider the effect of building a system with a coherent slit detector. In order to achieve this, we need simply to place (if the slit is infinitely narrow) a slit filter in the pupil plane at right angles to the direction of the effective coherent slit detector that we want to mimic.

If we look now at the form of the transfer function:

$$c(m) = P_1 \otimes P_{2eff}(\tilde{m}, \tilde{n}) \quad (3.23)$$

we see that the function will be very different for features running parallel and perpendicular to the slit. If we specialise to features which run perpendicular to the coherent slit detector, that is, parallel to the slit pupil plane filter, then it is simple to calculate the transfer function as:

$$c(m) = 1 - \tilde{m}/2; \quad \tilde{m} \leq 2 \quad (3.24)$$

whereas for the true confocal:

$$c(m) = \frac{2}{\pi}\left[\cos^{-1}\left(\frac{\tilde{m}}{2}\right) - \frac{\tilde{m}}{2}\sqrt{1 - \left(\frac{\tilde{m}}{2}\right)^2}\right]; \quad \tilde{m} \leq 2 \quad (3.25)$$

A plot of these two functions reveals that the straight line of equation (3.24) always lies above the gentle curve of equation (3.25), suggesting that the imaging might be better in this case. A further object of importance is the straight edge. It is easy to show that the gradient of the response at the edge is given by:

$$\frac{dI}{dx} \sim \int_0^2 c(m)\,d\tilde{m} \tag{3.26}$$

If we perform these integrals for the transfer functions of equations (3.24) and (3.25), we find that the gradient is 17.8% sharper in the coherent slit case. These predictions are borne out in practice [Wilson et al., 1990], where it is also found that the depth discrimination strength of the coherent slit system is better than that of an incoherent slit system, and quite similar to the traditional, true, confocal case, Figure 3.26.

A further advantage of the coherent slit technique, apart from its enhanced imaging of certain classes of line structure, is that it is possible to calculate, usually analytically, the images of a large class of objects. This is often not possible in the true confocal case. The reason lies, essentially, in the difference in complexity between equations (3.24) and (3.25). This has important implications for image analysis and interpretation.

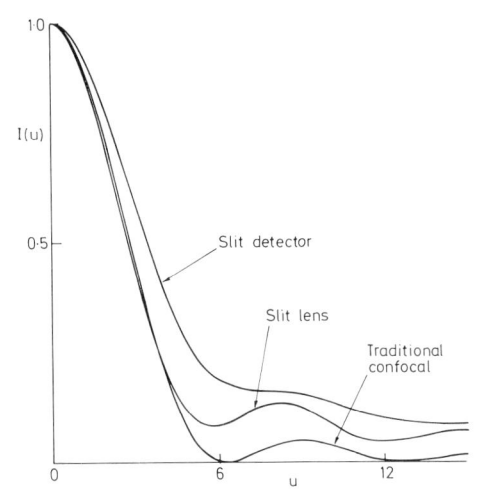

Figure 3.26. Theoretical $I(u)$ curves for the traditional confocal, slit lens and incoherent slit detector microscope systems.

V. Pupil function filters

We have just seen how a pupil plane filter may be used to mimic a coherent detector in a confocal system. An alternative approach is simply to regard pupil plane filters as a method of tuning the effective point spread function, $h_1 h_2$, of the confocal microscope. We shall not discuss this in detail here, as it is the subject of Chapter 5. However,

Optical Aspects of Confocal Microscopy

we will briefly consider the annular filter at this point, both because it has some interesting properties and because it may be thought of as the simplest "building block" from which certain more complicated, radially symmetric, filters may be constructed. The annular filter might be defined mathematically as:

$$F(\rho) = \begin{cases} 1 & \text{for } \epsilon \leq \rho \leq 1 \\ 0 & \text{otherwise} \end{cases} \quad (3.27)$$

In this case, the $V(z)$ response may be written as:

$$I(u) = \left\{ \frac{\sin[(1 - \epsilon^2)u/2]}{(1 - \epsilon^2)u/2} \right\}^2 \quad (3.28)$$

and here the optical sectioning is degraded by a factor of $1/(1 - \epsilon^2)$. This suggests that by choosing a high value of ϵ, we may be able to find the section of interest within a thick specimen more easily. Once it has been located, ϵ can be reduced to zero and high axial resolution imaging can be performed. In this way we can choose ϵ to determine the degree of sectioning we require.

Another attraction of the annular filter is that the lateral resolution is not dramatically changed as ϵ varies. The point spread function, h, for an in-focus system is given by [Born and Wolf, 1975]:

$$h(v) = \frac{1}{(1 - \epsilon^2)} \left[\frac{2J_1(v)}{v} - \epsilon^2 \left(\frac{2J_1(\epsilon v)}{\epsilon v} \right) \right] \quad (3.29)$$

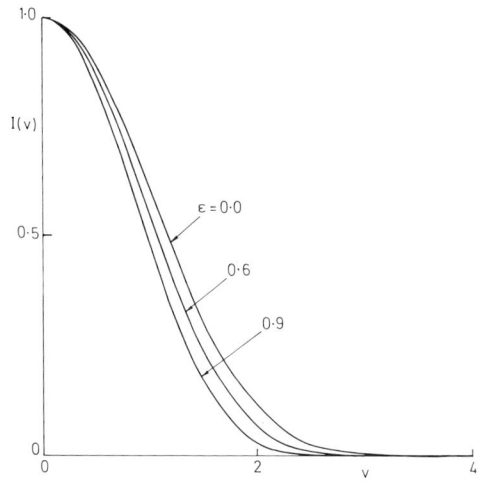

Figure 3.27. The image of a single point object in a confocal microscope employing one full and one annular lens for various values of ϵ.

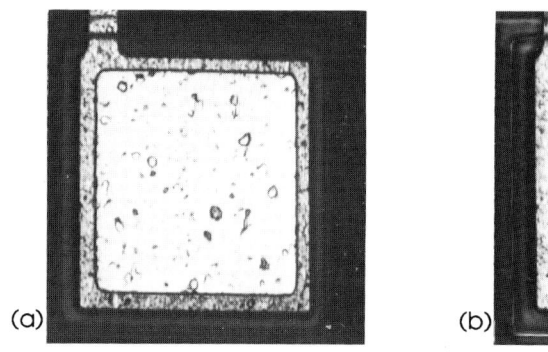

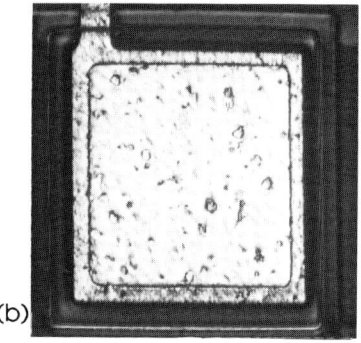

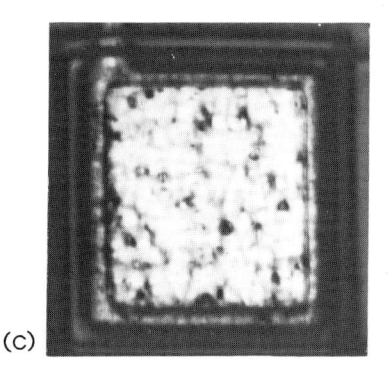

Figure 3.28. Confocal images of the metallisation on the bond pad of a transistor taken with (a) two full lenses, (b) one full and one annular lens and (c) two annular lenses. The annuli correspond to $\epsilon = 0.77$.

As a specific example, Figure 3.27 shows the image of a single point object in a confocal microscope employing one full and one annular lens for various values of ϵ.

Figure 3.28(a) shows a confocal image of the metallisation on a bond pad of a transistor. In this case, both lenses were unobscured and the microscope was focussed on the central region of the metal. Detail from all other parts of the specimen has been rejected by the optical sectioning property of the microscope. Figure 3.28(b) shows the corresponding image when a confocal system employing one full and one annular lens is used, and, as we would expect, the sectioning strength is reduced. The result of using two annular lenses is shown in Figure 3.28(c), which shows that this combination does not produce good images. The reason here is that although the image of a single point is improved, the image of extended objects is poor, as may be seen from the form of the transfer function [Wilson and Hewlett, 1990c], Figure 3.29.

Optical Aspects of Confocal Microscopy

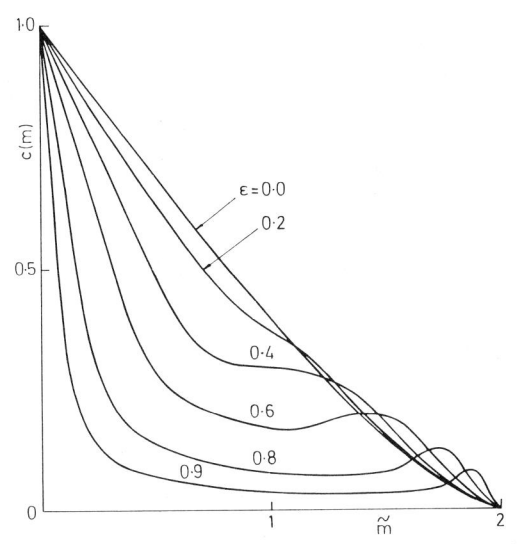

Figure 3.29. The transfer function of a confocal microscope employing two equal annular lenses.

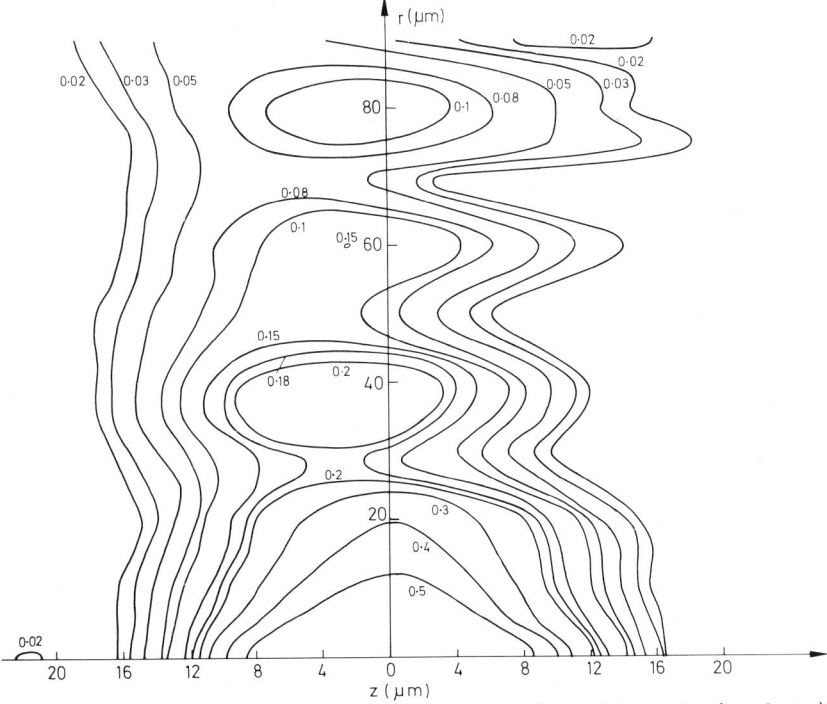

Figure 3.30. Experimentally determined contours of equal intensity (isophotes) of the function $|h(z,r)|^2$. The filter was such that $\epsilon = 0.77$ and the numerical aperture was 0.4.

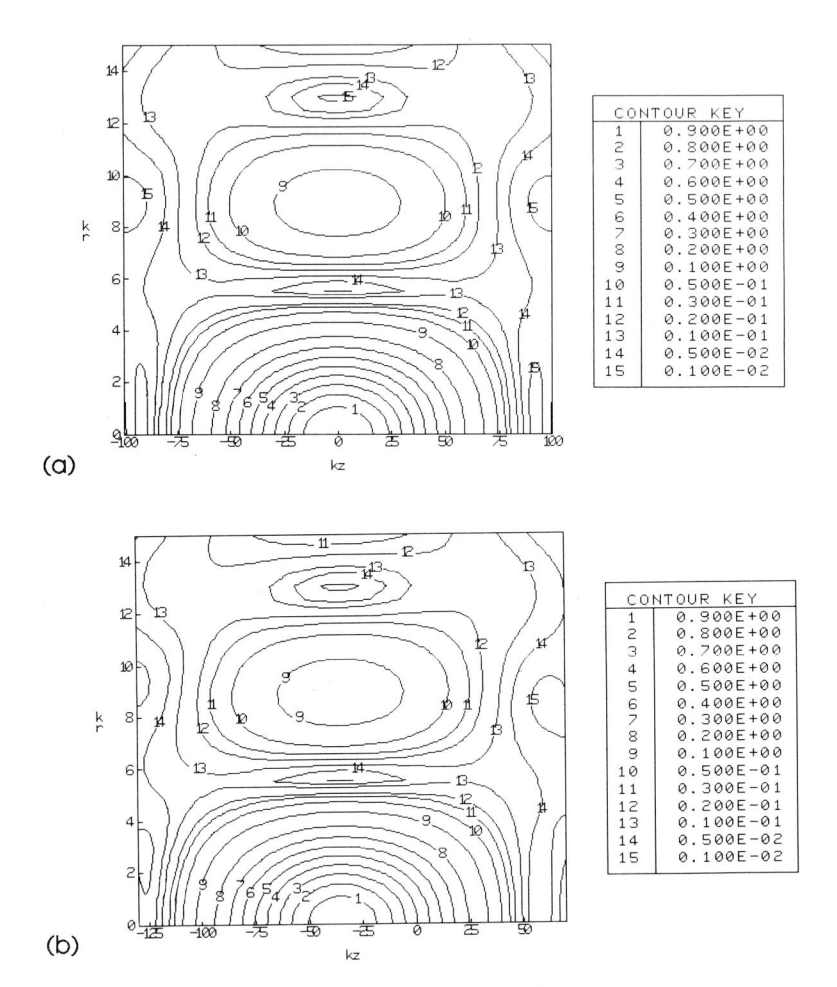

Figure 3.31. Theoretical isophotes for $\epsilon = 0.7$ and (a) no aberration, and (b) spherical aberration corresponding to $A = 0.5$, equation (3.12).

The question of the role of aberrations is an important one [Sheppard and Wilson, 1979], and it is found that when an annular pupil plane filter is employed, the effect of aberrations is reduced because the wavefront is partially obscured. This is confirmed in Figure 3.30, where we present experimentally measured isophotes for an annular lens with an obscuration corresponding to $\epsilon = 0.77$. These were obtained by a method similar to that used to obtain Figure 3.9. Theoretical comparisons are shown in Figure 3.31.

We note finally that pupil plane filters which are potentially suitable for brightfield imaging may not be so useful in fluorescence, because of the very different imaging properties of the two modes.

VI. Enhancing the optical sectioning

The use of an annular pupil plane filter, or of detector apertures which are not of ideal size, all lead (amongst other things) to a reduction in optical sectioning strength. We have not discussed any scheme, so far, in which the sectioning in a particular imaging mode can be made better than that obtained with an ideal, vanishingly small, point detector. An improvement can be made, however, if we form a composite image by combining two confocal images with different sectioning strengths. The principle is shown in Figure 3.32. The top figure shows two axial $V(z)$ responses obtained simultaneously, but using either different wavelengths, different numerical aperture objectives, or different-sized detector pinholes. If we denote the full curve by I_1 and the dashed curve by I_2, we can construct the composite image, I, as:

$$I = I_1 - \gamma I_2 \tag{3.30}$$

where γ is a constant. This is shown in the lower part of Figure 3.32, where we see that, although the response goes negative, the first zero, and also the half-width, defined when $I = 0.5$, is considerably reduced. In this way the sectioning strength can be improved. The method is equally applicable to fluorescence and brightfield imaging.

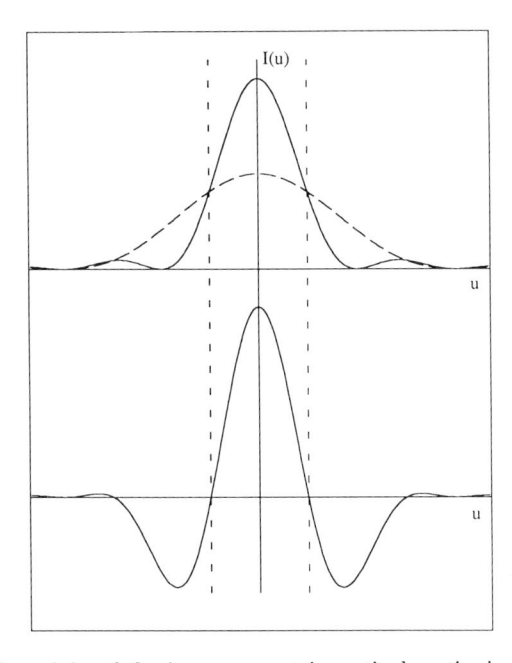

Figure 3.32. The origin of the improvement in optical sectioning strength which results from forming a composite image.

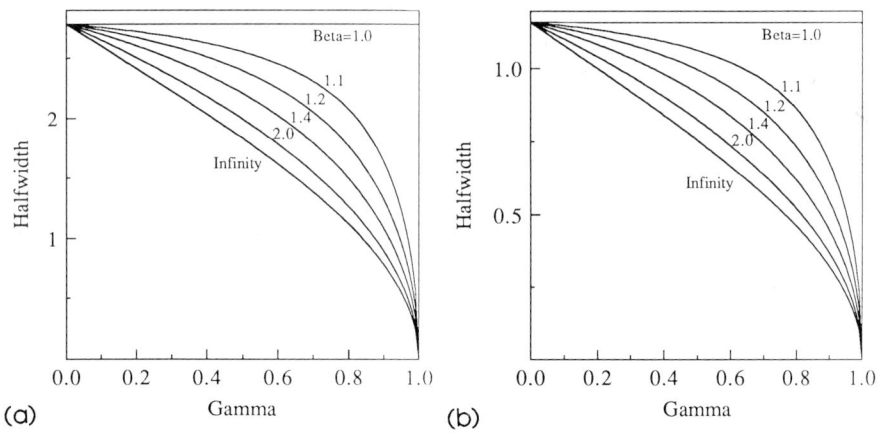

Figure 3.33. The improvement in (a) axial and (b) lateral resolution which becomes possible by forming a composite image. The parameter β denotes the wavelength ratio λ_2/λ_1.

If we now specialise to the situation in which we obtain two images at wavelengths λ_1 and λ_2, we can describe the optical sectioning by the combined $V(z)$ response as:

$$I(u) = \frac{1}{(1-\gamma)} \left\{ \left[\frac{\sin(u/2)}{u/2}\right]^2 - \gamma \left[\frac{\sin(u/2\beta)}{u/2\beta}\right]^2 \right\} \quad (3.31)$$

where u is defined relative to the shorter wavelength, λ_1 and we have introduced the wavelength ratio, $\beta = \lambda_2/\lambda_1$. We have also normalised the response such that $I(0) = 1$. As a measure of the strength of the optical sectioning which is available, we plot in Figure 3.33(a) the value of u at which $I(u) = 0.5$ as a function of γ for a variety of wavelength ratios, β. It is clear that considerable improvement is possible.

It is also important to see how the scheme affects the lateral resolution. We consider the in-focus image of a single point object, which is given by:

$$I(v) = \frac{1}{(1-\gamma)} \left\{ \left[\frac{2J_1(v)}{v}\right]^4 - \gamma \left[\frac{2J_1(v/\beta)}{v/\beta}\right]^4 \right\} \quad (3.32)$$

Figure 3.33(b) shows the variation of the value of v at which $I(v) = 0.5$ as a function of γ for a variety of values of β, and reveals that the lateral resolution, when measured in this way, may also be improved.

Optical Aspects of Confocal Microscopy

Although we have specifically considered the linear combination of images taken at two different wavelengths, the formal similarity in the sectioning expressions means that all the above comments apply equally well to the annular case. In particular, when an annular pupil image is subtracted from a confocal image, we simply replace the parameter β by $1/(1-\epsilon^2)$ in the composite sectioning equation, (3.31).

If we apply the same technique of subtracting two images with different-sized detectors [Sheppard and Cogswell, 1990, and Chapter 8] we find that the sectioning can again be improved. We will consider here only the specific cases of a true confocal image and one obtained with a system employing a finite-sized detector. The composite image may now be written as:

$$I = I(v_p = 0) - \gamma I(v_p) \qquad (3.33)$$

Figure 3.34 indicates the improvement in sectioning and lateral resolution that can be obtained for varying detector sizes [Wilson and Hewlett, 1990d]. A similar degree of improvement is available in fluorescence imaging. Figure 3.35 shows the improvement in sectioning in the special limiting case of equal excitation and fluorescence wavelengths.

Figure 3.36(a) shows two experimental $V(z)$ curves. The narrower one was obtained with a 5μm diameter pinhole which gave, essentially, true confocal behaviour, while the outer, broader curve resulted from using a 50μm-diameter ($v_p \sim 4$ optical units) pinhole. The composite curve in Figure 3.36(b) was obtained by subtracting half the "50μm curve" from the "5μm curve."

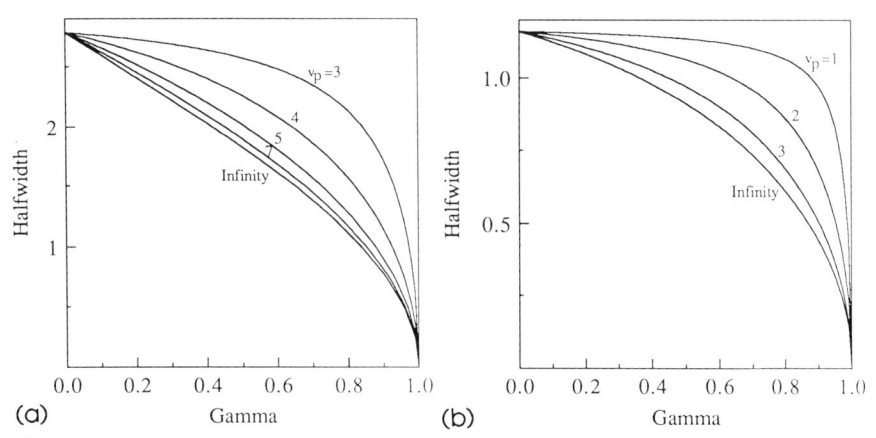

Figure 3.34. The improvement in (a) axial and (b) lateral resolution which becomes possible by forming a composite image by subtracting part of an image obtained with a finite-sized detector from a true confocal one.

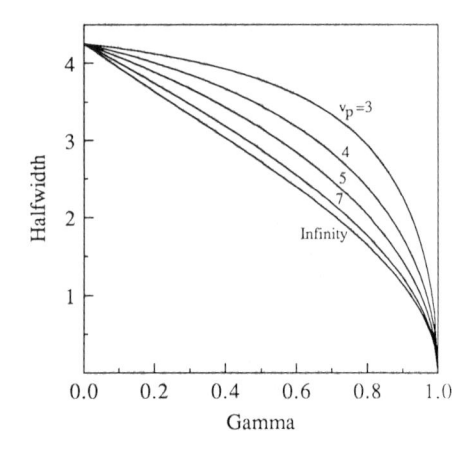

Figure 3.35. The improvement in axial resolution available in fluorescence imaging in the case of equal excitation and fluorescence wavelengths.

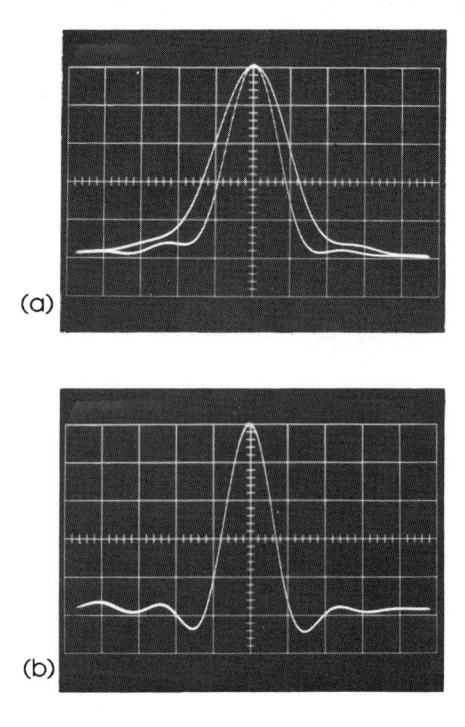

Figure 3.36. (a) $V(z)$ responses obtained using 5μm (inner curve) and 50μm (outer curve) diameter pinholes. (b) Composite $V(z)$ obtained by subtracting half the "50μm curve" from the "5μm curve." The full horizontal scale corresponds to an axial scan of 15μm, and a 32X, 0.5 numerical aperture objective was used with red 0.6328μm wavelength light.

Optical Aspects of Confocal Microscopy 129

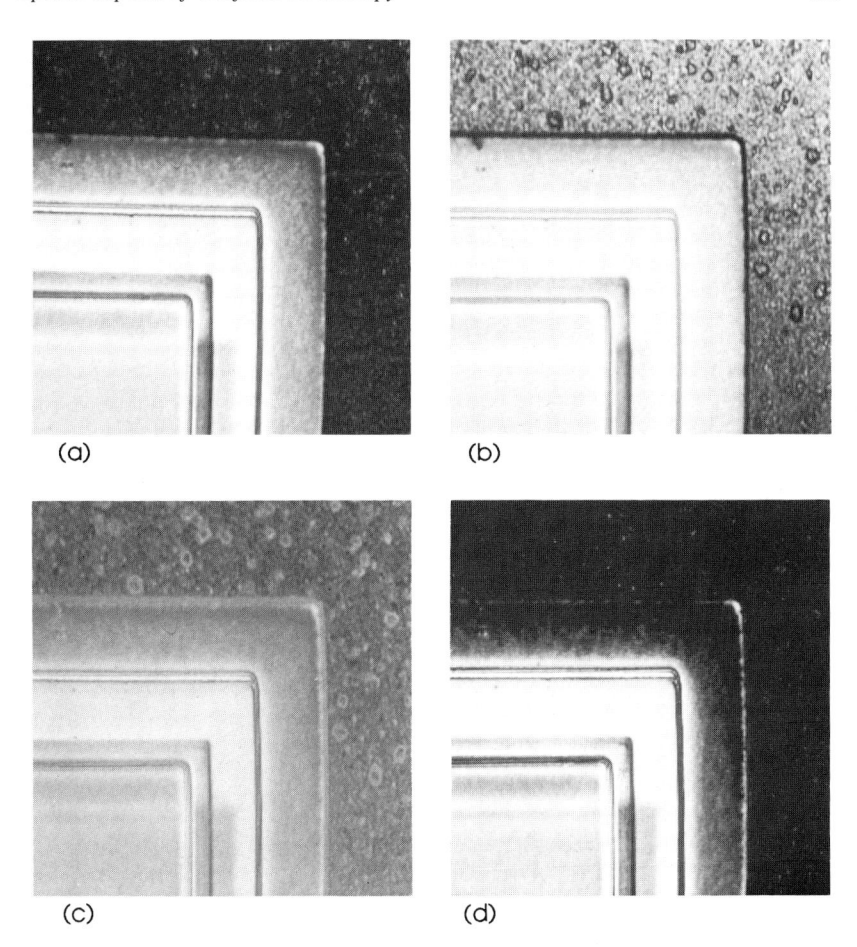

Figure 3.37. (a) Confocal image taken with a 5μm diameter pinhole and (b) taken with a 50μm diameter pinhole. (c) Composite image formed by subtracting half of the image of Figure 3.37(b) from that of Figure 3.37(a), but with an electronic offset added in such that the image intensity was always positive. (d) As (c), but without the electronic offset, and displaying only the positive parts of the composite image.

The different sectioning strength obtained with the two different-sized pinholes is shown in Figure 3.37(a) and (b). The microscope was focussed on the bright, right-angled portion of silicon, and the sectioning is considerably greater in the 5μm detector case.

If we now decide to form a composite image, we have to make a decision about what to do with the portions of the signal which go negative. If we simply choose to add in an electronic offset such that the composite signal is always positive, we obtain the rather washed-out image of Figure 3.37(c), which, because the negative-going signals have been included, does not exhibit improved sectioning. On the other hand, if we simply exclude all negative-going signals from the image, we obtain Figure 3.37(d), which shows a tremendous improvement in sectioning.

It should be emphasised that although the predicted improvement is confirmed by Figures 3.37(d), we have not taken into account, in the theory, variations in reflectivity across the specimen. This will produce variations in the optical sectioning strength in the composite image. Further images obtained with this technique are shown in Chapter 8.

We have simply discussed combining two images in this section. Naturally the method could be extended to forming a composite image from any number of individual images.

VII. The direct-view microscope

The imaging and design criteria of these instruments will not be discussed in great detail at this point, as this will be dealt with more fully in Chapters 9 and 14. We shall use the words "direct-view" and "tandem scanning" (microscopes) interchangeably in the following, as they are, unfortunately, two different names for the same instrument. We shall also restrict our discussion to a comparison of the optical sectioning properties of these instruments to those of single detector laser-based systems.

The operating principle of the tandem scanning microscope has already been discussed in Chapter 1, section VI and Figure 1.41. It essentially consists of a source aperture disc, S, and a detector aperture disc, D. The scanning is achieved by rotating the aperture discs. In the reflection systems which we will consider, S and D are of course identical.

We will begin by considering brightfield imaging, where we can write, again using the simple methods of Fourier optics, that the $V(z)$ response is given by [Wilson and Hewlett, 1990e]:

$$I(u) = \int Q(v)|h(2u,v)|^2 \, v \, dv \qquad (3.34)$$

where $Q(v) = S(v) \otimes D(v)$, and we have assumed, for the moment, that the apertures in the discs are circular, of normalised radius v_p, and that they are sufficiently far apart that there is no cross-talk between neighbouring pinholes. This permits us to write:

$$Q(v) = \frac{2}{\pi}\left[\cos^{-1}\left(\frac{v}{2v_p}\right) - \left(\frac{v}{2v_p}\right)\sqrt{1 - \left(\frac{v}{2v_p}\right)^2}\right] \qquad (3.35)$$

Optical Aspects of Confocal Microscopy 131

which is zero for $v \geq 2v_p$. If we compare this result to the single pinhole laser-based system, equation (3.7), we can see that the direct-view system is equivalent, but with an effective detector given by $Q(v)$ rather than $\text{circ}(v/v_p)$.

We plot in Figure 3.38(a) a series of responses for finite-sized apertures. The behaviour is broadly similar to that of the single pinhole system of Figure 3.2, although the curves have a different shape. This means that the sectioning, as measured by the half-width, is better in the direct-view case at higher values of v_p, Figure 3.38(b). On the other hand, if we measure the sectioning at the 10% width rather than the half-width, we find that the single aperture approach always gives better sectioning, Figure 3.38(c).

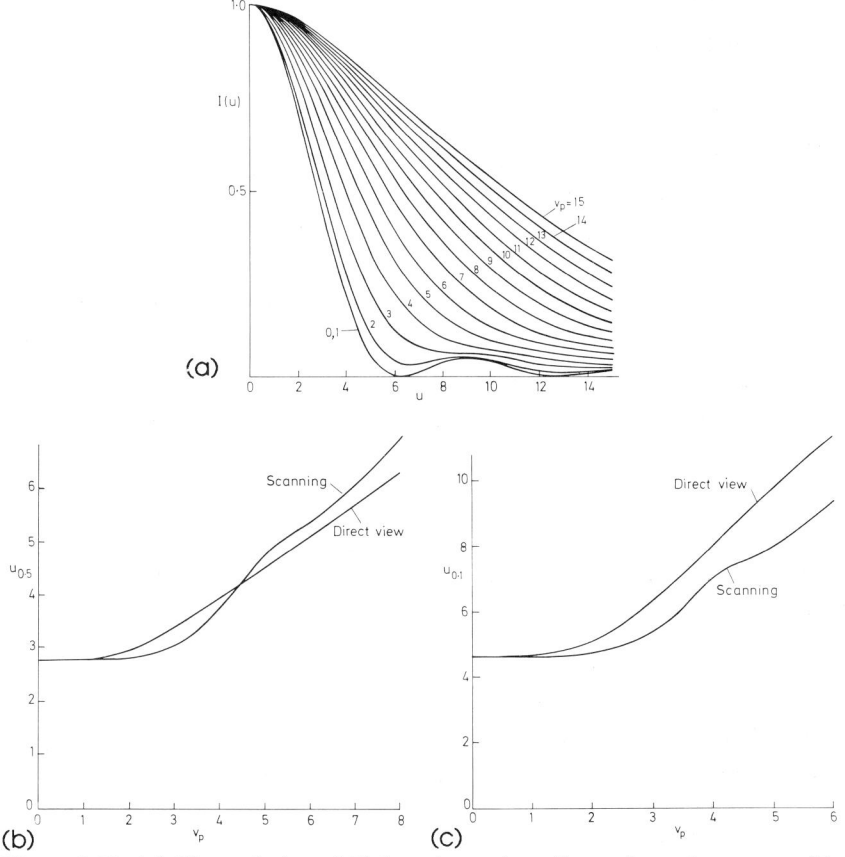

Figure 3.38. (a) The variation of $I(u)$ against u in a direct view microscope with various sizes of aperture pinhole. (b) The half-width of the curves of (a) as a function of aperture size. The scanning curve refers to the case of the single aperture laser-based system of Figures 3.2 and 3.3. (c) As (b), but now the 10% width is used as a metric of sectioning.

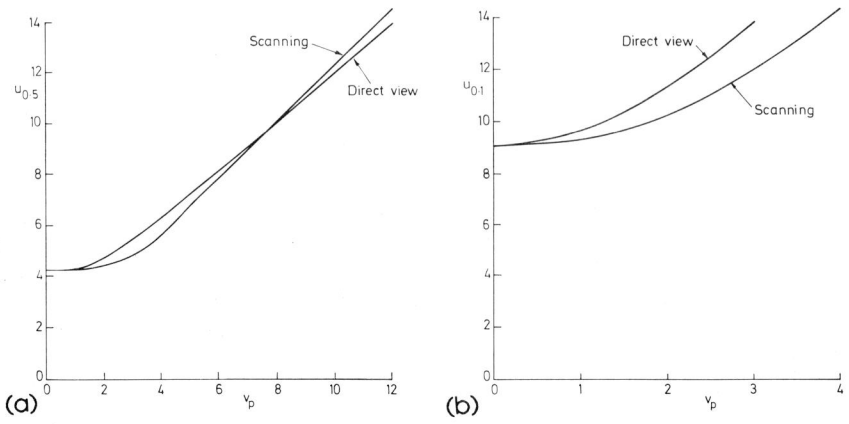

Figure 3.39. Axial resolution in fluorescent imaging in the direct-view and single aperture scanning microscopes for the case of equal excitation and fluorescence wavelengths. (a) The half-width resolution and (b) the 10%-width resolution.

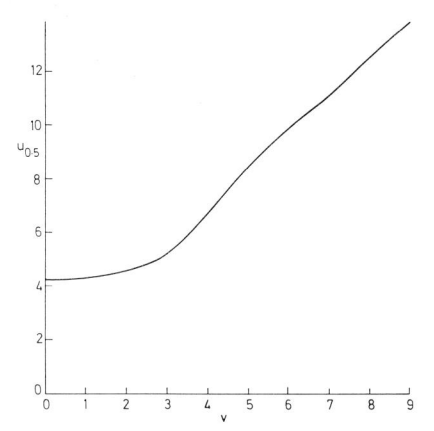

Figure 3.40. The half-width axial resolution in a direct-view fluorescence microscope with laser excitation for the case of equal excitation and fluorescence wavelengths.

Similar comparisons can be made in the fluorescence case, and again the shape of the curves is different. The direct-view response again falls off more quickly at first. Figure 3.39 shows the behaviour in the limiting case of equal excitation and fluorescence wavelength. In the case of fluorescence imaging, the signal level in the direct-view case is often low, and it has therefore been suggested that laser illumination might be used to advantage. If we adopt this approach, the curves of Figure 3.39 no longer apply, because they were based on incoherent illumination of the source apertures. If we repeat the calculations for coherent illumination, we find that the half-width of the fluorescence $V(z)$ varies as shown in Figure 3.40.

Optical Aspects of Confocal Microscopy

All our discussion so far has been concerned with the case in which the apertures are so far apart that neighbouring apertures do not affect each other. In a practical design, we must know how close together the apertures should be placed. We will illustrate the problem by considering the case in which the apertures are spaced on a rectangular grid, T optical units apart. In practice, however, the apertures are usually arranged along a series of Archimedean spirals to simplify the scanning. We will also consider only the incoherently illuminated brightfield case. It is intuitively clear that as the apertures come closer together, the sectioning will deteriorate. This is illustrated in Figure 3.41(a), in which we have considered infinitely small aperture pinholes, and Figure 3.41(b) where pinhole radii corresponding to $v_p = 4$ have been considered.

It is an important observation from these curves that $I(u)$ does not generally tend to zero as u becomes very large. In other words, the sectioning does not die away completely. We present this in Figure 3.42 for a variety of pinhole sizes. This suggests an alternative design criterion for direct view microscopes. We decide how small we want $I(u \to \infty)$ to be, and then use Figure 3.42 to determine suitable aperture sizing and spacing.

At this point, we have discussed only the strength of the optical sectioning by considering planar objects. The imaging of extended objects, which has been considered by Sheppard and Wilson (1981b) and Wilson and Sheppard (1984), will also be discussed in Chapters 9 and 14 of this book.

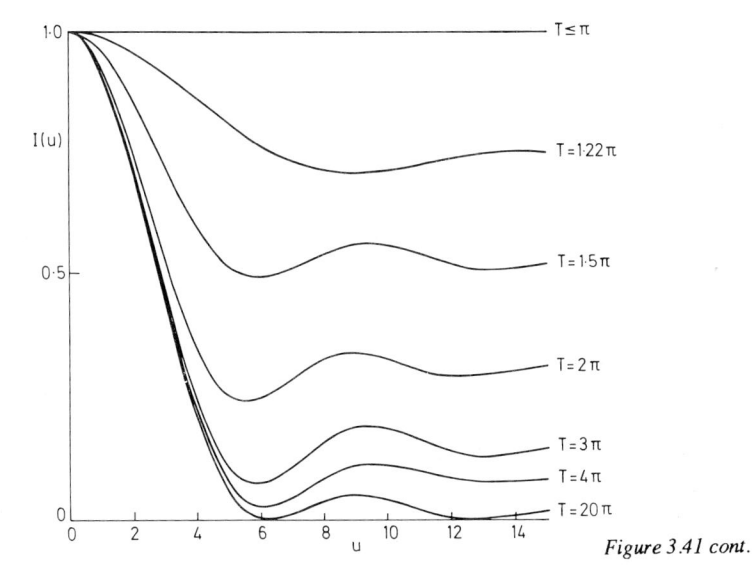

Figure 3.41 cont.

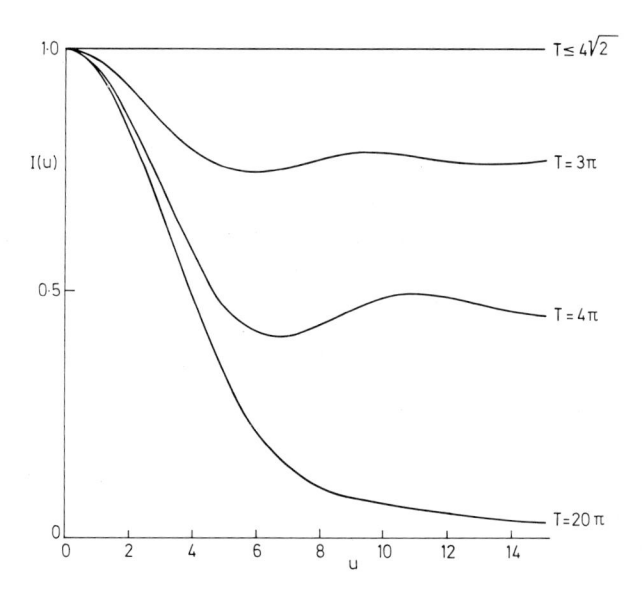

Figure 3.41. The variation in $I(u)$ for a brightfield direct view microscope in which the apertures are spaced on a rectangular grid, T optical units apart. (a) For the case of infinitesimally small apertures and (b) for apertures of normalised radius, $v_p = 4$.

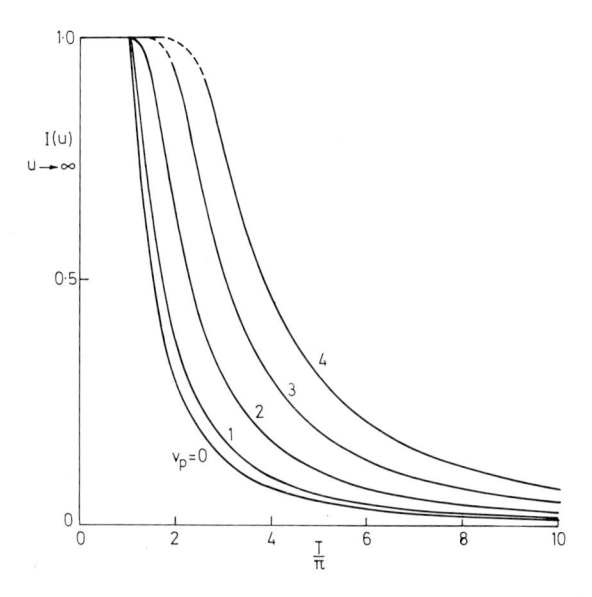

Figure 3.42. The variation of $I(u \to \infty)$ with aperture spacing for a variety of aperture sizes, for a brightfield direct view microscope.

VIII. Scanning mechanisms

We will conclude this chapter by discussing a few scanning methods and microscope modifications which lead to good confocal imaging. In our discussion so far, we have said that beam-scanning systems using vibrating galvanometer mirrors do not really give real-time imaging, and indeed, in fluorescence imaging, for example, this may be an advantage. However, in all beam-scanning systems we have to introduce an extra optical system in order to prevent vignetting, and to ensure that the important pupil apertures are always filled. In practice, this means that a transfer lens system such as that shown in Figure 3.43 must be used.

If we place a scanning element at the plane XX, this will cause deflected beams of the type shown in the bold and the dashed lines. As we require all of this light to enter the entrance pupil of a lens or the aperture of a scanning unit (i.e., a mirror), then we must arrange to place that unit at a "zero-deflection" plane. Such a plane is shown

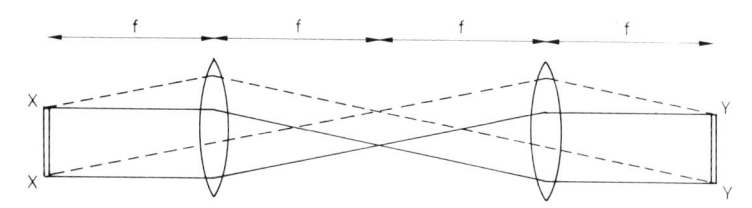

Figure 3.43. The layout of the $4 - f$ scan transfer lens system.

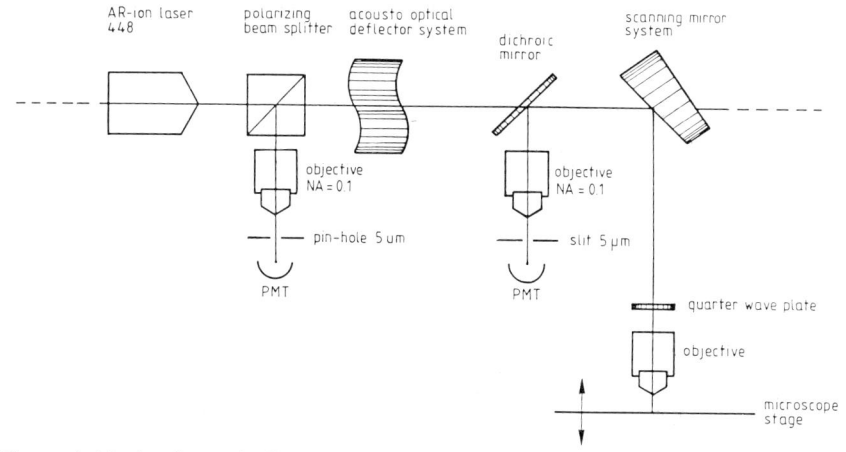

Figure 3.44. A schematic diagram of a beam-scanning microscope incorporating a slit-shaped detector for fluorescence imaging and a pinhole detector for brightfield imaging. (Courtesy of Dr. A. Draaijer, T. N. O., Delft, whose work was financially supported by Tracor Northern.)

as YY in Figure 3.43. The figure has been drawn for parallel light, but a similar scheme can be devised for use with diverging beams. Thus a system involving scanning mirrors might use two such relay systems, one for the x-scan and one for the y-scan, where the plane YY of the second unit coincides with the entrance pupil of the microscope. It may be possible, however, to dispense with the second set of lenses if the final scanning mirror is placed in the correct position with respect to the entrance pupil of the microscope.

If, however, we wish to obtain real-time scanning, then an acousto-optic deflection system may be used. These systems are fine for brightfield imaging, but care needs to be taken when imaging in fluorescence. This is because the fluorescence radiation will be at a different wavelength from the incident radiation, and hence will not be properly de-scanned by the acousto-optic deflector. In this case, a system such as that shown in Figure 3.44 might be used. Here the scanning is achieved with a combination of an acousto-optic beam deflector and a scanning mirror. The brightfield image, which can be de-scanned by the acousto-optic element, is detected by a stationary pinhole. The fluorescent light, however, is beam-split off after being de-scanned in one direction by the scanning mirror system. It is then allowed to scan over a slit, where it is detected. As an alternative, a single line CCD array may be used in place of the slit/photomultiplier combination. The fluorescence images of Figure 3.24 were obtained with the system of Figure 3.44, as was the brightfield image of Figure 3.45.

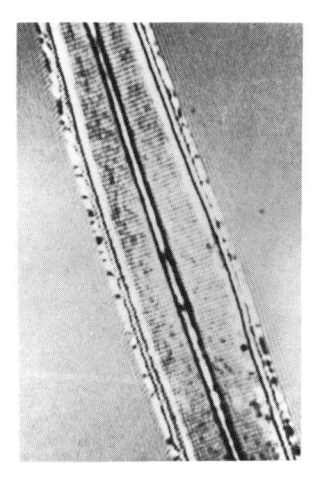

Figure 3.45. A reflected light image of the diatom Amphipleura pellucida, showing lines spaced $0.25\mu m$ apart and dots $0.17\mu m$ apart. The image was taken with a 1.25 numerical aperture achromatic objective. The horizontal field of view is $17\mu m$. (Courtesy of Dr. A. Draaijer, T. N. O., Delft, whose work was financially supported by Tracor Northern.)

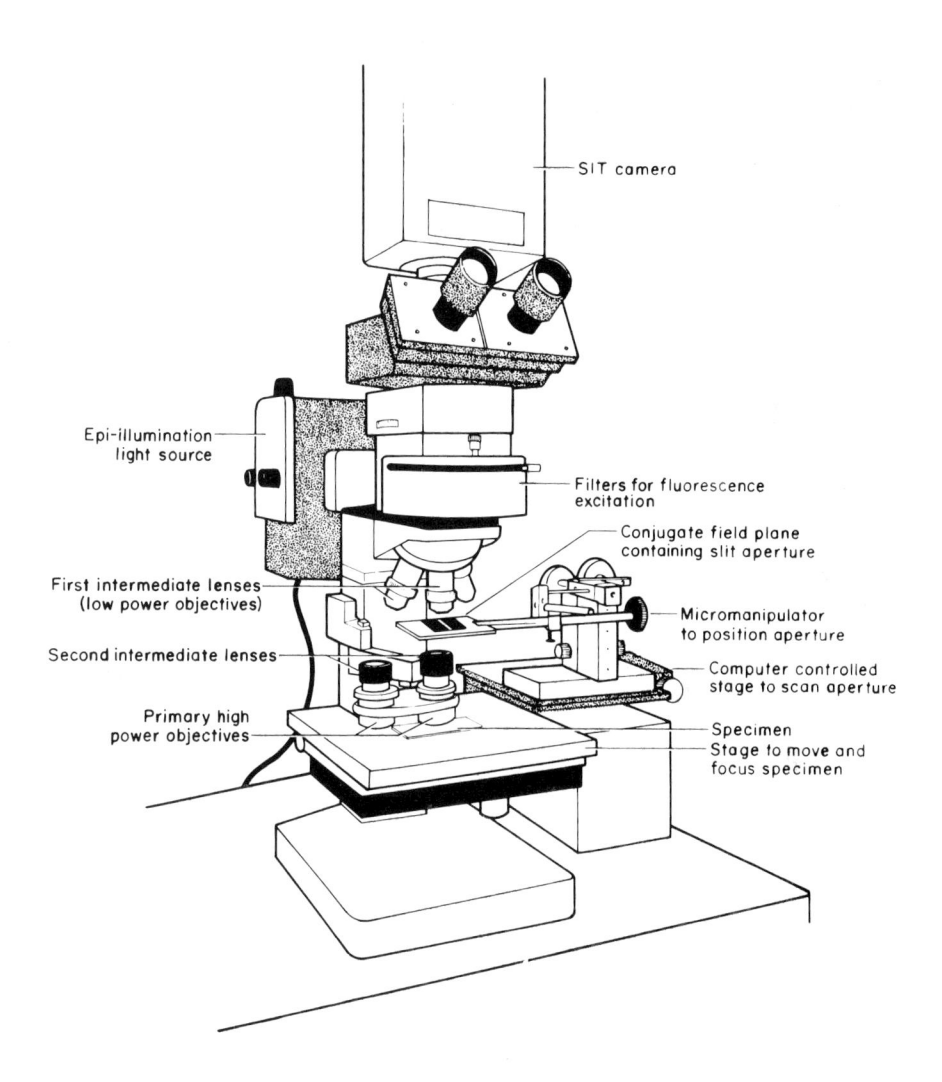

Figure 3.46. Illustration of a single-aperture confocal microscope. An upright epi-illumination fluorescence microscope (Leitz Laborlux D) was modified to permit confocal imaging. The two modifications are (i) lowering the focussing stage approximately 10 cm and reattaching it to the focussing mechanism, and (ii) adding a swivelling attachment to the microscope to hold the primary high-power objective(s). Next to the microscope is a computer-controlled stage upon which a micromanipulator has been placed to hold the aperture. A 75-watt xenon bulb provided fluorescence illumination using Leitz cubes (I_2 and N) for excitation and barrier filtration. (Courtesy of Dr. J. W. Lichtman.)

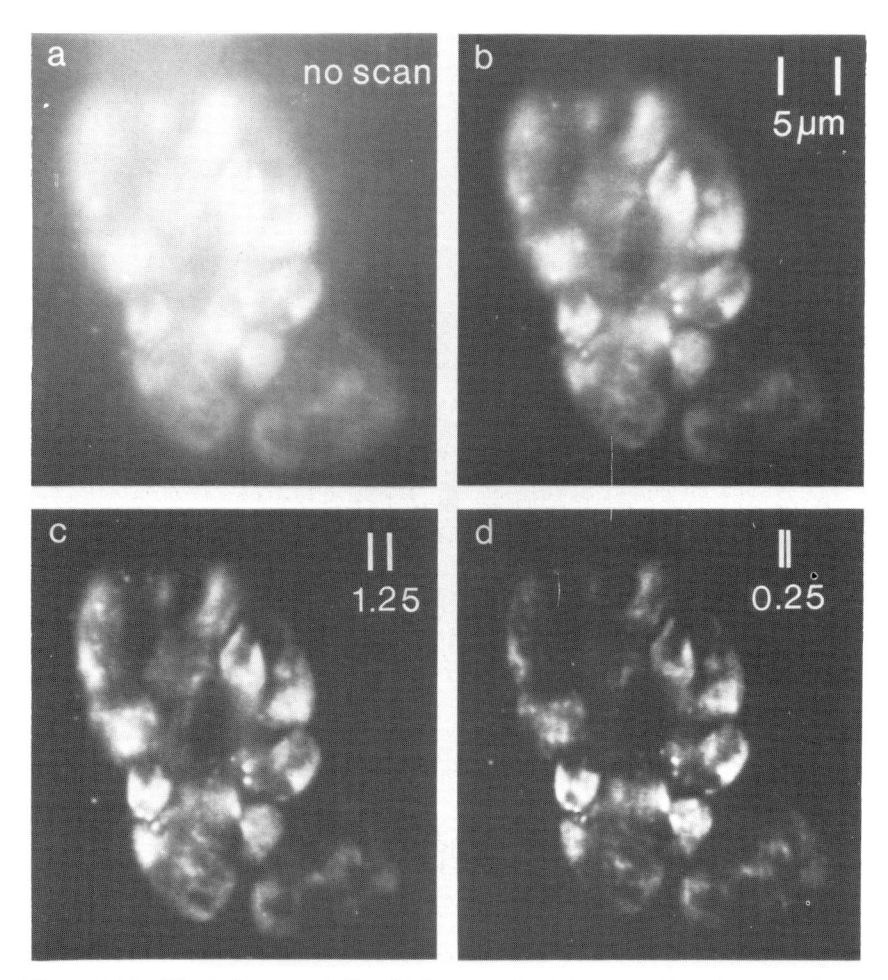

Figure 3.47. The influence of slit width on confocal images. Four images of one motor nerve terminal (snake transversus abdominis muscle) were stained with the fluorescent probe sulforhodamine 101. (a) Averaged conventional image (32 frames) digitised from a video camera through identical optics and magnification; as other panels except that the scanning slit was removed. Note background and out-of-focus fluorescence that obscure details of individual terminal boutons. (b-d) Confocal images scanned by slits of three different widths (slit width is shown to scale at upper right of each panel). (b) Significant improvement in background was obtained by using the widest slit (100μm, which illuminated 5μm in the specimen plane). (c) Reduction of slit width to 25μm (1.25μm at the specimen) further reduced the background. (d) Diffraction-limited slit (5μm, 0.25μm at the specimen) provided maximum optical sectioning capability (minimal depth of field). Note that some regions of the terminal are out of the plane of focus, and therefore vanish from the image. Images (b-d) were obtained by scanning at progressively slower rates proportional to the slit width, such that the total integrated light presented to the video camera was held constant. Slit widths serve as calibration bars. (Courtesy of Dr. J. W. Lichtman.)

One way of thinking of the tandem scanning or direct-view microscope is that it is a method of converting a conventional microscope into confocal operation without losing the ability to view the image directly by eye. Recently Lichtman [Lichtman et al., 1989] has suggested a straightforward modification to a conventional fluorescence microscope to permit confocal fluorescence imaging. It is shown in Figure 3.46, and involves scanning a slit aperture, in one direction only, in a conjugate field plane. Figure 3.47 shows a series of fluorescence images taken with this system, which also emphasises the importance of choosing the correct slit width.

IX. Conclusions

We have tried to highlight some of the optical problems which will be encountered in confocal microscopy. If optimum results are to be obtained, great care must be taken to ensure that the system is properly aligned. This is particularly important if further post-processing is envisaged, especially if this processing relies on an accurate knowledge of the microscope's point spread function. The simple $V(z)$ response is a particularly powerful test of alignment which is very easy to use.

References

Awamura, D., Ode, T. and Torezawa, M., (1987). Colour laser microscope. Proc. S.P.I.E., **765**, 53-60.

Born, M., and Wolf, E., (1975). *Principles of optics*. Pergamon Press, Oxford.

Corle, T. R., Chou, C.-H. and Kino, G. S., (1986). Depth response of confocal optical microscopes. Opt. Lett., **11**, 770-7702.

Hellmuth, T., Seidel, P. and Siegel, A., (1988). Spherical aberration in confocal microscopy. Proc. S.P.I.E., **1028**, 28-32.

Hopkins, H. H., (1953). On the diffraction theory of optical images. Proc. Roy. Soc. Lond., **A217**, 408-432.

Keller, H. E., (1989). Objective lenses for confocal microscopy. In: *The handbook of biological confocal microscopy*. Pawley, J. (Editor), 69-77. I.M.R. Press, Madison, Wisconsin.

Koester, C. J., (1980) A scanning mirror microscope with optical sectioning characteristics: applications in ophthalmology. Appl. Opt., **19**, 1749-1757.

Lichtman, J. W., Sunderland, W. J. and Wilkinson, R. S., (1989). High-resolution imaging of synaptic structure with a simple confocal microscope. The New Biologist, **1**, 75-82.

Matthews, H. J., (1987). D. Phil. Thesis, University of Oxford.

Schutt, W., Reinholz, F., Grummer, G., Kuhlmann, F. and Kraft, S. K., (1988). Proc. Int. Congress on Optical Science and Engineering (S.P.I.E.), Paris.

Sheppard, C. J. R. and Cogswell, C. J., (1990). Confocal microscopy with detector arrays. J. Mod. Opt.

Sheppard, C. J. R. and Mao, X. Q., (1988). Confocal microscopes with slit apertures. J. Mod. Opt., **35**, 1169-1185.

Sheppard, C. J. R. and Wilson, T., (1978). Image formation in scanning microscopes with partially coherent source and detector. Opt. Acta, **25**, 315-325.

Sheppard, C. J. R. and Wilson, T., (1979). Effect of spherical aberration on the imaging properties of scanning optical microscopes. Appl. Opt., **18**, 1058-1063.

Sheppard, C. J. R. and Wilson, T., (1981a). Effects of high angles of convergence on V(z) in the scanning acoustic microscope. Appl. Phys. Lett., **38**, 858-859.

Sheppard, C. J. R. and Wilson, T., (1981b). The theory of the direct-view confocal microscope. J. Microsc., **124**, 107-117.

Stamnes, J. J., (1986). *Waves in focal regions*. Adam Hilger, Bristol.

van der Voort, H. T. M. and Brakenhoff, G. J., (1990). Three-dimensional image formation in high-aperture fluorescence confocal microscopy: a numerical analysis. J. Microsc., **158**, 43-54.

Wilson, T., (1989a). Depth response of scanning microscopes. Optik, **81**, 113-118.

Wilson, T., (1989b) Optical sectioning in confocal fluorescent microscopes. J. Microsc., **154**, 143-156.

Wilson, T., (1990). Detector arrays in confocal microscopy. Scanning Microscopy, **4**, 21-24.

Wilson, T. and Carlini, A. R., (1987). Size of the detector in confocal imaging systems. Opt. Lett., **12**, 227-229.

Wilson, T. and Carlini, A. R., (1988). Three-dimensional imaging in confocal imaging systems with finite-sized detectors. J. Microsc., **149**, 51-66.

Wilson, T. and Carlini, A. R., (1989). The effect of aberrations on the axial response of confocal imaging systems. J. Microsc., **154**, 243-256.

Wilson, T. and Hewlett, S. J., (1990a). Imaging in scanning microscopes with slit-shaped detectors. J. Microsc.

Wilson, T. and Hewlett, S. J., (1990b). Coherent detection in scanning microscopes. Inst. Phys. Conf. Ser., **98**, 629-632.

Wilson, T. and Hewlett, S. J., (1990c). The use of annular pupil plane filters to tune the imaging properties in confocal microscopy. J. Mod. Opt.

Wilson, T. and Hewlett, S. J., (1990d). Imaging strategies in three-dimensional confocal microscopy. Proc. S.P.I.E., **1245**.

Wilson, T. and Hewlett, S. J., (1990e). Optical sectioning strength in direct-view scanning microscopes employing finite-sized pinhole arrays.

Wilson, T., Hewlett, S. J., and Sheppard, C. J. R., (1990). The use of objective lenses with slit pupil functions in the imaging of line structures.

Wilson, T. and Sheppard, C. J. R., (1984). *Theory and practice of scanning optical microscopy.* Academic Press, London.

Zernike, F. and Nijboer, B. R. A., (1949). In: *La theorie des Images Optiques.* (Paris, Revue d'Optique).

4. Three-dimensional Imaging in Confocal Microscopy

C. J. R. SHEPPARD AND C. J. COGSWELL

I. Introduction

Perhaps the most important property of confocal microscopy is its ability to form a three-dimensional image of an object possessing appreciable depth. In biological microscopy, for example, this allows the investigation of structure inside a cell. In materials science or semiconductor device technology it allows the measurement of surface topography. This ability to form a three-dimensional image is extremely powerful, especially when coupled with the improved resolution and contrast of confocal microscopy.

The three-dimensional imaging capability results from the strong optical sectioning property of confocal microscopy. This is of a completely different nature from the restricted depth of field in conventional microscopy, the difference being that in a conventional microscope out-of-focus information is merely blurred, whilst in the confocal system it is actually detected much less strongly: light scattered at some place axially separated from the focal plane is defocused at the detector plane and hence fails to pass efficiently through a mask placed there.

Using a confocal microscope, a series of image slices through a thick object can be obtained. For example, Figure 4.1 shows four sections of a specimen of rat cerebellum. An alternative is to generate an x-z image, which is a vertical slice through the object. This is illustrated in Figure 4.2.

The optical sectioning property also allows the use of three further imaging techniques: the extended focus [Wilson and Hamilton, 1982; Sheppard et al., 1983] and auto-focus [Cox and Sheppard, 1983a] techniques result in vastly increased depth of field whilst retaining high resolution, whereas the surface profiling technique [Caulfield and Kryger, 1978; Hamilton and Wilson, 1982; Cox and Sheppard, 1983b] allows non-invasive investigation of surface topography.

If the object is scanned axially in a confocal microscope, the distance moved from some datum for which the maximum intensity is detected is a measure of the surface height of the specimen at this point. This is the basis of the surface profiling technique, which can be used to produce one- and two-dimensional plots of surface topography with a sensitivity of better than 100nm. The results can be presented in the form of an isometric plot [Hamilton and Wilson, 1982] or by modulating the image brightness [Cox and Sheppard, 1983b] or colour [Sheppard, 1986a]. The maximum intensity can be determined either using analogue [Hamilton and Wilson, 1982], or digital, [Cox and Sheppard, 1983b] techniques. Alternatively a white-light source and a photodiode array may be used to avoid the necessity for axial scanning [Molesini et al., 1984].

As well as recording the surface height, the maximum intensity, which gives a measure of local surface reflectivity, may also be displayed. The depth of focus of this auto-focus technique is in principle

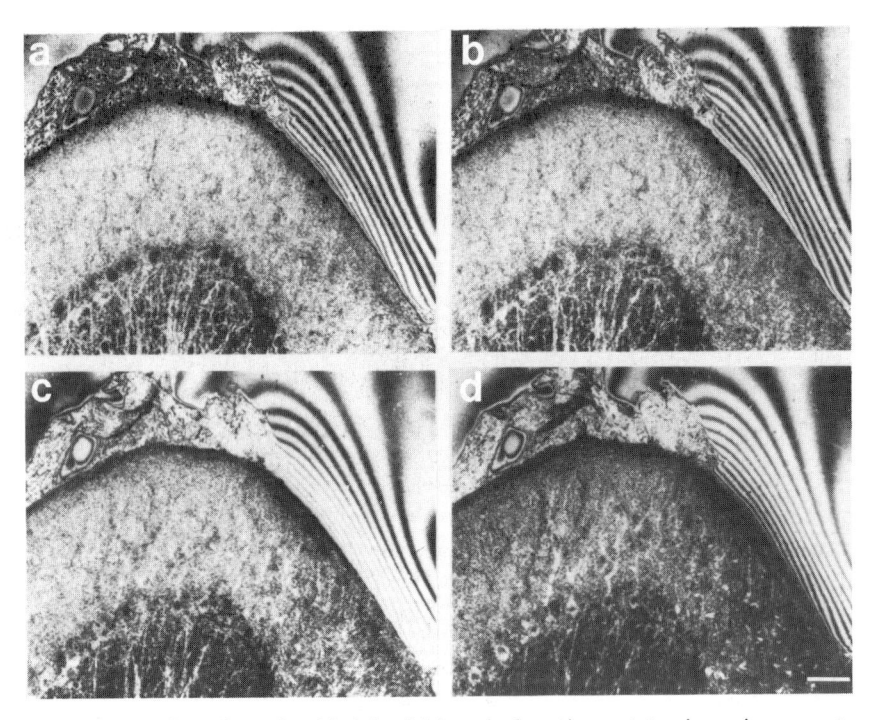

Figure 4.1. Reflected confocal bright field optical sections at 1 micron increments of focus from a 30 micron thick section of rat cerebellum. In (a) and (b), unstained axons are clearly visible in the lower regions of the images (granule cell layer) under the semicircular row of Purkinje cells. In (d), which is focused 3 microns below (a), the axons are no longer visible but now dendrites can be seen to radiate upwards into the molecular layer. Scale = 50 microns. (Preparation courtesy of Dr. R.A.J. McIlhinney, University of Oxford.)

unlimited. The overall appearance of an auto-focus image is very similar to that in the confocal mode, except that absolute height information is suppressed. Thus high resolution diffraction-limited imaging may be achieved with a depth of field vastly greater than in a conventional microscope. The auto-focus technique is highly non-linear and hence imaging cannot be expressed in terms of a transfer function. One advantage of auto-focusing is that the surface profile can be recorded simultaneously. Thus, reflectivity and surface height information can then be processed to produce three-dimensional stereoscopic image pairs. Figure 4.3 illustrates an alternative display technique where an isometric view of the object is constructed.

As an alternative to the auto-focus technique for producing a large depth of focus, the confocal microscope may also be used in the extended focus mode. In this technique the object is again scanned axially but, instead of recording the maximum intensity, an integral over all axial positions is performed. The confocal system produces an optical slice through a thick object and these slices are then summed to form the extended focus image, which is the projection of the thick object in a direction parallel to the optical axis. As the object is scanned axially, each point on the surface at some time travels through the focal point and, because out-of-focus information is detected only weakly in the confocal system, this does not greatly affect the image.

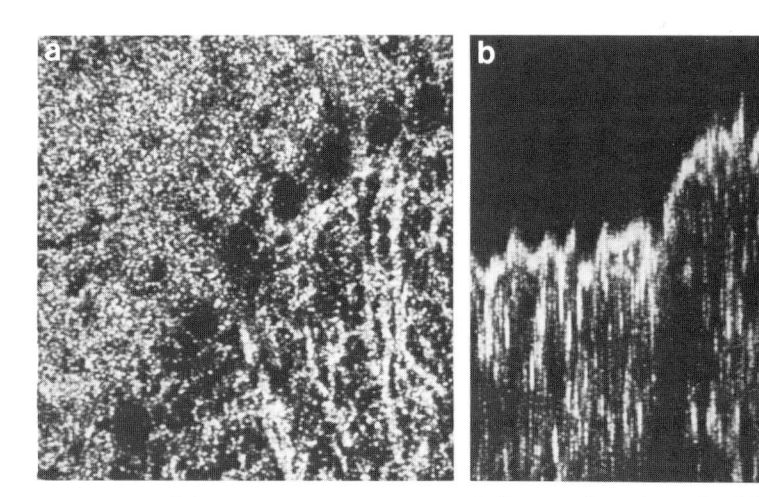

Figure 4.2. (a) Enlarged view of rat cerebellum, reflected confocal bright field. The dark circles are Purkinje cells. (b) x-z image created by scanning in depth (z-axis) across a single horizontal line through the centre of image (a). This produces a side view image in which variations in specimen height appear as bright peaks while the location of a Purkinje cell below the surface is visible as a dark hole. Note: horizontal scale = 20 microns (both images), vertical scale in image (b) = 1.5 microns which produces an exaggerated depth appearance.

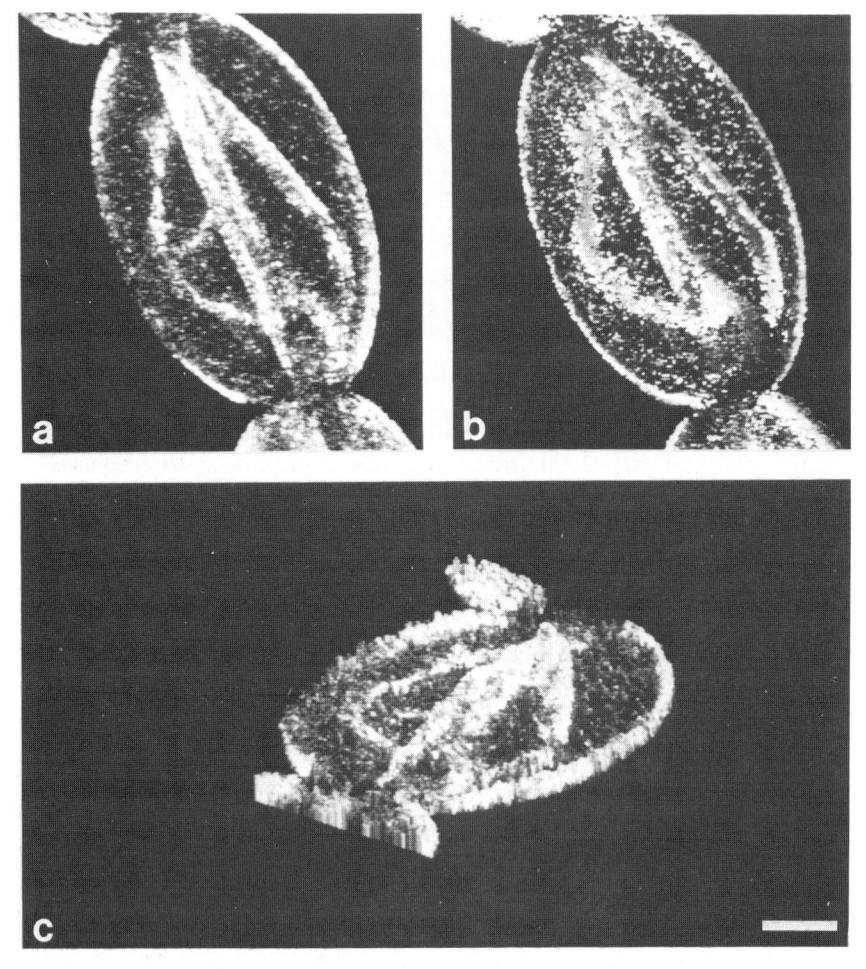

Figure 4.3. *Tradescantia* stamen hair cell. (a) Auto-focus image. (b) Surface profile image where a grey scale has been used to code the significant bits in the image according to their location in depth (light grey = top, dark grey = bottom of cell). (c) Isometric projection using information from both (a) and (b). Scale=20μm.

If on the other hand the axial scanning and integration are performed with a conventional system the out-of-focus information produces a blurred final image. The integration may be performed very simply in an analog manner by photographic recording, or by digital techniques.

The extended focus method, as in the auto-focus method, results in a depth of focus which is in principle unlimited. In practice it is restricted by the available amplitude of the axial scan and by the limited working distance of the objective but, nevertheless, depths of field of several hundred microns have been achieved [Sheppard et al., 1983].

The images produced using the auto-focus method and the extended focus method are visually similar. Both result in greatly increased depth of focus, whilst maintaining a lateral resolution usually better than in conventional microscopy. In practice, auto-focus images usually exhibit better contrast but, on the other hand, the extended focus method is superior in filtering out noise in the image. A further difference between the auto- and extended focus modes is that the latter technique, for objects with slowly varying height, results in partially coherent imaging for which the transfer function has been derived [Sheppard et al,. 1983]. This has some advantage for numerical image calculations and processing.

Image projections, produced using either the auto-focus or extended focus techniques, can be used as a basis for forming stereo pairs. Figure 4.4 shows an example of this method. In practice the stereo pairs can be generated during scanning or, alternatively, optical sections can be stored for subsequent processing.

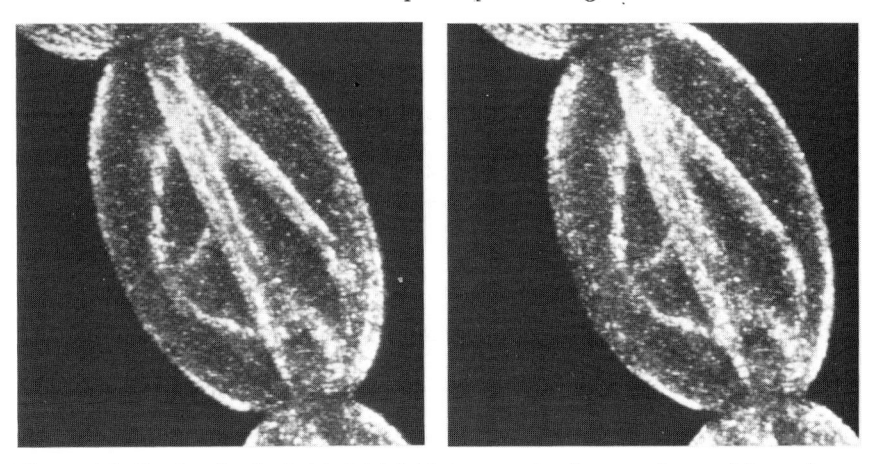

Figure 4.4. Confocal reflected bright field stereo pair of seven planes of focus from the *Tradescantia* stamen hair cell of Figure 4.3. Location of the brightly reflecting cytoplasmic channels in depth can be clearly distinguished from the surrounding, darker-appearing vacuoles.

II. Imaging in confocal systems: a single point object

We continue with a discussion of image formation in confocal systems. In a confocal microscope light from a laser is focused by an objective lens onto the specimen, and the reflected (or transmitted, or fluorescent) light collected by a second lens, the collector, and refocused onto a detector, the area of which is limited by a small pinhole. In a reflection system the objective and collector lenses are physically the same lens.

Let us introduce a normalised radial coordinate ρ in the pupil plane of the first objective, defined by:

$$\rho = r/a_1 \qquad (4.1)$$

where a_1 is the radius of the pupil and r is the actual radial co-ordinate. Then the point spread function of the lens, h_1, is given by the Hankel transform of the pupil function:

$$h_1(v,u) = 2\int_0^\infty P_1(\rho)\exp\left(\frac{1}{2}ju\rho^2\right)J_0(v\rho)\,\rho\,d\rho \qquad (4.2)$$

Here v and u are cylindrical and axial optical coordinates defined by:

$$v = kr\sin\alpha_1 \qquad (4.3)$$

and

$$u = 4kz\sin^2(\alpha_1/2) \qquad (4.4)$$

in which k is the wave number ($= 2\pi/\lambda$), λ is the wavelength and α_1 is the angular aperture of the objective. The term $P_1(\rho)$ is the pupil function of the lens, assumed radially symmetric, which is complex in general to account for aberrations. These aberrations exclude defocus, for which the parabolic phase error is introduced explicitly in the term $\exp(ju\rho^2/2)$. The point spread function is normalised so that its value is unity at the focal point for an aberration-free, unapodized pupil. In an analogous fashion, the point spread function for the second lens, the collector, is:

$$h_2(v,u) = 2\int_0^\infty P_2(\rho)\exp\left(\frac{1}{2}ju\rho^2\right)J_0(v\rho)\,\rho\,d\rho \qquad (4.5)$$

for a reflection system and

$$h_2(v,u) = 2\int_0^\infty P_2(\rho)\exp\left(-\frac{1}{2}ju\rho^2\right)J_0(v\rho)\,\rho\,d\rho \qquad (4.6)$$

for a transmission system. The difference between these equations appears because the two defoci are equal in the reflection system and equal and opposite in transmission. The confocal system behaves as a coherent system with an effective point spread function given by the product of the individual point spread functions for the two lenses [Sheppard and Choudhury, 1977]. Thus, the intensity in the image of a single point in a confocal system is given by:

$$I(v,u) = |h_1(v,u)h_2(v,u)|^2 \qquad (4.7)$$

For an aberration-free, in-focus system with two equal pupils we then have for the intensity in the image of a single point:

$$I(v) = \left(\frac{2J_1(v)}{v}\right)^4 \qquad (4.8)$$

This is shown in Figure 4.5 in which it is seen that the central peak is sharpened up by a factor of 1.4 relative to the image in a conventional system (at half the peak intensity). The side-lobes are also drastically reduced (from 1.7% to 0.03%, or by 18dB) so we would expect a marked reduction in the presence of optical artifacts in the resultant images.

If the two lenses are not of the same aperture, the images of a single point can be approximately compared by expanding the point spread function as a power series and retaining just the first two terms:

$$h(v) \simeq 1 - av^2 \qquad (4.9)$$

where a is a constant. Then if the second lens has a relative aperture

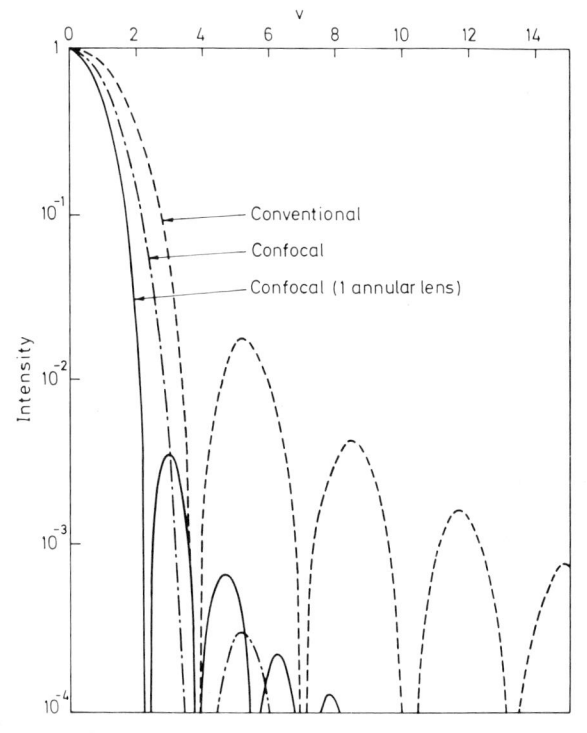

Figure 4.5. Image of a single point in conventional and confocal microscopes.

s, again keeping just the first two terms, the image of the single point in the confocal system is sharpened up by a factor $\sqrt{1+s^2}$, which is independent of a. This approximate behaviour is exactly true for the case of gaussian pupils, which was considered by Sheppard and Wilson (1978a). For the particular case of s equal to unity, the image is sharpened up by a factor of $\sqrt{2}$ which is very close to the exact figure for circular apertures. Another particular value of s which may be of interest occurs when the first zero of the point spread function of one lens coincides with the first maximum of that of the other, at $s = 0.745$ for circular pupils. This gives a sharpening of about a factor of 1.25 with very low side-lobe levels.

Figure 4.5 also gives the image of a single point in a confocal system employing one lens with an annular aperture (the width of the annulus being very small). For one circular and one annular lens of equal aperture the intensity is given by:

$$I(v) = \left(\frac{2 J_0(v) J_1(v)}{v} \right)^2 \quad (4.10)$$

so that the central peak is now even narrower (a factor 1.67 narrower compared with a conventional system) and the side-lobes are weak as the maxima of the spread function of the annulus coincide with the zeros of that of the circular lens. Although the side-lobes are weak compared with the conventional microscope, they are in fact stronger than in the confocal microscope.

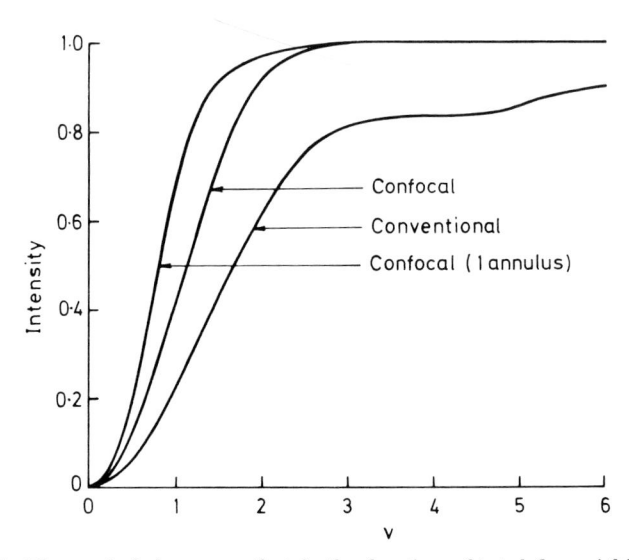

Figure 4.6. The encircled energy, that is the fraction of total flux within a disk of normalised radius v.

3-D Imaging in Confocal Microscopy

For large values of v the intensity in the image of a single point in a conventional system falls off as the third power of v. In a confocal system with two circular lenses it falls off as the sixth power, whereas with one annular lens it falls off as the fourth power. With two infinitely narrow annuli, however, it falls off only as the second power and the total normalised power does not converge; such an arrangement is clearly unusable.

The encircled energy, that is the fraction of the total flux within a disk of normalised radius v, can be computed for the confocal system and is shown in Figure 4.6. The radius containing 90% of the total energy is reduced from about $v = 6$ for a conventional system to less than 2 for a confocal system with two equal circular pupils (that is by a factor of 3.1), and to less than 1.5 for a confocal system with one annular pupil (by a factor of 4.2). The total energy in the point image is found to be [Sheppard and Wilson, 1978b]:

$$\int_0^\infty \left(\frac{2J_1(v)}{v}\right)^4 2\pi v\, dv = 5.78 \qquad (4.11)$$

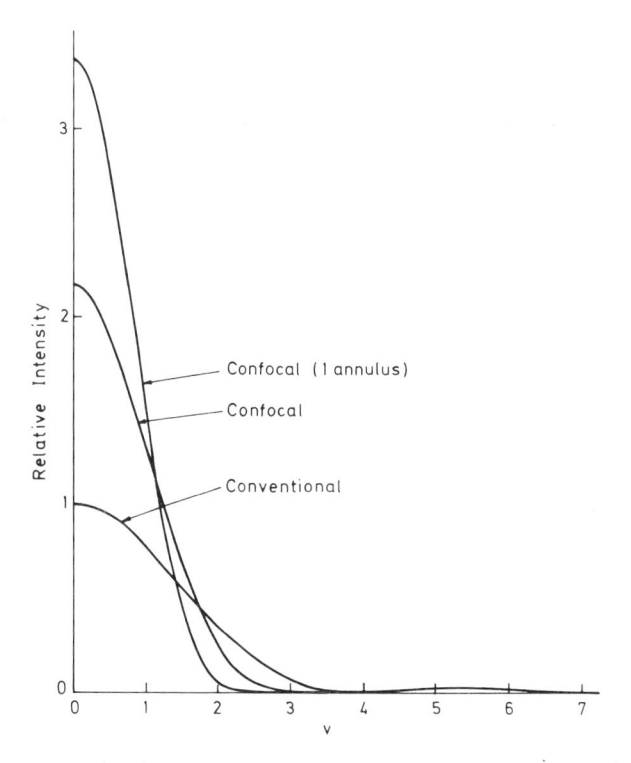

Figure 4.7. Image of a single point after renormalisation to equal total flux.

and

$$\int_0^\infty \left(\frac{2J_0(v)J_1(v)}{v}\right)^2 2\pi v\, dv = 3.73 \tag{4.12}$$

respectively for these two cases, which compares with the conventional figure of:

$$\int_0^\infty \left(\frac{2J_1(v)}{v}\right)^2 2\pi v\, dv = 4\pi = 12.57 \tag{4.13}$$

These allow the point images to be renormalised so that the total flux is constant, as shown in Figure 4.7. This plot demonstrates very strongly the improved performance in confocal imaging.

We now turn to the effects of defocus. If the pupils are unshaded, aberration-free apertures, Equations (4.2)-(4.4) can be written as:

$$h_{1,2}(v,u) = C(v,u) \pm jS(v,u) \tag{4.14}$$

where

$$C(v,u) = 2\int_0^1 \cos\left(\frac{1}{2}u\rho^2\right) J_0(v\rho)\, \rho\, d\rho \tag{4.15}$$

and

$$S(v,u) = 2\int_0^1 \sin\left(\frac{1}{2}u\rho^2\right) J_0(v\rho)\, \rho\, d\rho \tag{4.16}$$

Thus, the confocal image of a single point object, whether in transmission or reflection, is:

$$I(v,u) = \left[C^2(v,u) + S^2(v,u)\right]^2 \tag{4.17}$$

for two equal circular pupils. For a system with one circular pupil and one narrow annular pupil the corresponding image is:

$$I(v,u) = \left[C^2(v,u) + S^2(v,u)\right] J_0^2(v) \tag{4.18}$$

as the annular pupil results in an infinite depth of focus for this lens. The intensity in a plane through the axis for these cases is shown in Figure 4.8. The region of space for which the intensity is greater than one per cent of the maximum is indicated. It should be noticed that for the confocal system this is much smaller than the corresponding region for the conventional microscope which is large and irregularly shaped. This accounts for the superior imaging of thick objects in confocal microscopy. For the confocal system with one annular lens the focal spot is stretched out along the axis and some satellite peaks are evident.

3-D Imaging in Confocal Microscopy

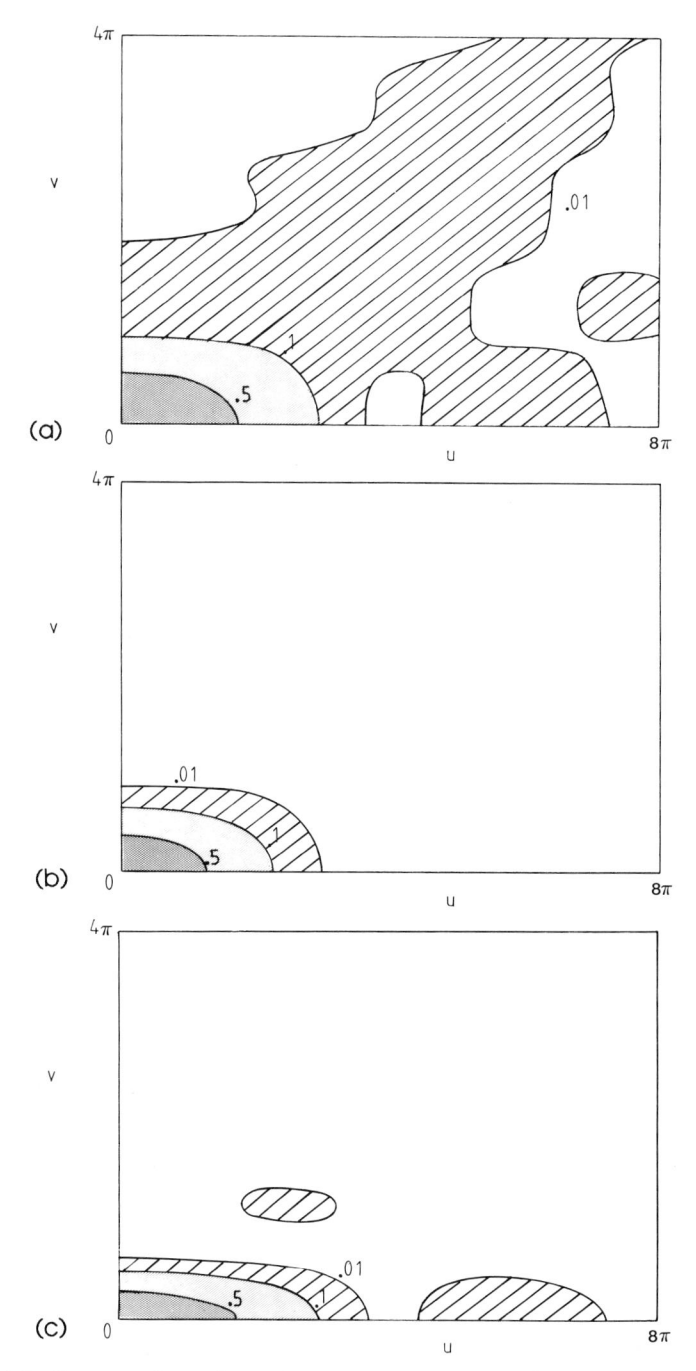

Figure 4.8. Intensity in a plane through the optic axis. (a) Conventional. (b) Confocal. (c) Confocal with one annular pupil.

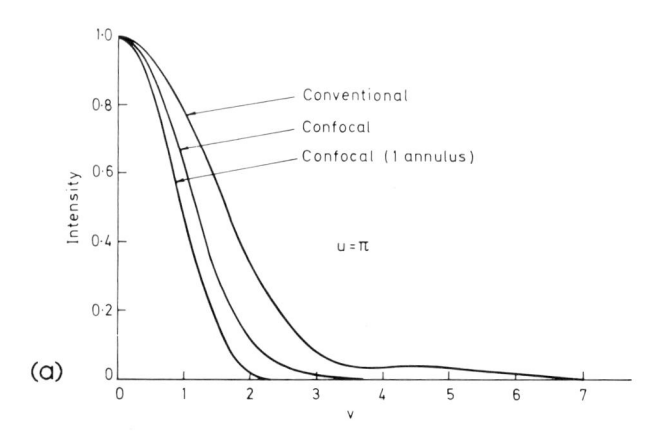

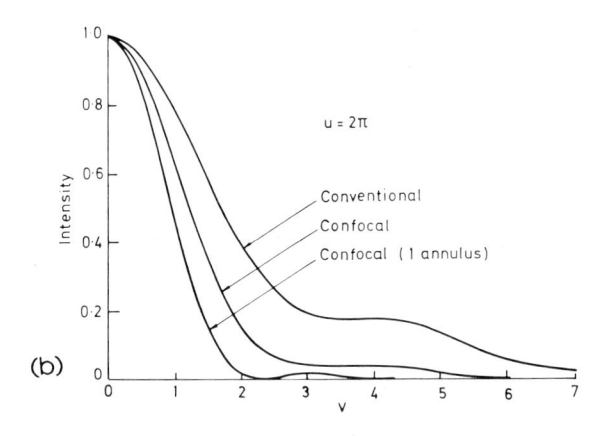

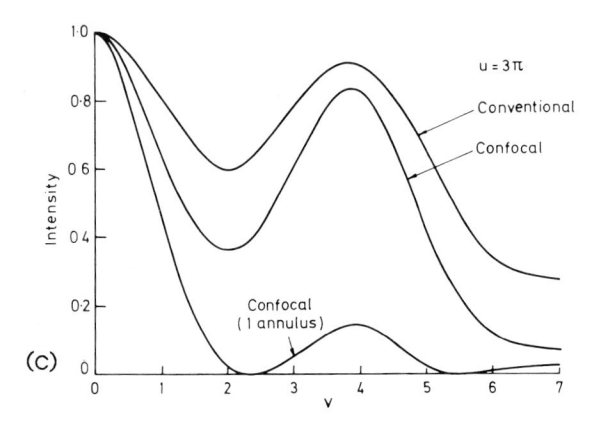

Figure 4.9. Intensity in planes perpendicular to the axis. (a) $u = \pi$. (b) $u = 2\pi$. (c) $u = 3\pi$.

The intensity along the axis is given by:

$$I(u) = \left[\frac{\sin(u/4)}{u/4}\right]^2 \qquad (4.19)$$

for the conventional system, whilst for the confocal systems it is:

$$I(u) = \left[\frac{\sin(u/4)}{u/4}\right]^4 \qquad (4.20)$$

so the axial side-lobes are extremely weak (the first side-lobe is reduced from 0.83% to 0.007% or by 20 dB).

In Figure 4.9 we show the intensity in planes perpendicular to the axis after normalization to unity on the axis. By $u = 2\pi$ the central spot is not broadened appreciably but for a conventional system the intensity is slow to decay below a value of about a quarter. For the confocal system the intensity decays quickly to a value smaller than 5% of the axial intensity, which itself is only 16% of the focal intensity. For the confocal system with one annular pupil, the point spread function is still reasonably sharp by $u = 3\pi$.

As a conventional microscope is defocused, the image of a single point spreads out, the peak intensity reducing and the minima becoming less pronounced. The net result is that, by conservation of energy, the total power in the image is independent of defocus. This is no longer the case in the confocal microscope as conservation of energy no longer applies because some energy is obstructed by the detector pinhole. The total power has been computed [Sheppard and Wilson, 1978b] and falls off monotonically, reaching the half-power point at a distance of 0.35λ from the focal plane for a numerical aperture of 1.4. For large values of u, the total energy falls off according to an inverse square law, as can be shown from geometrical optics. Thus the total energy in the three-dimensional image is finite, unlike in a conventional system. These results apply equally for reflection and transmission systems.

III. Three-dimensional image formation

For the calculation of the images of more complicated objects, or for general comparison of different imaging systems, it is convenient to use a method based on the spatial frequency response of the object. The object transmittance (or reflectance) can be expanded in a series of harmonics each represented by a spatial frequency component. The imaging system behaves as a filter with a certain transfer function. In this way, the properties of the system are separated from those of the object. The two-dimensional transfer function for a confocal

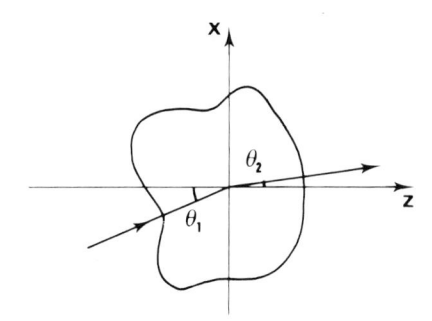

Figure 4.10. The scattering function, $S(\theta_1,\theta_2)$, for a two-dimensional system.

reflection system with defocus present is given by Sheppard et al. (1983). However, here we choose to generalize this approach to cover the imaging of a complete three-dimensional object.

Consider a thick microscopic object. If the object is two-dimensional in the $x-z$ plane and is illuminated in this plane with a plane wave at an angle θ_1, to the z axis, Figure 4.10, the scattering can be described by a scattering function, $S(\theta_1,\theta_2)$, giving the amplitude of the plane wave component scattered at an angle θ_2. For the full three-dimensional case, it is convenient to introduce direction cosines μ,ν for the waves so the scattering function becomes $S(\mu_1,\nu_1;\mu_2,\nu_2)$.

Consider now an object in which the scattering is weak so that the first Born aproximation is valid. The object is assumed to extend from z_1 to z_2 with an homogenous medium outside. The object is described by a three-dimensional function of position $t(x,y,z)$. In a non-fluorescent system, the total light field is equal to the sum of the incident light and the scattered light. The object function can be written as [Wolf, 1969]:

$$t(x,y,z) = \delta(z) - \frac{jk_0}{2}\left[n^2(x,y,z) - n_0^2\right] \quad (4.21)$$

where the first term represents the unscattered light. Here k_0 is the wave number $2\pi/\lambda_0$, n_0 is the refractive index in the region outside of the object, and n is the refractive index, in general complex, at any point inside the object. For a fluorescent object, t represents simply the fluorescent strength as a function of position.

If the object is illuminated with a plane wave, the scattered wave in any direction can be calculated by summing over the contributions from each point of the objects so that:

$$S(\mu_1,\nu_1;\mu_2,\nu_2) = \int\int\int_{-\infty}^{\infty} t(x,y,z)$$

$$\times \exp\{jk_0[(\mu_1 - \mu_2)x + (\nu_1 - \nu_2)y$$

$$+ \left(\sqrt{1 - \mu_1^2 - \nu_1^2} - \sqrt{1 - \mu_2^2 - \nu_2^2}\right)z]\} \, dx\, dy\, dz \quad (4.22)$$

Here we assume that the incident beam is not depleted by the scattering and also that the waves travel through a medium of constant refractive index n_0 in getting to and from the scattering point. Now any object can be expressed as a sum of constituent sinusoidal gratings inclined at various angles. This is a three-dimensional generalization of the Fourier series. Suppose a particular grating has a grating vector **K**, that is the wavelength of the grating is $\Lambda = 2\pi/K$ and the direction of fastest change of the scattering strength is in the direction of **K**. We can also introduce the spatial frequency of the grating, defined as the reciprocal of the grating wavelength, which can be resolved into components m, n, and r. Let the strength of the particular elemental grating be $T(m,n,r)$. Then, summing over all possible sinusoidal (or rather complex exponential) gratings, we get:

$$t(x,y,z) = \int\int\int_{-\infty}^{\infty} T(m,n,r) \exp\left[-2\pi j(mx + ny + rz)\right] \, dm\, dn\, dr \quad (4.23)$$

This expression can be inverted to give:

$$T(m,n,r) = \int\int\int_{-\infty}^{\infty} t(x,y,z) \exp\left[2\pi j(mx + ny + rz)\right] \, dx\, dy\, dz \quad (4.24)$$

Comparing equation (4.22) with equation (4.20), we can put:

$$\lambda m = \mu_1 - \mu_2 \quad (4.25)$$
$$\lambda n = \nu_1 - \nu_2 \quad (4.26)$$

and

$$\lambda r = \sqrt{1 - \mu_1^2 - \nu_1^2} - \sqrt{1 - \mu_2^2 - \nu_2^2} \quad (4.27)$$

so that a single grating component results in a beam scattered in that particular direction μ_2, ν_2. For given illumination and scattering angles we observe one point m, n, r in the spatial frequency space.

This development is the well-known Ewald sphere construction, Figure 4.11, of x-ray diffraction. The incident wave vector plus the grating vector is equal to the scattered wave vector. As the wavelength of the incident and scattered light are equal (for the non-fluorescence

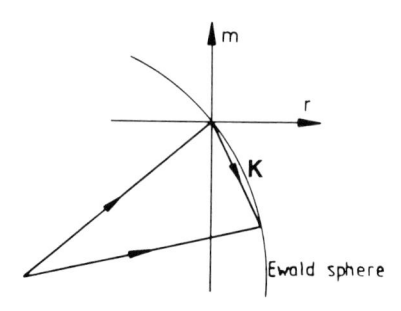

Figure 4.11. The Ewald sphere construction for diffraction by a grating, vector **K**.

case) the locus of all points in spatial frequency space which can be observed lie on the surface of a sphere through the origin. For a coherently illuminated microscope the range of scattering angles is also limited by the aperture of the objective and we can only image spatial frequencies on the cap of a sphere. Figure 4.12 shows the range of imaged spatial frequencies for coherent imaging systems for various apertures, in both transmission and reflection. It is seen that in reflection the spatial frequencies imaged have a high r component which results because of the phase term $\exp(2jkz)$. It should also be noticed that a small increase in aperture results in a comparatively large increase in the range of axial spatial frequencies imaged.

In order to image the three-dimensional structure of the object, we require to observe as much of the spatial frequency space as is possible. The range of spatial frequencies can be increased by illuminating at more than one angle. For example with two illuminating beams we now image spatial frequencies on two caps of spheres.

At this point we should mention that a projection of the three-dimenisonal object in a particular direction depends only on the spatial frequencies on a plane perpendicular to that direction through the origin of spatial frequency space. So a stereo pair contains only those frequencies on a pair of planes. Thus, only a limited range of spatial frequencies are imaged in a stereo pair and, indeed, only a limited range of spatial frequencies are needed to form a stereo pair.

For our purpose it is convenient to introduce a sphere each for the illuminating and scattered wave. These spheres, which are the three-dimensional Fourier transforms of the three-dimensional point spread functions, have their centre at the origin and radius of $1/\lambda_0$. For a fluorescent system of course, the spheres have differing radii. For a real system we do not have a complete sphere but only a cap, and the strength and phase may vary across the surface of the second sphere to account for apodization and aberrations. In a conventional microscope, for the first sphere there is only variation from apodization

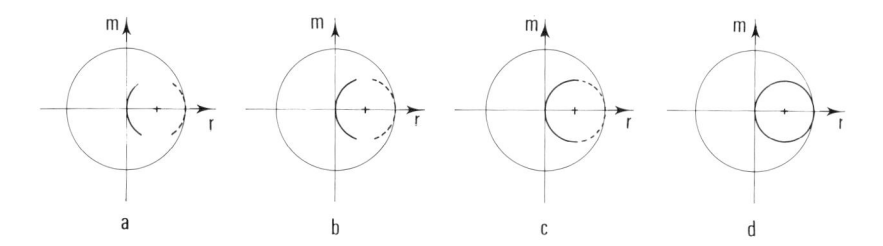

3-D Imaging in Confocal Microscopy

Figure 4.12. Range of spatial frequencies imaged for coherent imaging systems of different apertures. (a) $\alpha = \pi/3$. (b) $\alpha = 5\pi/12$. (c) $\alpha = \pi/2$. (d) $\alpha = \pi$.

because the phase effects vanish for an incoherent source. We denote these three-dimensional pupil functions Π_1, Π_2. We can show that, exactly as for two-dimensional conventional microscope, the image intensity can be written:

$$I(x) = \int\int\int\int\int\int_{-\infty}^{\infty} T(\mathbf{m})T^*(\mathbf{m}')C(\mathbf{m};\mathbf{m}')$$
$$\times \exp\left[2\pi j(\mathbf{m}-\mathbf{m}')\cdot\mathbf{x}\right]\,d\mathbf{m}d\mathbf{m}' \qquad (4.28)$$

where $C(\mathbf{m};\mathbf{m}')$ is the partially coherent transfer function (transmission cross-coefficient) given by:

$$C(\mathbf{m};\mathbf{m}') = \int\int\int_{-\infty}^{\infty} \Pi_1(\xi)\,\Pi_2(\xi+\mathbf{m})\,\Pi_2^*(\xi+\mathbf{m}')\,d\xi \qquad (4.29)$$

For a non-fluorescent confocal system on the other hand the image amplitude can be written:

$$U(x,y,z) = \int\int\int_{-\infty}^{\infty} T(\mathbf{m})c(\mathbf{m})\exp(2\pi j\mathbf{m}\cdot\mathbf{x})\,d\mathbf{m} \qquad (4.30)$$

where the coherent transfer function is given by:

$$c(\mathbf{m}) = \int\int\int_{-\infty}^{\infty} \Pi_1(\xi)\,\Pi_2(\xi+\mathbf{m})\,d\xi \qquad (4.31)$$

in which Π_1 *can* now have variations associated with aberrations. This is simply the convolution of the two spherical shells.

The resulting transfer function [Sheppard, 1986b, 1986c] thus has

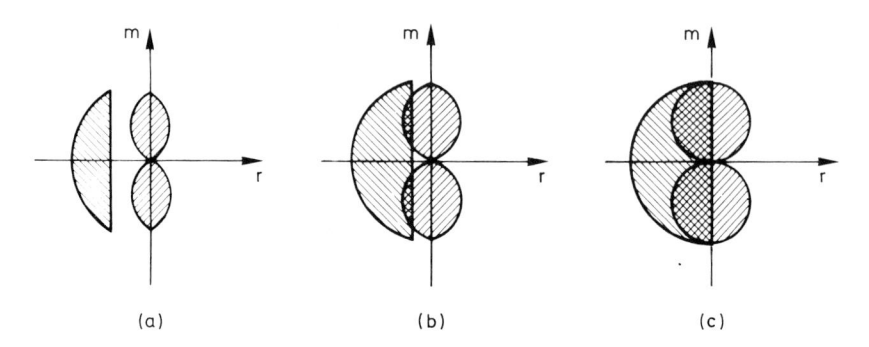

Figure 4.13. The cut-off in spatial frequency response for the three-dimensional transfer function of a confocal system. (a) $\alpha = \pi/3$. (b) $\alpha = 5\pi/12$. (c) $\alpha = \pi/2$.

the cut-off as shown in Figure 4.13. The complete three-dimensional transfer function is a volume of revolution about the r-axis. For a transmission system if $m = n = 0$ then the transfer function is zero unless $r = 0$. Thus, an absorbing plane perpendicular to the axis with no transverse variations in transmission is *not imaged*: we cannot tell where it is positioned in the axial direction. We can only image depth if transverse structure is present. For high apertures the spatial frequency bandwidth in the axial direction tends to exactly one half that in the transverse direction. The boundaries of the various regions are spheres but, in the paraxial approximation (for small apertures), they can be approximated to parabolas. For this paraxial approximation, the value of transfer function is given by:

$$c(\tilde{m},\tilde{n},\tilde{r}) = \frac{2}{\sqrt{\tilde{m}^2+\tilde{n}^2}}\text{Re}\left\{\sqrt{1-\left[\frac{|\tilde{r}|}{\sqrt{\tilde{m}^2+\tilde{n}^2}}+\frac{\sqrt{\tilde{m}^2+\tilde{n}^2}}{2}\right]^2}\right\}$$
(4.32)

where the normalised spatial frequencies $\tilde{m}, \tilde{n}$ and $\tilde{r}$ have been introduced.

In reflection a range of values of r *are* imaged when $m = n = 0$ but, as in the coherent system described earlier, the response is displaced along the r-axis. This means that the system behaves as dark field in the axial direction because the direct unscattered light is not detected. Thus, phase information in the object (i.e. refractive index variations) are imaged. Actually the system behaves similar to a Schlieren system in the axial direction so that the phase gradients as well as amplitude variations are imaged. Again, as in the coherent

3-D Imaging in Confocal Microscopy

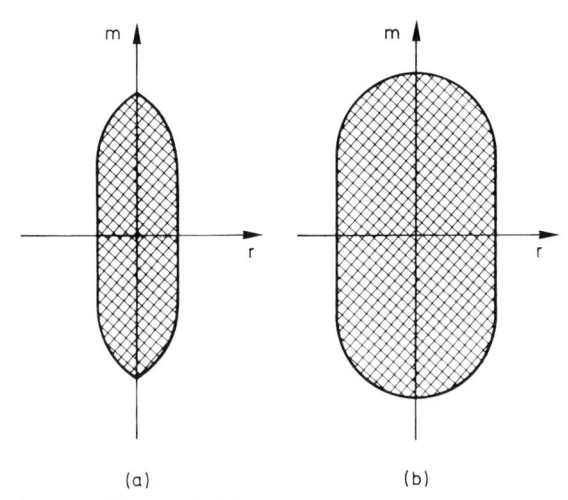

Figure 4.14. The cut-off in spatial frequency response for a confocal fluorescence system for equal incident and fluorescent wavelengths. (a) $\alpha = \pi/3$. (b) $\alpha = \pi/2$.

case considered earlier, a small increase in aperture results in greatly improved depth imaging and, in the limiting case of unity numerical aperture, the spatial frequency bandwidth in the axial direction approaches one half that in the transverse direction. However, unlike the transmission case, the cut-off frequency in the axial direction is $\lambda_0/2$ whatever the aperture so that higher spatial frequencies can be imaged in the axial direction than in the transverse directions.

For a confocal fluorescence system, the transfer function is given by:

$$C = \Pi_1 \otimes \Pi_1^* \otimes \Pi_2 \otimes \Pi_2^* \tag{4.33}$$

where $\otimes$ denotes a three-dimensional convoluton. For the particular case of equal incident and fluorescent wavelengths (which is of course not exactly realisable in practice) the cut-off for the transfer function is shown in Figure 4.14. The bandwidths in both the transverse and axial directions are exactly twice those for the confocal non-fluorescence case but the transfer function falls off rapidly with distance from the origin of spatial frequencies, so that resolution in the axial direction is not as good. This does, however, suggest that there is opportunity for improving the high spatial frequency response either by digital processing or pupil plane filters.

IV. Axial response

For an object which exhibits only axial variations, the image amplitude in a confocal microscope is given by:

$$U(z) = \int_{-\infty}^{\infty} T(r)c(0,0,r)\exp(2\pi j r z)\, dr \tag{4.34}$$

so that the image is completely characterised by the behaviour of the transfer function for m and n zero. The resulting intensity we call the axial or defocus response. We must therefore calculate the volume of overlap of two intersecting spherical caps. Consider first the volume of overlap of two intersecting spherical shells, which results in a transfer function $2/r\lambda$. This must now be adapted to take into account the pupil functions of the lenses. We obtain for a reflection system:

$$c(0,0,r) = \frac{1}{\pi r \lambda} \int_0^{2\pi} P_1(r,\phi) P_2(r,\phi)\, d\phi \tag{4.35}$$

where P_1 and P_2 are expressed as functions of the spatial frequency r. Now as the light on reflection passes throught the opposite side of the lens, we have:

$$P_2(r,\phi) = P_2(r, \pi - \phi) \tag{4.36}$$

so that the symmetry dictates that all aberrations which have odd-fold rotation symmetry (such as coma) do not affect the axial response of the system. For an aberration-free system which obeys the sine condition [Richards and Wolf, 1975] the apodization of the lenses falls off exactly to cancel out the $2/r\lambda$ term in equation (4.31) and we obtain very simply:

$$c(0,0,r) = \begin{cases} 1 & -2/\lambda \le r \le -(2\cos\alpha)/\lambda \\ 0 & \text{otherwise} \end{cases} \tag{4.37}$$

Substituting equation (4.31) into equation (4.30) we have:

$$U(z) = \int_0^{2\pi} \int_{-\infty}^{\infty} \frac{T(r)}{\pi r \lambda} P_1(r,\phi) P_2(r,\phi) \exp(2\pi j r z)\, dr\, d\phi \tag{4.38}$$

Comparing this with the usual expression for the axial response in terms of pupil functions expressed as functions of angle [Sheppard and Wilson, 1981]:

$$V(z) = \int\int R(\theta,\phi) P_1'(\theta) P_2'(\theta,\phi) \exp(2jkz\cos\theta) \sin\theta\, d\theta\, d\phi \tag{4.39}$$

we see that the spectrum of the object is related to the angular reflectivity:

$$T(r) \doteq R(\theta)\cos\theta \tag{4.40}$$

3-D Imaging in Confocal Microscopy

Note that in equation (4.35) the usual $\cos\theta$ term is incorporated in the $\cos^{1/2}$ terms for each of P_1' and P_2' and that equation (4.35) is not normalised in the same way as equation (4.34). For a single scattering plane oriented perpendicular to the axis we have:

$$T(r) = 1 \qquad (4.41)$$

and thus

$$R(\theta) = 1/\cos\theta \qquad (4.42)$$

The reflectivity of a single scattering plane was discussed by Darwin (1914) in connection with x-ray diffraction.

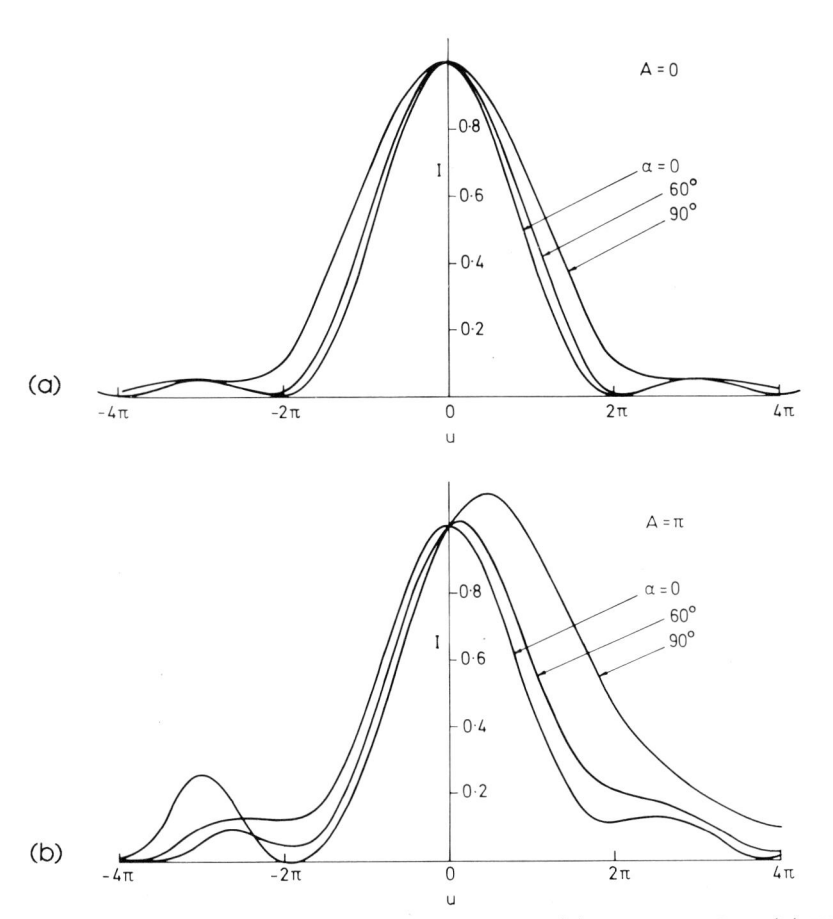

Figure 4.15. Axial response for a confocal system. (a) Aberration-free. (b) With primary spherical aberration of $\lambda/2$ (maximum wavefront error $= \lambda/8$ at mid-focus).

Alternatively, for a perfect reflector:

$$R(\theta) = 1 \qquad (4.43)$$

and we must take

$$T(r) = \lambda r/2 \qquad (4.44)$$

so that the perfect reflector is equivalent to a dipole layer in the Born approximation. In this case equation (4.34) takes the simple form:

$$U(z) = \frac{1}{2\pi} \int \int P_1(r,\phi) P_2(r,\phi) \exp(2\pi j r z) \, dr \, d\phi \qquad (4.45)$$

that is the Fourier transform of the product of the pupil functions averaged around a circle. This expression is valid even for high apertures, the effective apodization for a system obeying the sine condition being incorporated in the pupil function.

In the presence of aberrations, the axial response can be calculated from equation (4.41) [Sheppard, 1988]. The response for a system exhibiting primary spherical aberration is shown in Figure 4.15. It is seen that the response is asymmetric for a high aperture system, even though it is symmetric in the low aperture case. This emphasises that the low aperture treatment does not predict all the important aspects of the behaviour. In the presence of spherical aberration the response is broader and exhibits stronger side lobes. It should be noted that the aberration introduced in Figure 4.15 is quite weak so that *transverse* imaging is degraded only slightly. It seems generally true that the axial imaging properties are more sensitive to the presence of aberrations than those in the transverse direction.

V. The axial response for a low aperture system

An alternative treatment for the axial response is based on consideration of well-known results on the intensity in the focal region of a lens. In a confocal reflection system, with a perfect reflector as object, the light travels throught the objective twice and is then focused on to the detector plane where the confocal pinhole is situated. The intensity distribution in this region is thus exactly as has been described for a lens, and is shown in Figure 4.8(a) for an aberration-free system. A point-like pinhole simply samples this distribution so that, as the system is focused at different depths, the response is a section through the diagram. If the pinhole is aligned on-axis, the repsonse is given by the value along the u-axis, where u is the optical coordinate, equation (4.3).

Consider now a point-like pinhole offset by a distance ρ from the axis, where the normalised offset is, from equation (4.3):

$$v_d = k\rho \sin \alpha_d \qquad (4.46)$$

with α_d the aperture in the detector space. Then the response is given by the intensity along a line parallel to the u-axis, offset by v_d. The resulting response is shown in Figure 4.16. The central peak is reduced in intensity, the relative height of the side-lobes increases and they also get slightly closer to the origin. For v_d of about 3, a minimum appears at the centre, and the response consists of two equal peaks, symmetrically placed. The system would therefore result in a double image in the axial direction. Increasing the offset further, the intensity at $u = 0$ rises to a maximum for $v_d = 5.13$ but, as there is a saddle point, the axial response still consists of two peaks [Wilson and Carlini, 1988].

The intensity in the focal region for a lens with spherical aberration has been given, for example, by Born and Wolf (1975 p. 477). Note that as the beam passes through the objective twice this doubles the effective amount of aberration. For a low aperture system, the response for an on-axis pinhole is symmetrical but, as the pinhole is displaced, the side-lobes do become asymmetric. The central peak moves sideways and a new peak appears on the other side of the origin which moves further from the origin with larger displacement. For larger aberrations and no offset, a minimum may occur at the origin in u. Again one side-lobe becomes dominant with offset, with sometimes a series of peaks on one side decreasing in amplitude. Figure 4.17 illustrates the measured response with defocus and pinhole offset for an objective of numerical aperture 0.5. It is very similar to the theoretical results in Born and Wolf (1975) for a lens with weak spherical aberration and is discussed in more detail in chapter 3.

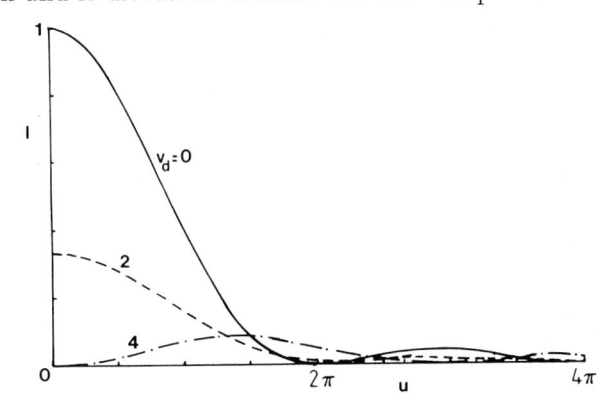

Figure 4.16. Axial response for a detector offset by a normalised distance v_d.

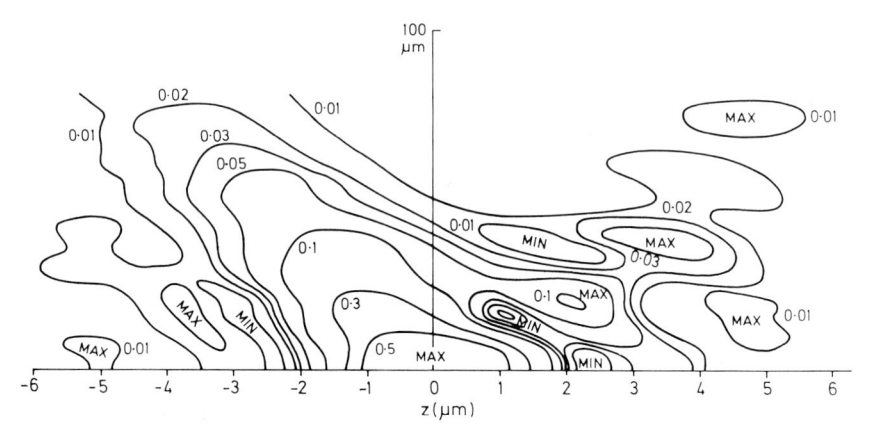

Figure 4.17. Measured response with defocus and pinhole offset for a lens of numerical aperture 0.5 (data recorded by S. J. Hewlett).

With a small amount of astigmatism present, one central peak occurs at the diffraction focus. For an offset of the pinhole in one direction (say along the x-axis) the peak moves to the sagittal focus, whereas for offset in the perpendicular direction the peak moves to the tangential focus. For stronger aberration, a double peak occurs when the pinhole is on-axis, one or other becoming stronger with offset according to the direction of offset. Thus, by observation of the behaviour of the axial response with pinhole offset, it is possible to recognize the presence of particular aberrations.

The axial response also changes as the reflector is tilted from the normal position, as shown in Figure 4.18. The peak intensity drops off and the response broadens. It should be noted that although coma does not affect the response from a normal reflector it does affect the response if the object is tilted. This then may be a method of investigating the presence of coma.

For a system with an annular lens pupil, the axial response can again be ascertained from a study of the intensity variations in the focal region. The response is stretched out in the z direction and, for larger obscurations, the behaviour with pinhole offset is rather different in that there is still a single central peak when the pinhole is offset to the first secondary maximum.

Also given in the optics texts, for example Born and Wolf (1975), is the encircled energy in the focal region, from which the axial response for a finite (non-zero) pinhole diameter can be determined. Obviously the peak signal is smaller from a smaller pinhole and the response is broader for a larger pinhole. In practice we choose to use the smallest pinhole consistent with giving an adequate signal level.

3-D Imaging in Confocal Microscopy 167

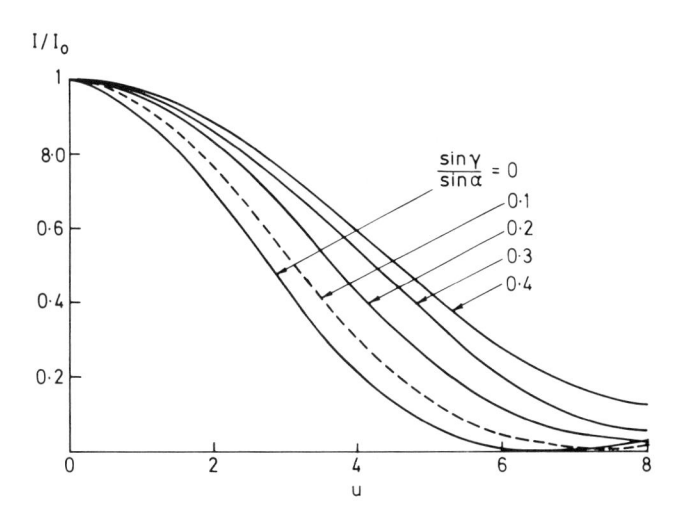

Figure 4.18. Effect of surface slope, γ, on the axial response (normalised) of a confocal microscope.

We have observed the axial response for a wide range of objectives in different systems and at different points in the field. Conditions such as pinhole size and position, and the presence of aberrations have also been investigated. The axial response is a convenient indicator of the three-dimensional imaging capabilities. It is found that it is important to use objectives of the correct tube length and coverslip correction as these greatly affect imaging in the axial direction. We have found that some commercial microscopes tend to display a double peak consistent with the presence of astigmatism or strong spherical aberration in the system. The choice of appropriate objectives is dependent on the design of microscope. In object-scanning systems, because of their on-axis geometry, fluorite objectives give adequate correction; in beam-scanning systems planapochromat objectives are preferred because of their freedom from field curvature and off-axis aberrations. Overall, we have found that fluorite objectives in an object-scanning system give a sharper axial response with weaker side-lobes.

VI. Conclusions

Confocal microscopy can be used to produce three-dimensional images, which can be generated and displayed using a variety of different techniques. Three-dimensional image formation can be considered in terms of an approach based on three-dimensional transfer functions. The axial imaging properties can be explored, both theoretically and experimentally, by investigation of the image of a plane mirror.

References

Born, M. and Wolf, E., (1975). *Principles of Optics*. Pergamon Press.

Caulfield, H. and Kryger, D. L., (1978). The use of microdensitometers as a basis for accurate metrology. Proc. S.P.I.E., **153**, 23-26.

Cox, I. J. and Sheppard, C. J. R., (1983a). Scanning optical microscope incorporating a digital framestore and microcomputer. Appl. Opt., **22**, 1474-1478.

Cox, I. J. and Sheppard, C. J. R., (1983b). Digital image processing of confocal images. Image and Vision Comp., **1**, 52-56.

Darwin, C. G., (1914). The theory of X-ray reflexion. Phil. Mag. **27**, 315-333, 675-690.

Hamilton, D. K. and Wilson, T., (1982). Three-dimensional surface measurement using the confocal scanning microscope. Appl. Phys., **27**, 211-213.

Molesini, G., Pedrini, G., Poggi, P. and Quercioli, F., (1984). Focus-wavelength encoded optical profilometer. Opt. Comm., **49**, 229-233.

Richards, B. and Wolf, E., (1959). Electromagnetic diffraction in optical systems. II. Structure of the image field in an aplanatic system. Proc. R. Soc. Lond. Ser. A., **253**, 358-379.

Sheppard, C. J. R., (1986a). Scanning methods in optical microscopy. Endeavour, **10**, 17-19.

Sheppard, C. J. R., (1986b). The spatial frequency cut-off in three-dimensional imaging. Optik, **72**, 131-133.

Sheppard, C. J. R., (1986c). The spatial frequency cut-off in three-dimensional imaging II. Optik, **74**, 128-129.

Sheppard, C. J. R., (1988). Aberrations in high aperture conventional and confocal imaging systems. Appl. Opt., **27**, 4782-4786.

Sheppard, C. J. R. and Choudhury, A., (1977). Image formation in the scanning microscope. Opt. Acta, **24**, 1051-1073.

Sheppard, C. J. R. and Wilson, T., (1978a). Image formation in scanning microscopes with partially coherent source and detector. J. Microsc., **25**, 315-325.

Sheppard, C. J. R. and Wilson, T., (1978b). Depth of field in the scanning microscope. Opt. Lett., **3**, 115-117.

Sheppard, C. J. R. and Wilson, T., (1981). Effects of high angle of convergence on $V(z)$ in the scanning acoustic microscope. Appl. Phys. Lett., **38**, 858-859.

Sheppard, C. J. R., Hamilton, D. K. and Cox, I. J., (1983). Optical microscopy with extended depth of field. Proc. Roy. Soc. Lond., **A387**, 171-186.

Wilson, T. and Hamilton, D. K., (1982). Dynamic focusing in the confocal scanning microscope. J. Microsc., **128**, 139-143.

Wilson, T. and Carlini, A. R. (1988). Effect of detector displacement in confocal imaging systems. Appl. Opt., **27**, 3791-3799.

Wolf, E., (1969). Three-dimensional structure determination of semitransparent objects from holographic data. Opt. Comm., 1, 153-156.

5. Pupil Filters in Confocal Imaging

Z. S. HEGEDUS

I. Introduction

A reduction of the size of the detector leads to significant changes in the imaging characteristics of a scanning system. This fact was first recognised by McCutchen (1967) who stated that if, in a scanned imaging system, the final image is passed through a pinhole which is much smaller than the diffraction disk produced by the imaging lens, the bandwidth of the imaging system increases.

The theory of this imaging mode, which became known as confocal imaging, was developed by Sheppard and Choudhury (1977). In the confocal scanning microscope the image becomes critically dependent on both the illumination and imaging components of the system. One of the advantages of this change is that the optical performance of the total instrument can be modified by the two independent optical systems. We investigate in detail the effects of altering the pupil transmittance of the component optical systems on the point spread function and on the shape of the transfer function of the confocal microscope.

II. Composite point spread and transfer functions

The conditions for confocal imaging are satisfied only if the spatial response of the detector approximates a delta function. This requirement, as shown by Wilson and Carlini (1987), is achieved when the size of the detector is reduced to about half the extent of the central diffraction spot, this latter being measured at its half intensity. To highlight the significance of a point detector, we summarize the main results of the imaging theory developed by Wilson and Sheppard (1984).

The amplitude a at the detector plane is expressed as a convolution of the amplitude transmitted by the object a_1 and the amplitude point spread function of the imaging system h_2. Denoting position vectors in the object and image (detector) planes by $\underline{o}$ and $\underline{i}$, this relationship is:

$$a(\underline{i},\underline{s}) = \int a_1(\underline{o},\underline{s})h_2(\underline{i}-\underline{o})\,d\underline{o} \tag{5.1}$$

where $\underline{s}$ corresponds to a given scan position of the object. The effect of the point detector positioned on-axis is clearly seen from equation (5.1). At the image plane the pinhole obstructs all points except the one at $\underline{i} = 0$, so the convolution integral becomes zero for all other values of $\underline{i}$. Thus, equation (5.1) is reduced to:

$$a(\underline{s}) = \int a_1(\underline{o},\underline{s})h_2(\underline{o})\,d\underline{o} \tag{5.2}$$

Here we have assumed that h_2 is an even function: $h_2(-\underline{o}) = h_2(\underline{o})$.

Further, if the illuminating source also approximates a delta function, the amplitude transmitted by the scanned object a_1 is described as the product of the point spread function of the illuminating system h_1 with the amplitude transmittance t of the object:

$$a_1(\underline{o},\underline{s}) = h_1(\underline{o})t(\underline{o}-\underline{s}) \tag{5.3}$$

If we substitute equation (5.3) into equation (5.2), we can see that a pinhole source and detector pair, positioned on-axis, transforms the image amplitude into a convolution between the object transmission t and the product of the two point spread function $h = h_1 h_2$:

$$a(\underline{s}) = \int t(\underline{o}-\underline{s})h_1(\underline{o})h_2(\underline{o})\,d\underline{o} \tag{5.4}$$

or

$$a = t \otimes (h_1 h_2) \tag{5.5}$$

where $\otimes$ denotes the convolution operation. Equation (5.5) states that:

> Confocal imaging is linear in amplitude. Hence, it is a coherent process, but the amplitude point spread function of the system is a composite one, formed by the product of two amplitude point spread functions which are independent of each other.

If we transform equation (5.5) into the spatial frequency domain, we obtain for the amplitude spectrum A behind the pinhole:

$$A = T(P_1 \otimes P_2) \tag{5.6}$$

where T is the amplitude spectrum of the object and P_1 and P_2 are the pupil functions of the optical systems. Equation (5.6) may be interpreted as follows:

> A confocal scanning instrument is a coherent imaging system whose pupil is a composite one, formed by the convolution of two independent pupil functions.

Pupil Filters in Confocal Imaging 173

Finally, we show the image intensity as measured through the pinhole:

$$I = |t \otimes (h_1 h_2)|^2 \tag{5.7}$$

or

$$I = |T(P_1 \otimes P_2)|^2 \tag{5.8}$$

III. Imaging with modified pupils

The effects of a variation of pupil transmittance on the image have been studied extensively. Reviews by Barakat (1961) and Jacquinot and Dossier (1964) have summarized the history and diffraction theory of pupil modification for incoherent and partially coherent imaging. Pupil modification in coherent imaging systems has been investigated in detail by Thompson and Krisl (1977) and more recently by Mills and Thompson (1986a; 1986b).

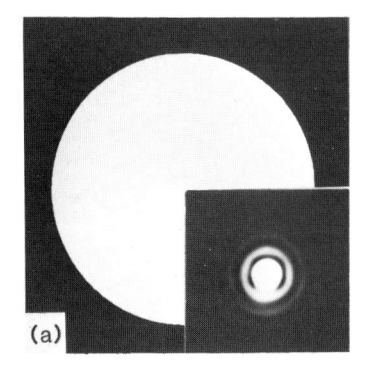

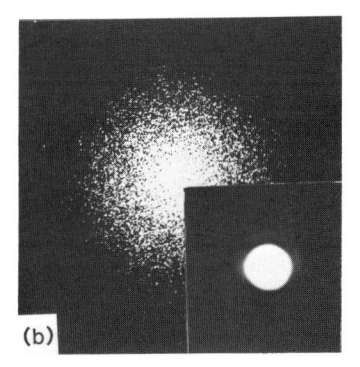

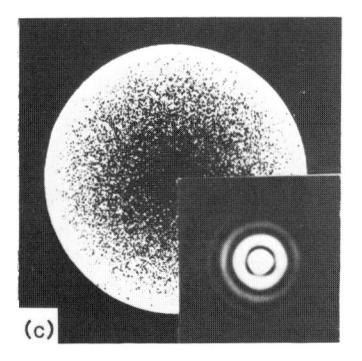

Figure 5.1. Far-field diffraction patterns (inserts with the corresponding pupils): (a) Clear pupil. (b) Quadratic apodising pupil. (c) Quadratic superresolving pupil.

The subject of pupil modification and its effect on an image is generally called apodisation. Apodisation, however, is only one aspect of this subject, namely, the weighting of the pupil function to suppress high order diffraction rings. Other pupil modification techniques cause an opposite effect: they increase energy in the higher order diffraction rings. For reasons explained below, such pupils are associated with the theme of superresolution.

Apodising and superresolving pupils cause complementary changes in diffraction. The redistribution of diffracted energy in the far field caused by these two types of pupils is illustrated in Figure 5.1. The inserts in the three photographs in Figure 5.1 show the diffraction patterns corresponding to the three pupils shown in the same figure. The pupil on the left is the fully transmitting, unmodified pupil with the familiar Airy pattern. All discussions on the energy distribution in diffraction patterns will be made with reference to this pattern. To indicate the relation of the energy distribution between the patterns, the exposures have been adjusted according to the total transmittance of the pupils. The exposure time for both modified pupils is 2τ, where τ is the exposure time used with the clear pupil. (The two pupil modifying functions in Figure 5.1 are: $1 - r^2$ and r^2, with r denoting the normalised pupil radius.)

The pupil in the centre is graded towards its edge, and its diffraction pattern shows that the energy in the high order diffraction rings is lower than that observed in the Airy pattern. We also notice that the lateral extent of the central spot increases. These two are major features of all apodising pupils.

The next pupil is graded towards its centre and, because the size of the central diffraction spot is reduced, we refer to such pupils as superresolving. This name, which is somewhat misleading, stems from the traditional Rayleigh criterion for resolution applied to two-point objects. In general, because of the inevitable increase of energy in the diffraction rings next to the centre, such pupils are not useful in the imaging of extended objects.

IV. Pupils in confocal imaging

IV.1. Amplitude point spread functions

We found that in the confocal microscope the effective point spread function is the product of two independently definable functions, each associated with the two components of the system.

If these components are identical, the resulting confocal amplitude point spread function will be positive; that is, it appears that the system behaves as if it were incoherent. This, however, is not exactly

Pupil Filters in Confocal Imaging

true because the imaging is not linear in intensity.

Indeed, equation (5.5) shows that while the confocal system is linear in terms of the object amplitude, its response is characterised with an intensity point spread, provided its illuminating and imaging components are identical:

$$a = t \otimes h^2 \qquad (5.9)$$

This is a very important feature of confocal imaging, since the disappearance of negative components of the amplitude response reduces the oscillations commonly observed in coherent image formation.

When the pupils of the components differ, the following general observations can be made:

(1) The number of diffraction rings in the resulting amplitude spread function will increase (as a result of non-coinciding zeros within h_1 and h_2).

(2) The envelope of the magnitude of the rings will be smaller than the envelope associated with either h_1 or h_2.

These properties are illustrated in Figure 5.2 for the two functions introduced in the previous section. In each drawing the dotted curve corresponds to the composite point spread of a confocal system with

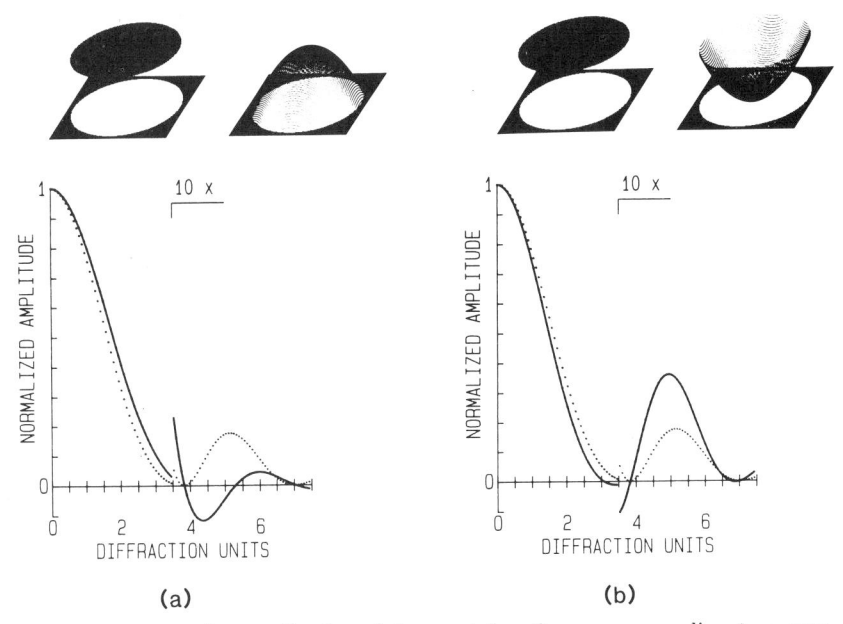

Figure 5.2. Composite amplitude point spread functions corresponding to a combination of a clear pupil with: (a) Apodising pupil. (b) Superresolving pupil. The drawings on top show the pupil functions involved. The dotted curve shows the composite amplitude point spread for the case of identical clear pupils.

identical and unobstructed pupils. These curves are therefore equivalent to the intensity distribution in the Airy pattern.

Figure 5.2(a) shows the resulting composite point spread when the clear pupil is used with the quadratic apodiser. We notice that the lateral extent of the central disk (the position of the first zero) remains unchanged, but that its shape is somewhat fuller. The absolute value of the amplitude corresponding to the first ring is reduced in comparison to the unmodified confocal system; that is to say, the apodising filter will behave similarly to the conventional imaging case. It should be noted that it now has a negative value. At the same time, the widening of the central spot is not apparent, suggesting that an apodiser will not noticeably change the two-point resolution of a confocal microscope.

The diagram on the right results when the clear pupil is used with the quadratic superresolving pupil. The first zero has shifted considerably towards the centre and the central disk is slimmer than the Airy disk.

The amplitude of the first ring is very small: it has almost disappeared between the first zeros of the diffracted amplitudes of the two pupils. The amplitude of the second ring is higher than that associated with the unmodified confocal system, but it is still around eight times smaller that the amplitude generated by the same filter in a conventional imaging system. The drawings at the top indicate the transmittances of the pupils in question.

In summary, in the confocal microscope the effect of pupil modifications on the composite amplitude point spread function is similar to that observed in conventional imaging. However, while in conventional imaging the action of these filters is accompanied by undesirable side effects, in confocal systems these appear to be minimised.

IV.2. Transfer functions

In a coherent imaging system, the transfer function is equivalent to the pupil function. As we have seen from equation (5.7), the pupil function in the confocal system is obtained by convolving the pupil functions of the illuminating and light-collecting optics of the instrument. We again have an additional degree of freedom in shaping the transfer function of the total system.

The convolution operation in two dimensions between two functions $P_1(x,y)$ and $P_2(x,y)$ can be interpreted as a volume of the product of P_1 and P_2 as a function of their relative position, while one of the functions is folded about each axis.

Since we are dealing with radially symmetric functions, the volume of intersection is independent of the particular direction of displacement between the functions. Hence the calculation of the convolution integral is greatly simplified. For our three cases, the integrals can be calculated analytically.

Figure 5.3 defines the major parameters of the integration. The two circles indicate the domain of the functions which are displaced by $2d$. The angle $\theta = \cos^{-1} d$ is a parameter in which the integrals are easily expressed.

For the first case, when both pupil functions are unobstructed — that is, $P_1 = P_2 = 1$ — the convolution $C(\theta)$ is well known. It is equivalent to the modulation transfer function for conventional incoherent systems:

$$C_1(\theta) = 2(\theta - \sin\theta \cos\theta) \qquad (5.10)$$

When one of the pupils is modified to a quadratic apodiser $P_1 = 1 - r^2$, and the other is unchanged — $P_2 = 1$ — the convolution becomes:

$$C_2(\theta) = \theta(\theta - 4\cos^2\theta) + \sin\theta \cos\theta(1 + 2\cos^2\theta) \qquad (5.11)$$

Similarly, for the quadratic superresolving pupil, $P_1 = r^2$, and with $P_2 = 1$ we have

$$C_3(\theta) = \theta - 3\sin\theta \cos\theta + 2\cos^2\theta(2\theta - \sin\theta \cos\theta) \qquad (5.12)$$

These functions have been normalised and then plotted against the displacement d in the second row of Figure 5.4. The first row shows the transmittance function of the first pupil, the transmittance of the second pupil being equal to 1 in every case. The normalising factors for the transfer functions are: $1/\pi$ for C_1 and $2/\pi$ for C_2 and C_3.

The bottom row of the same figure shows the difference between the transfer functions with respect to the transfer function of the confocal system with clear pupils; that is to say, the plots show $C_i - C_1$, where $i = 1, 2$ and 3.

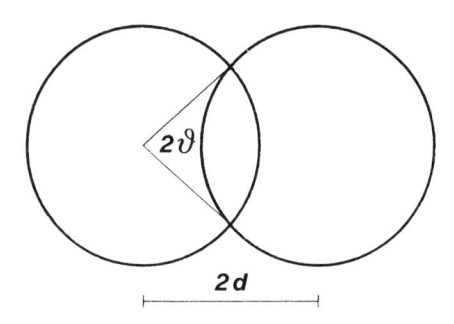

Figure 5.3. Derivation of the pupil function of a confocal system.

(a)   (b)   (c)

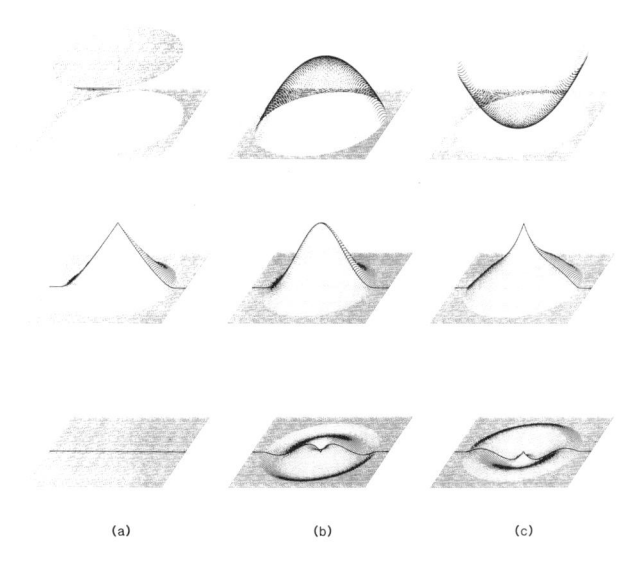

Figure 5.4. Examples of composite pupil functions of a confocal system. In each column the top drawing illustrates the transmittance of the first pupil: (a) Clear. (b) Apodising. (c) Superresolving. (The second pupil is clear for all cases.) The drawings in the middle row show the composite transfer functions and the bottom row corresponds to the difference between the given transfer function and the reference transfer function shown in the first column.

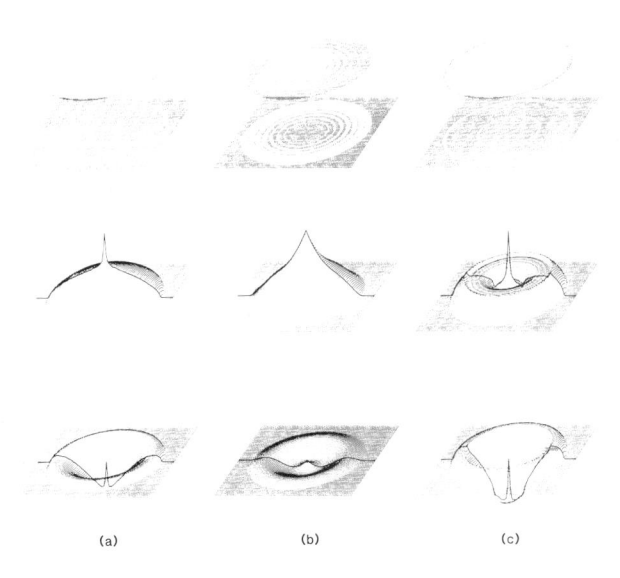

(a)   (b)   (c)

Figure 5.5. Same as in Figure 5.4 but now the first pupil functions are: (a) A finite annulus. (b) An inverse Gaussian approximated by an annular array. (c) Two finite annuli in anti-phase.

As expected, the apodiser will improve the transmission of spatial frequencies in the lower range, but above a certain frequency the transmittance is reduced. Exactly the opposite tendency is observed with the superresolving filter which, as detailed analysis shows, improves the response of the system at high spatial frequencies by almost 100%.

We must emphasise the importance of normalisation in these discussions. While in conventional imaging the energy efficiency is one of the most important optimising factors in analysing performance, in confocal imaging this becomes less important (except in the case of fluorescent imaging). With a laser source, it is relatively easy to increase the input illumination to compensate for the inevitable losses when pupil filters are used.

IV.3. General remarks and some experimental results

Pupil modification in confocal instruments leads to changes in image formation which are very similar to those predicted in conventional imaging. However, there is one major change in the confocal system which extends the possibilities of imaging near the diffraction limit. The so-called superresolving filters are no longer useless for imaging extended objects.

As we have seen in Figure 5.2, the amplitude of oscillations in the high order diffraction rings is very much reduced, which indicates that extended images can be collected with the advantages offered by the narrower central peak, but without the detrimental effects of strong ringing observed when these filters are placed in a conventional microscope. One of the simplest superresolving filters, the annulus, has been suggested by Sheppard (1977) for use in the confocal microscope. The transfer function of such an instrument with a finite annulus (obstruction ratio $\epsilon = 0.95$) is shown in the first column in Figure 5.5. The arrangement of this figure is identical to that of Figure 5.4. In the second column, another superresolving filter is shown. It is an inverse Gaussian function: $T(r) = \exp(-5r^2)$, which is approximated with an array of concentric annuli.

The last column illustrates the transfer function of a complex filter where the two rings are phase-shifted with respect to each other by $\pi/2$. These two filters were used in experiments whose results are shown in Figures 5.6 and 5.7. Figure 5.6 shows three images taken at the performance limit of optical microscopy. The oil-immersion lens had a numerical aperture of 1.36. The first image corresponds to the so-called type-1 imaging mode, in which the scanning microscope has a large detector. The second picture corresponds to a confocal image, in which both pupils are identical. The third image was obtained when a modest superresolving filter, shown in Figure 5.5(b), was placed in the illuminating arm of the instrument. Special attention is drawn to the top right corner of each of the images. The confocal image taken with the filter shows very clearly the diagonal lines which are barely distinguishable on the confocal image without the filter and totally absent on the first picture.

Figure 5.7 indicates the potential of a complex superresolving filter. The top scan is included for comparison, to show the object in more detail. The set of scans in the middle two show that while the type-1 scanning microscope does not resolve the thin line near the edge, in the confocal scan the presence of this detail is obvious. Finally, the scan on the extreme right shows that the complex two-ring filter produces an image at N.A. = 0.1 which is comparable to a confocal scan taken with N.A. = 0.2 without the filter (bottom scan).

The superresolving filters extend the effective range of spatial frequencies transmitted by the confocal system. Since the value of the transfer function is increased towards the cut-off frequency, spatial frequencies otherwise lost in noise are transmitted by the confocal system.

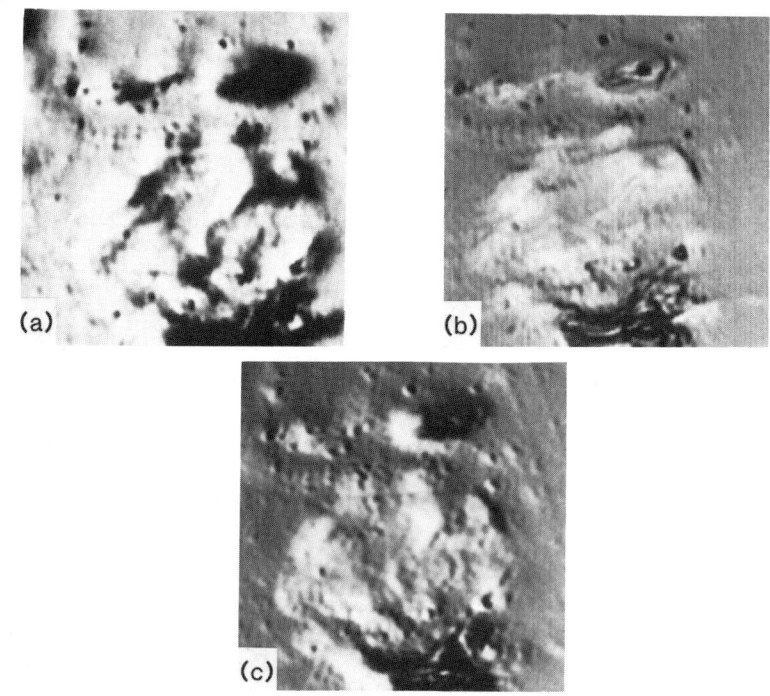

Figure 5.6. Reflection images of a thin Al coating: (a) Scanned image taken with a large detector. (b) Confocal image when both pupils of the system are clear. (c) Confocal image when the illumination part of the system includes a superresolving filter. The filter used is described in Figure 5.5(b). All images were obtained with an oil-immersion lens with N.A. = 1.36.

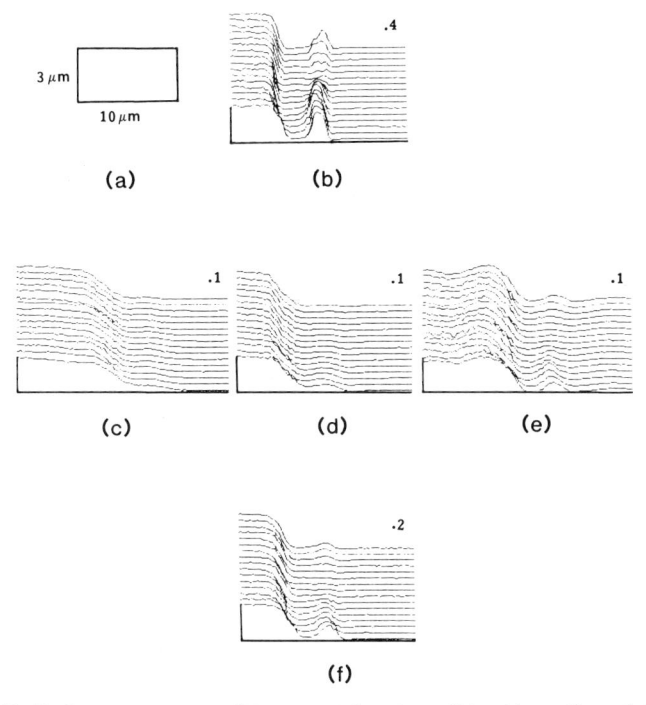

Figure 5.7. Reflectance scans of two scratches in a thin Al coating: (a) Scanned area. (b) Confocal scan with clear pupils (N.A. = 0.4). Middle row, scans with N.A. = 0.1. (c) Scan with large detector clear pupils. (d) Confocal scan with clear pupils. (e) Confocal scan with one pupil replaced with a complex superresolving pupil shown in Figure 5.4(c). Bottom: (f) Confocal scan with clear pupils (N.A. = 0.2).

When discussing the role of pupil filters, their influence on the longitudinal imaging performance must also be mentioned. The confocal microscope has a most striking three-dimensional imaging feature which manifests itself in excellent depth discrimination. Indeed, this property is more important in practical microscopy than the improvement in lateral performance. Depth discrimination is an easily observable effect: it occurs at pinhole sizes which are not so critical as in the case of true confocality.

The optical sectioning property is used to collect full, three-dimensional images by axial scanning of the object. Pupil filters also modify the axial distribution of diffraction and thus can play an important role in three-dimensional imaging. Sheppard and Hegedus (1988) have highlighted the relationship between axial and lateral performance of pupil filters; and it is suggested that, with appropriate filter combinations, the imaging performance of three-dimensional confocal microscopy can be extended significantly.

V. Pupil generation

There are a large variety of methods of producing pupils with varying transmittance. In practice, the realisation of continuous pupil functions is a complicated process. A recent overview of various methods of pupil generation was given by Hegedus (1985).

Filters used in confocal imaging can be produced as simple approximations to a continuous function. Coarse sampling of a continuous function leads to large oscillations in the high order diffraction. However, because of its multiplicative character, the confocal system attenuates these oscillations so that the oscillations in the composite point spread function are reduced to a tolerable level. This leads to a smooth transfer function, and, indeed, the transfer function shown in Figure 5.5(b) is very smooth, even though the corresponding pupil has been approximated only with ten annuli.

Transmittance variation may be simulated by approximating the original function with a large number of fine dots. The result is achieved by two possible processes, either by varying the dot size on a fixed grid matrix (as in traditional halftone techniques) or by the use of the same size dot with varying distribution. This latter method is easy to apply in digital halftone techniques as described, for example, by Anastassiou and Pennington (1982).

We found that the most convenient form of such dot printing technique is the variation of the density of dots of equal size on a random carrier. For example, the negatives for pupils illustrated in Figure 5.1 were generated by the following simple algorithm:

```
for every point within the pupil
    of RND P
        then draw a dot
        else move to the next point
```

Here RND is a random number and $P = P(r)$ is the desired pupil function. Both quantities are normalised.

In general, in the pupil-approximating methods the original is drawn on large scale and the final filter is obtained by photographic reduction. The original negatives corresponding to the filters on Figure 5.1 were plotted on a light plotter directly on film, which was then reduced to one-tenth of its size onto a very high-resolution film. Needless to say, during the reduction process care should be taken to ensure that the ratio of the dot size to the diameter of the pupil remains constant.

Production of complex filters is more critical. Polarising masks have been suggested for simultaneous phase and transmittance control (see, for example, Gupta, 1983). The two-ring superresolving pupil shown in Figure 5.5(c), which was used in our experiments, Figure 5.7, and described in detail earlier [Hegedus and Sarafis, 1986], was manufactured by a thin film deposition method.

Acknowledgements

The author is indebted to V. Sarafis, who took the images shown in Figure 5.6 at the Department of Engineering Science, University of Oxford, and to F. Torok who assisted in generating many of the drawings.

References

Anastassiou, D. and Pennington, K. S., (1982). Digital halftoning of images. IBM J. Res. Develop., **26**, 687-697.
Barakat, R., (1961). *Progress in optics.* Vol. I. Wolf, E. (Editor). North Holland.
Gupta, S. D., (1983). Performance evaluation of an apodized optical system with polarization masks. Opt. Acta, **30**, 607-621.
Hegedus, Z. S., (1985). Annular pupil arrays. Application to confocal scanning. Opt. Acta, **32**, 815-826.
Hegedus, Z. S. and Sarafis, V., (1986). Superresolving filters in confocally scanned imaging systems. J. Opt. Soc. Am., **A3**, 1892-1896.
Jacquinot, P. and Roizen-Dossier, B., (1964). *Progress in optics.* Vol. III. Wolf, E. (Editor). North Holland.
McCutchen, C. W., (1967). Superresolution in microscopy and the Abbé resolution limit. J. Opt. Soc. Am., **57**, 1190-1192.
Mills, J. P. and Thompson, B. J., (1986a). Effects of aberrations and apodization on the performance limit of coherent optical systems. I. The amplitude impulse response. J. Opt. Soc. Am., **A3**, 694-703.
Mills, J. P. and Thompson, B. J., (1986b). Effects of aberrations and apodization on the performance limit of coherent optical systems. II. Imaging. J. Opt. Soc. Am., **A3**, 704-716.
Sheppard, C. J. R., (1977). The use of lenses with annular aperture in scanning optical microscopy. Optik, **48**, 329-334.
Sheppard, C. J. R. and Choudhury, A., (1977). Image formation in the scanning microscope. Opt. Acta, **24**, 1051-1073.
Sheppard, C. J. R. and Hegedus, Z. S., (1988). Axial behaviour of pupil-plane filters. J. Opt. Soc. Am., **A5**, 643-64.
Thompson, B. J. and Krisl, M. E., (1977). Photogr. Sci. Eng., **21**, 109-114.
Wilson, T. and Carlini, A. R., (1987). Size of the detector in confocal imaging systems. Opt. Lett., **12**, 227-229.
Wilson, T. and Sheppard, C. J. R., (1984). In: *Theory and practice of scanning optical microscopy.* 47-48. Academic Press, London.

6. Three-dimensional Image Representation in Confocal Microscopy

G. J. BRAKENHOFF, H. T. M. VAN DER VOORT AND J. L. OUD

I. Introduction

The origin of the ideas on which confocal microscopy is based can be traced to the mid-fifties. Although they used different terminology, Minsky (1961) and Petráň et al. (1968) described systems in which the important properties of confocal microscopy, such as improved resolution and optical sectioning, were first introduced and demonstrated. The first demonstration of the resolution improvement in confocal microscopy at high numerical apertures, 1.3 and 1.4, was published by Brakenhoff et al. (1978, 1979).

After a slow beginning, this type of microscopy has seen a rapidly increasing acceptance by biologists in particular. In their field, it turns out that the main value of confocal microscopy is its sectioning capability, thanks to which it is able to rapidly provide high-resolution three-dimensional information from the specimen, which in principle may still reside in its natural, watery environment. Although the resolution is not as high as that provided by electron microscopy, the loss is often compensated for by more reliable morphology. This is because the fixation, dehydration and other biochemical preparation steps needed for electron microscopy often cause severe shrinkage and deformation of the object, or may even create artefacts [Woldringh et al., 1976].

For a basic treatment of confocal imaging we refer to other chapters in this book and to Wilson and Sheppard (1984). In this chapter we will concentrate on two interrelated topics. First, a discussion of factors which may affect the imaging (i.e., three-dimensional data collection) in real confocal microscopy, and, second, the representation of the biological data thus collected. We emphasize that for a correct interpretation of the confocal data, account should be taken of how the data are obtained — an aspect which is crucial if a quantitative analysis is required. A short description of the data-collection instrumentation is included.

II. Aspects of confocal image formation

Image formation in confocal microscopy can be viewed as the interaction of the object with the product of the illumination and detection sensitivity distributions. In the case of ideal lenses, in the paraxial approximation, and an ideal point detector, this approach results in the lateral confocal point response being described by the well-known square of the Airy disc, equation (1.21) [Sheppard and Choudhury, 1977; Brakenhoff et al., 1979; Wilson and Sheppard, 1984]. In practical confocal microscopy we may meet optical conditions which significantly depart from the ideal, such as not satisfying the paraxial conditions, use of polarized light and presence of aberrations. Sheppard and Wilson (1982) showed that high angle effects result in a broadening of the confocal response by about 10% when high numerical aperture objectives are used. We have investigated [van der Voort et al., 1990] the influence of polarization on image formation. We found, for instance, an appreciable azimuthal asymmetry of the energy distribution, Figures 6.1(a) and 6.1(b), if polarized laser light is used for the illumination. In order to avoid orientation effects on the imaging, it therefore seems advisable to use circularly polarized light for confocal imaging [Brakenhoff et al., 1979].

In real-life systems, however, the imaging situation may be more complicated. Of the factors affecting confocal imaging in practical

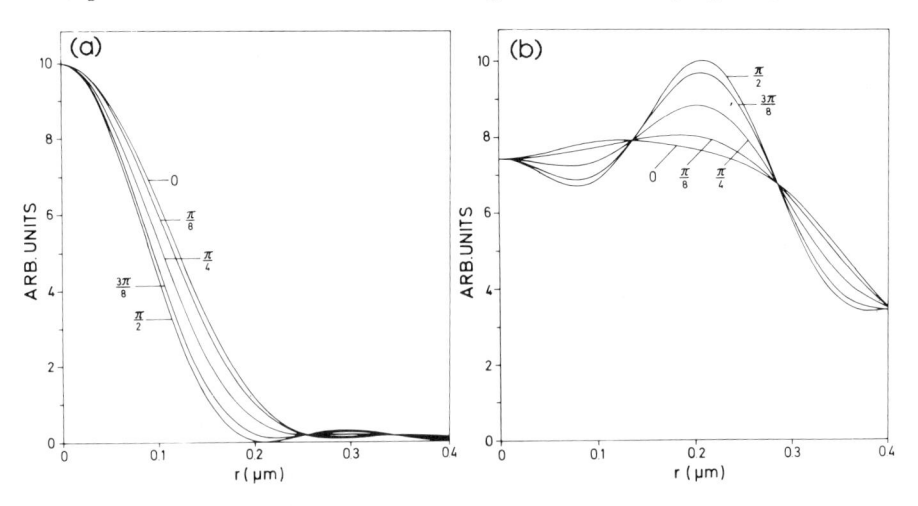

Figure 6.1. Effect of polarization on the electrical energy density distribution near the focus for a high numerical aperture immersion system N.A.$= n \sin \alpha$ and $n = 1.51$ for light with a wavelength of $0.5 \mu m$. Shown is the energy distribution (a) in focus and (b) at an axial position $0.44 \mu m$ before focus as a function of angle (in radians) between the x-axis and the direction of the electrical vector of the polarized light.

situations we would like to mention the following:

(1) The actual three-dimensional shape of the illumination and detection sensitivity distributions is basically determined by the optical properties of the lenses used, and hence the aberrations inherent in them. It is therefore important to use the lenses under the precise optical conditions for which they were designed. Use of the correct refractive index of the imaging medium is of particular importance. In practice, in biological specimens, the refractive index may not be precisely known and may also vary along the optical path. If we use as a criterion that optimal confocal imaging will be affected by wavefront aberrations greater than $\lambda/8$, we can calculate that, at a numerical aperture of 1.3 and a wavelength of 500nm, this limit will already be surpassed at penetration depths of 10 micrometers in a medium with a refractive index differing by 0.01 from the design value.

(2) The superposition of the illumination and detection sensitivity distributions in specimen space is determined by the relative positioning of the illumination and detection pinholes. Ideally, these should be at conjugate points with respect to the confocal point. We found in practice that, in addition to the lateral fit, the axial fit is also quite critical. For instance, at a numerical aperture of 1.3 and wavelength of 500nm, we found that the axial fit must be better than 50nm in specimen space for the system to have good detection sensitivity and high resolution. In fact, in an optimal or near-optimal confocal system, these two latter properties go together. If systems are used at various wavelengths and/or in fluorescence, account should be taken of this aspect in order to evaluate the amount of chromatic aberration permitted in the optical system.

(3) The type of interaction with the specimens is very important for the evaluation of the imaging. Confocal imaging systems are most often used in the reflection, transmission and fluorescence modes. In the transmission mode the imaging is relatively straightforward. If no major refractive index changes occur along the optical path, the detected intensities will be related directly to the absorption characteristics of the specimen. In fluorescence, owing to the incoherence of the emitted radiation, the detected fluorescence will be, in the first approximation, a faithful representation of the local fluorescence distribution. In brightfield reflection, the imaging is complicated by the coherent nature of the radiation, which can lead to unwanted interference effects between features separated axially. This is unlike the situation in the incoherent imaging of fluorescence microscopy. A case where problems might arise is, for instance, with specimens labeled with gold beads of 20 or 50nm size, where the scattering of the remaining tissue is low in comparison to the scattering of these gold beads. Of course, if the gold labels are used at sufficiently low concentration, mutual interference will be rare.

In view of the very global treatment given above, it is clear that the fluorescence mode is a good choice for reliable three-dimensional confocal imaging of biological specimens. Although brightfield reflection imaging may suffer from problems of interpretation, it has the definite advantage over fluorescence that no bleaching problems are encountered. In principle, the transmission mode makes absorption stains available for specimen marking and characterization. A drawback is that in transmission it is considerably more difficult to maintain conjugate positioning of the illumination and detection pinholes in comparison to confocal reflection or fluorescence microscopy in the usual epi-illumination arrangement.

III. Confocal data collection

The key to the success of a confocal microscope as a three-dimensional imaging instrument is its ability to sample just one volume element, in first approximation, independently of its surroundings. A *three-dimensional image* is then created by scanning this sample element through the volume of the specimen. By analogy with *pixel* for an element in a two-dimensional picture, the term *voxel* is used for an element from a three-dimensional or volume image. When coupled to a computer system, this three-dimensional image will reside in a three-dimensional memory array. These data are the starting point for the subsequent three-dimensional image visualization and analysis operations presented later in this chapter. Figure 6.2 shows the basic arrangement of the confocal instrument on which the three-dimensional data used below were collected. It is based on mechanical scanning of the specimen through the confocal point. This *on-axis* scanning approach has several advantages over the *off-axis* approach, where the scanning is realized by sweeping the beam optically over the specimen. The main ones are:

(1) The lens is used on-axis, thus avoiding off-axis aberrations.

(2) The area to be imaged by the microscope is not limited by the field of view of the optics employed but by the scan amplitude of the mechanical scan, which may be up to several millimeters wide.

(3) The imaging properties for each point of the scanned specimen are — apart from the abovementioned specimen-associated aberrations — identical, a property of significance in view of subsequent image processing and analysis.

Specimen-scanning seems, at first, to be inherently slower than data collection by beam-scanning methods [Carlsson *et al.*, 1985] or the Nipkow disc scanning approach [Petráň *et al.*, 1985]. However, in fluorescence microscopy especially, it is the signal-to-noise ratio in the three-dimensional image which is the limiting factor in determining the

3-D Image Representation in Confocal Microscopy

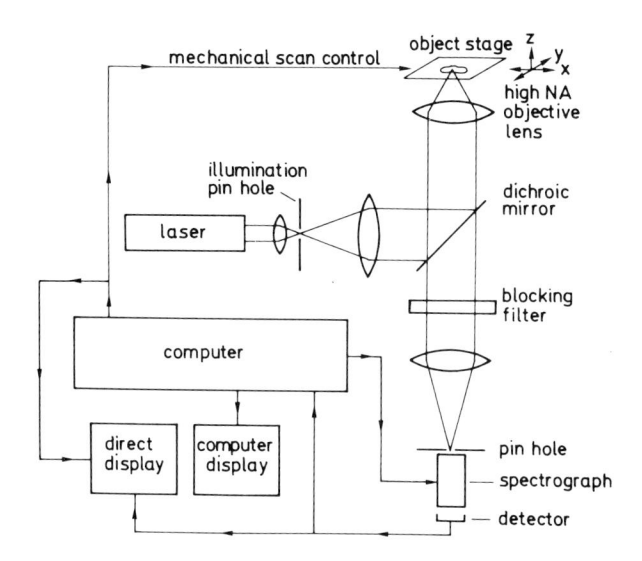

Figure 6.2. The confocally-collected data are directly stored in the computer system, which also controls the actual x, y, z scan movement and the sampling of the data. The optical arrangement of the dichroic mirror/blocking-filter is as for standard epi-fluorescence microscopy. Operation at various fluorescence detection channels is possible with the help of the spectrograph, the wavelength and bandwidth of which can be set by the computer system.

image acquisition time. Then, because of the necessary integration time, image collection times with typical laser light sources of 20-100mW power become comparable to the time it takes to build up an image by object scanning. This is because the amount of radiation that has to be collected from each point in the specimen to obtain a certain signal-to-noise level is identical in both methods. If the fluorescence emission is at or near saturation, higher excitation levels will not increase the photon flux (they are actually detrimental to image formation), and the only way to improve the signal-to-noise level will be to increase the observation time per voxel.

The sampled volume element in fluorescence (excitation wavelength 480nm, detection above 520nm) is typically 200×200nm laterally and 800nm axially when high numerical aperture immersion optics (N.A. = 1.3) are used along with a reasonably-sized pinhole (half the Airy disk diameter projected in specimen space). Data are collected automatically. With a line scan frequency of 50Hz (up to 150Hz is possible) it takes about 80 seconds to collect the data with 8-bit precision for a 16 section 256×256 data set. The images shown below are based on such data sets.

IV. Three-dimensional data sets and representation techniques

Data in a confocal scanning microscope obtained either by the object-scanning or the beam-scanning technique are collected on a three-dimensional sampling grid. Care should be taken that no unintentional orientation-dependent filtering occurs during the collection process. If the usual scanning method is followed, data are obtained x-y plane by x-y plane at various height or z-positions in the specimen. The fact that the detection channel has a limited bandwidth (for reasons concerning signal-to-noise levels) an orientation-dependent smoothing filter may have been applied over the data set during the collection process. This can most easily be understood by realizing that in the x-direction (the fast scan direction) the change in voxel value is limited by the detection bandwidth, while in the other directions no limitations apply. A second aspect is that the dimensions of the voxel volume do not necessarily match those of the three-dimensional grid on which the data have been collected. With a voxel shape determined by the numerical aperture, the alignment of the confocal optics, and the size of the pinholes used, it may occur that, depending on magnification, the sample is either over- or under-sampled. The processing of confocal data sets and the choice of filtering techniques should take account of these facts. The data presented below are collected under the conditions indicated, while the sampling grid can be derived from the magnification used.

An essential aspect of confocal microscopy is to develop techniques to make the three-dimensional information accessible to users in such a way that they are able to draw the (biological) conclusions necessary for their work. A number of approaches have been developed. We will demonstrate these on three-dimensional data sets representing the spatial chromosomal structures of two different organisms. In the software environment in which we process and analyze our three-dimensional images, we have a number of filters (Gaussian, median, Sobel) and image correction routines available, all directly operating on three-dimensional data sets, which can be used for specific purposes.

IV.1. Sections and cross-sections

The simplest method of presenting the contents of a three-dimensional data set is in the form of a series of images taken sequentially at various heights in the specimen. Such a set contains the basic, unprocessed information and provides direct information about the lateral and depth positioning of the various elements in the specimen volume. An example is given in Figure 6.3(a). This image set also demonstrates an effect which occurs in objects of appreciable density. Owing to absorption and effects associated with the phenomena discussed above, we see that a signal returning from greater depths is progressively attenuated by the intervening layers. A simple correction of this effect can be realized by multiplying each subsequent layer by an empirically determined quantity, chosen such that the apparent intensity from each layer is about equal. Figure 6.3(b) shows the result of this operation, and further applications will be demonstrated on this corrected data set. The various sections have been multiplied by a proportionally increasing factor such that data from the deepest lying section was multiplied by a factor of 2.4.

The data can also be presented in cross-sections through data sets along the x-z plane, Figure 6.4. Such cross-sections (x-y, x-z and y-z) provide a direct insight into the axial extension of structures. If the data are stored in a reasonably fast computer system, one can also travel forward and backward through the various cross-sections in real time. This facility, as implemented on our system, turns out to be particularly useful for a first examination of a three-dimensional data set.

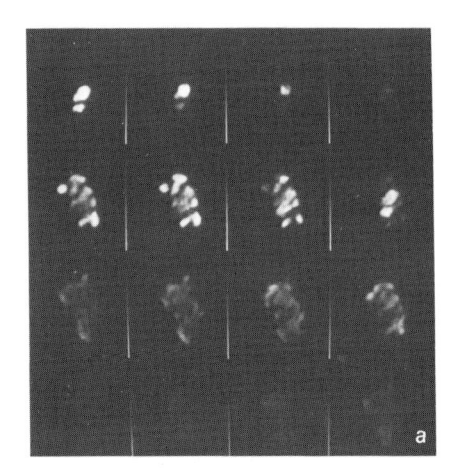

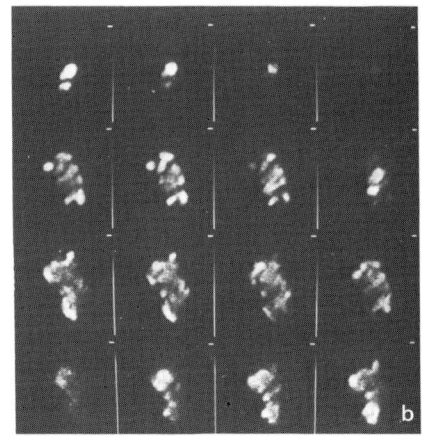

Figure 6.3. (a) Confocal microscopy of Allium, triquetium metaphase I chromosomes in pollen mother cells. Shown are 16 lateral sections taken at mutual distances of 2μm. Section width 25.8μm. The sections were taken starting from above the material — top right image — going down into the material to the deepest lying section — bottom left image. Because of absorption along the optical path, the signal from the deeper-lying sections is greatly attenuated. The fluorescence images were obtained by specific Feulgen staining of the DNA (excitation 528nm, detection 580nm). (b) Data set of Figure 6.3(a) after correction for absorption effects by multiplying the subsequent sections with a proportionally increasing factor (see text). Specimen kindly provided by G. K. Richards, Wellington, NZ.

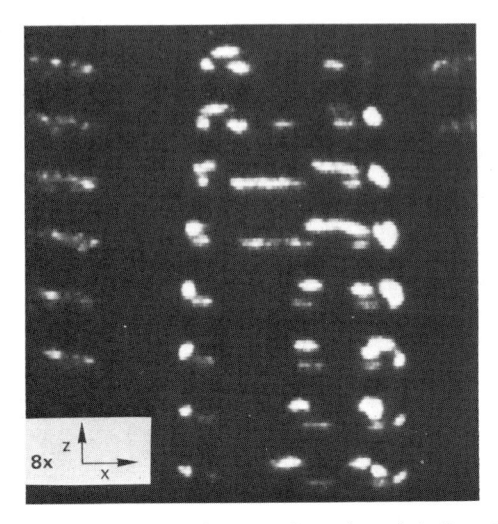

Figure 6.4. 8 cross-sections along the x-z plane (z axial direction) through the corrected data set shown in Figure 6.3(b). The cross-sections are evenly spaced in the y-direction through the chromosomal structure.

IV.2. Superposition and stereoscopic images

The three-dimensional information in a data set can be projected [van der Voort *et al.*, 1985] on an image plane along a certain direction. Each pixel in the image-to-be is built up from the data along a projection through the data set ending at that pixel. Such an image we call a superposition image. To determine the value to be assigned to that pixel from the voxel values encountered along the projection line, one can use various algorithms, such as addition (the extended-focus image), retaining the maximum or minimum along the line (the autofocus image), etc. Superposition images, apart from being informative in themselves, can be used in various ways. A sequence of such images obtained at various projection angles can be presented in rapid succession on a display monitor. The viewer then gets the impression of observing a rotating object, and is able, because of the movement, to perceive depth relations in the three-dimensional object shown. It is more suitable for publication purposes, however, to present these images in an array with each image constructed by projection along the appropriate projection angle. Any pair of these images can then be viewed stereoscopically. An example is shown in Figure 6.5. Larger arrays of such images are possible which enable the viewer to observe the structure stereoscopically from various angles, as each pair in such an array under any angle provides a stereoscopic view [Brakenhoff *et al.*, 1988].

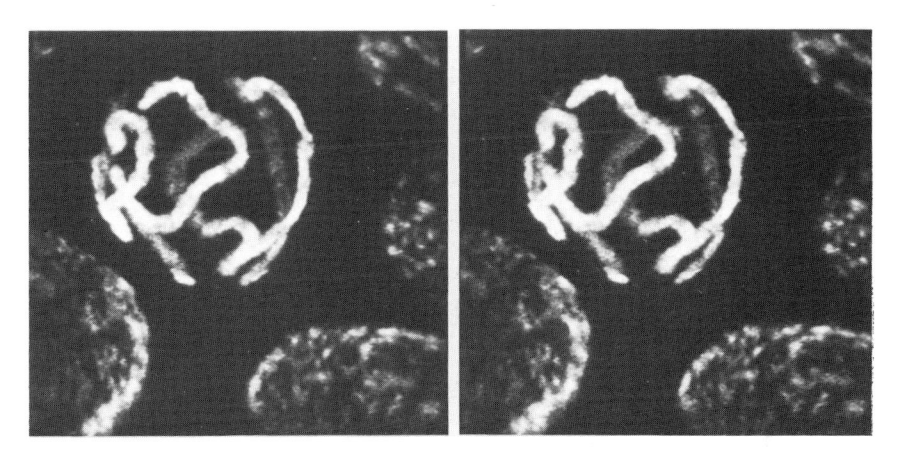

Figure 6.5. A set of two superposition images (see text) at different angles provides a stereoscopic view of the arrangement of the chromosomes during prophase of a *Crepis capillaris* root top cell, as observed in fluorescence after DNA-specific Feulgen staining. The image field is 18.4μm square. The original confocal data set contained 16 sections collected at 1μm spacing.

V. Representation by simulated fluorescence processing (SFP)

We have developed an alternative way of presenting spatial information from stereoscopy. This is useful because not everybody is able to perceive depth relations from stereo images [Richards, 1970]. The idea behind the SFP method [van der Voort et al., 1989] is to simulate, in the computer, the excitation and the emission steps of the fluorescence process by operating directly on the voxel values of the three-dimensional image. The computer simulation method consists of two independent steps. First the data set is *illuminated* from a simulated light source which is placed at infinity (parallel projection). In this excitation step, an exciting ray is made to travel through the data set. When it encounters a non-zero voxel value, the SFP algorithm sets the value of this voxel point equal to the triple product of the excitation intensity times the original voxel value times an absorption factor. The intensity of the exciting light ray will also be reduced at this point to account for absorption by a factor determined by the original voxel value times the absorption factor. Absorption at higher levels will thus reduce the degree of excitation of voxels deeper in the data set.

The emission phase is the second step in the SFP algorithm. Here the voxel is made to radiate in the observation direction, which may be different from the illumination direction. The amount of light received

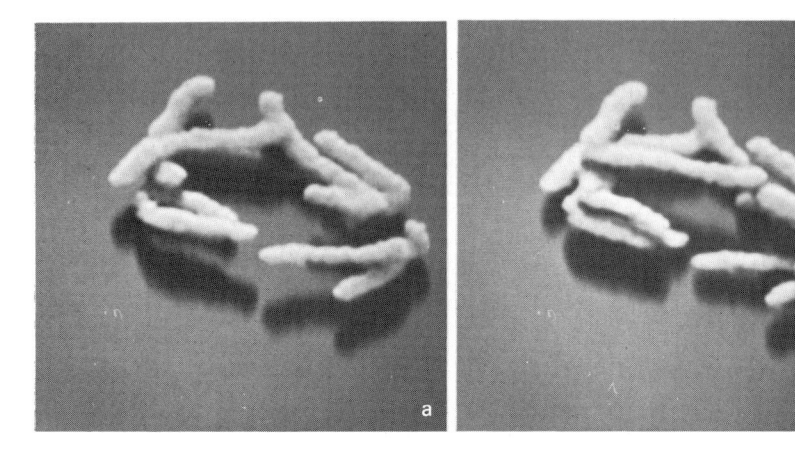

Figure 6.6. (a) Image created by the simulated fluorescence method (see text) of the chromosome arrangement during anaphase of a *Crepis capillaris* root top cell. The image field is 26µm square. (b) Same procedure, only the data set is illuminated and viewed from the direction opposite the one used to create Figure 6.6(a).

by an observer — again placed at infinity — from each voxel will, just as in the excitation phase, be influenced by absorption proportional to the voxel values of the intervening pixels. By adding below the deepest layer of the three-dimensional data set a layer of voxels with a uniform, intermediate value before starting the SFP process, a background for the structure to be imaged will be created. Figure 6.6(a) demonstrates the result of this operation with a very life-like representation of the object, thanks to the natural way in which structural elements are highlighted and shadowed. It is also very easy to change the observation point so that, for instance, the object may be viewed from the back by choosing appropriate illumination and observation points for the SFP process as shown in Figure 6.6(b). By varying the absorption and emission factors, the apparent transparency of the object may be controlled. A direct extension of this procedure is to place the observer not at infinity but rather closer, or possibly even inside the data set. Now we have to take account of the effects of perspective on SFP image formation. This turns out, however, to be very calculation-intensive, and we have therefore not developed this procedure into an operational form.

VI. Graphic techniques

For three-dimensional data representation and manipulation, we can take advantage of graphic techniques. By this term we mean the type of image representation tools provided for the display, on workstations, of three-dimensional objects designed in CAD-CAM environments. These programs are very versatile, and enable one to perform operations such as changing the point of observation, i.e., rotating the object, dissecting the object, showing internal views, etc., all in real time. The starting point for these programs is the surface data of the objects to be displayed, which have to be supplied to the workstation in an appropriate form. This means that before such routines can be used, the surface data of the object embedded in the three-dimensional data set have to be derived. This step is often non-trivial and also involves biological judgement. In the segmentation and surface definition process, one has to include the criteria which determine where and at which signal level an object will be different from the background or other objects with which it is embedded in the three-dimensional data set. These are essentially biological decisions. Nevertheless, this approach has definite promise for certain applications. As an example we show the chromosomal structure of *Crepis capillaris*, Figure 6.7. These images were constructed by first determining the central axis of each chromosome by a skeletonizing operation, and then creating the apparent surface from cylindrical elements of a certain diameter centered on the axis. These are joined and provided with a capping element. The final result is presented as a wireframe. It is clear therefore that the surfaces as shown no longer have any relation to the actual chromosomal surfaces. They are, however, eminently suitable for showing relative chromosome positions within the nucleus of this plant cell, and making these visible on a workstation from various observation points in real time.

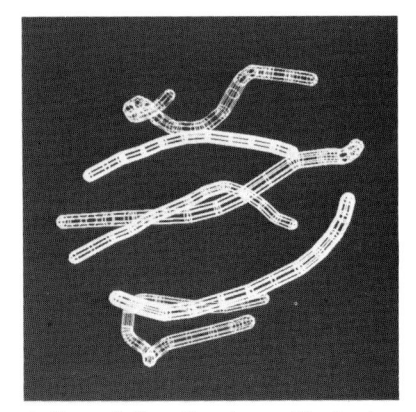

Figure 6.7. Representation of the *Crepis capillaris* chromosomal structure by graphic techniques. Such images can be manipulated in real time on a suitable workstation. Shown is a wireframe representation of a *Crepis capillaris* chromosome arrangement.

VII. Conclusions and comments

We have not taken account in this chapter of the relation between object orientation and structure and the apparent intensity of image elements in the three-dimensional confocal image. For instance, take the imaging of a fluorescent sphere with a uniform fluorescence distribution over its surface. If we image such an object by confocal microscopy, we will observe that the apparent intensity coming from the top and the bottom (viewed along the optical axis) is considerably lower than the signal detected from areas along the sphere's equator. The reason for this is basically the non-sphericity of the effective confocal probing volume with which the specimen is sampled. See, for a further discussion, van der Voort et al. (1989), where the imaging of a number of object structures is investigated as a function of orientation. We are of the opinion that this intensity dependence on object and orientation should be taken into account for faithful image representation and analysis of confocal data.

The results presented in this chapter should be seen as a qualitative demonstration of a number of possibilities in the area of image representation. It is by no means exhaustive, and rather demonstrates the techniques which are most frequently used at present. In other areas, especially in the processing of data acquired by X-ray computer tomography, sophisticated data representation and handling methods have been developed, and some of these could be used for confocal microscopy data.

In this chapter we have considered some aspects of confocal imaging and representation. We are of the opinion, however, that optimal use of the three-dimensional data acquired by this technique is made when they form the basis for the spatial analysis of material of either biological or industrial origin. This can be combined with multi-parameter imaging through the detection of different substances at specific fluorescence wavelengths. Instruments with such optical and image-processing facilities will be of great use for the study and quantitative description of the three-dimensional organization and structure of the specimen.

Acknowledgements

Development was supported by Stichting Technische Wetenschappen (STW) and by the Foundation for Biological Research (BION), which is subsidized by the Netherlands Organization for Scientific Research.

References

Brakenhoff, G. J., Blom, P. and Bakker, C., (1978). Confocal scanning light microscopy with high aperture optics. Proc. ICO Conf., Madrid, Spain, 215-218.

Brakenhoff, G. J., Blom, P. and Barends, P., (1979). Confocal scanning light microscopy with high aperture immersion lenses. J. Microsc., **117**, 219-232.

Brakenhoff, G. J., van der Voort, H. T. M., Baarslag, M. W., Mans, B., Oud, J. L., Zwart, R. and van Driel, R. W., (1988). Visualization and analysis techniques for three-dimensional information acquired by confocal microscopy. Scanning Microsc., **2**, 1831-1838.

Carlsson, K., Danielsson, P. E., Lenz, R., Liljeborg, A., Majlöf, L. and Åslund, N., (1985). Three-dimensional microscopy using a confocal laser scanning microscope. Opt. Lett., **10**, 53-55.

Minsky, M., (1961). Microscopy apparatus. United States Patent No. 3,013,467. Filed Nov. 7, 1957; granted Dec. 19, 1961.

Petráň, M., Hadravský, M., Egger, M. D. and Galambos, R., (1968). Tandem-scanning reflected-light microscopy. J. Opt. Soc. Am., **58**, 661-664.

Petráň, M., Hadravský, M., Benes, J., Kucera, R. and Boyde, A., (1985). The tandem scanning reflected light microscope. Part 1 - The principle, and its design. Proc. Roy. Microsc. Soc., **20**, 125-129.

Richards, W., (1970). Stereopsis and stereo blindness. Exp. Brain. Res., **10**, 380-388.

Sheppard, C. J. R. and Choudhury, A., (1977). Image formation in the scanning microscope. Opt. Acta, **24**, 1051-1073.

Sheppard, C. J. R. and Wilson, T., (1982). The image of a single point in microscopes of large numerical aperture. Proc. Roy. Soc. Lond., **379**, 145-158.

van der Voort, H. T. M. and Brakenhoff, G. J., (1990). Three-dimensional image formation in high aperture fluorescence confocal microscopy: a numerical analysis. J. Microsc.

van der Voort, H. T. M., Brakenhoff, G. J. and Baarslag, M. W., (1989). Three-dimensional visualization methods. J. Microsc., **153**, 123-132.

van der Voort, H. T. M., Brakenhoff, G. J., Valkenburg, J. A. C. and Nanninga, N., (1985). Design and use of a computer-controlled confocal microscope. Scanning, **7**, 66-78.

Wilson, T. and Sheppard, C. J. R., (1984). *Theory and practice of scanning optical microscopy*. Academic Press, New York.

Woldringh, C. L., de Jong, M., van den Berg, W. and Koppes, L., (1976). Morphological analysis of the division cycle of two *Escherichia coli* substrains during slow growth. J. Bacteriol., **131**, 270-279.

7. Optical Cell Splicing with the Confocal Fluorescence Microscope: Microtomoscopy

E. H. K. STELZER AND
R. W. WIJNAENDTS-VAN-RESANDT

I. Issues in confocal fluorescence microscopy

Modern biological fluorescence microscopy of mammalian cells is mainly concerned with samples in which specific structures have been labeled with a fluorophore. Parameters of interest are the spatial distribution of the fluorescing dye and the intensity of the fluorescence response, although many other parameters can be observed (depolarization, intensity ratios, etc.). The technique that will make a specific target (e.g. lipids, proteins, DNA, organelles) visible depends on the biochemical properties of the cell system, the number of target sites, the biological question in mind, the cellular developmental stage, etc., and on the type of fluorescence microscopy (conventional, video-enhanced, confocal) that is to be performed on this sample. It is important to note that any kind of fluorescence microscopy measures the spatial intensity distribution of the dye, and this, in turn, is interpreted to reveal the spatial distribution of the target. Not all targets are equally accessible, and independent verification techniques are necessary to confirm that the labeling has been successful. Confocal fluorescence microscopy is one of the tools available [Bacallao and Stelzer, 1989].

Conventional fluorescence microscopy has the handicap of a limited depth discrimination capability. Thick and densely packed samples will not allow an observer to focus into a specific plane of interest without observing out-of-focus contributions from structures several micrometers above and below the focal plane. This problem can be overcome in various ways:

(1) Using cells that are relatively flat (e.g., fibroblasts).
(2) Flattening and spreading the cells during the preparation.
(3) Labeling only parts of the target.
(4) Improving the specific labeling of the target.
(5) Removing parts of the target.

These methods are used extensively by fluorescence microscopists. Many samples observed in the confocal fluorescence microscopes at the European Molecular Biology Laboratory had an uneven staining (in three dimensions) or were only a fraction of the expected thickness. Techniques that cause such results evolve because the criterion for quality is usually the visibility of certain structures in the conventional fluorescence microscope. Labeling only parts of a cell, e.g., either the basal or the apical domain in polarized epithelial cells, decreases out-of-focus contributions and increases the contrast. It is important to realize that standard techniques for conventional fluorescence microscopy will in many cases be rated inappropriate by confocal fluorescence microscopy. Preparation and fixation methods must be critically reviewed before they can be used to draw conclusions concerning three-dimensional spatial distributions. This, in many instances, will force the development of new labeling and fixation methods.

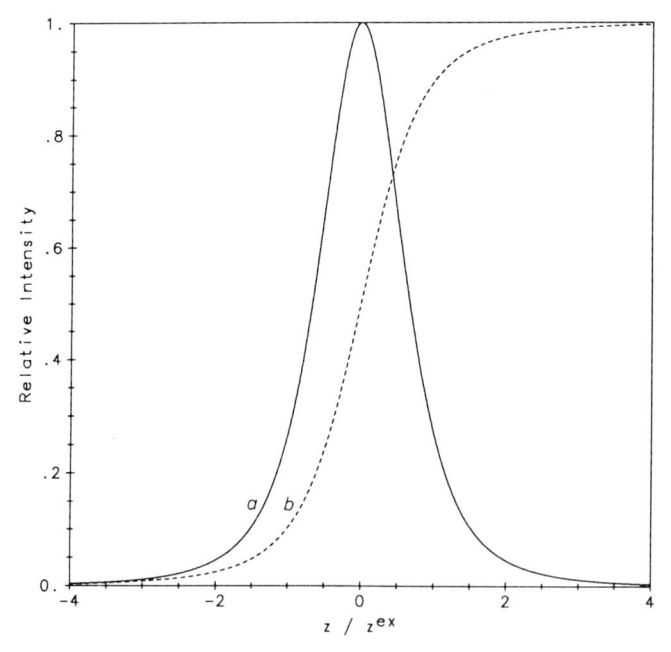

Figure 7.1. Depth discrimination of a confocal fluorescence microscope. Displayed are normalized intensities as a function of the normalized unit u along the optic axis. Curve (a) shows the fluorescence intensity as a function of the distance from the geometrical center of a fluorescent point source (e.g. a bead). Curve (b) shows the integral over the signal displayed in curve (a) [Wilson and Sheppard, 1984, pp. 72].

Many samples which have been previously unobservable, such as thick (10μm - 60μm) densely packed cells [Rosa et al., 1989] or cells growing on filters, can now be observed in more detail [van Meer et al., 1987]. Since conventional fluorescence microscopy is unsuitable, confocal fluorescence microscopy, with its optical cell-slicing capability, [Wijnaendts et al., 1985; Stelzer and Wijnaendts, 1986] has become the ideal instrument with which to analyze these specimens.

II. Depth discrimination in a confocal fluorescence microscope

Discrimination against out-of-focus light (depth discrimination) is explained in detail elsewhere in this book. In short, the *field pinhole* discriminates against light emitted from outside the common focal volume of the source and the detector pinhole. In a first order approximation, for large z it can be shown that the photon flux falls off with the square of the distance from the focal volume. Fluorophores that are at distance z from the center of the focal volume receive incident light at an intensity of I_0/z^2 (I_0 is the maximum intensity). Since the image in front of the pinhole is also enlarged by a factor proportional to z, the intensity discrimination can be said to be roughly proportional to $1/z^4$. The graph in Figure 7.1 plots the intensity as a function of the distance of the focal volume from an infinitely small fluorescent bead and the integral of this signal [Wilson and Sheppard, 1984, p. 72; and chapter 1 of this book]. While signal (a) cannot be measured easily, it is simple to measure the response to a step function, and hence the integral over signal (a), by scanning into an arbitrarily thick layer of fluorescent dye [Wijnaendts et al., 1985], Figure 7.2. The response function can be interpreted to reveal the depth discrimination properties of a confocal fluorescence microscope, Figure 7.3. The full-width-half-maximum (FWHM) presents a rough estimate of the size of the focal volume along the optic axis. Approximately 80% of the total energy penetrates a fraction of the total volume whose boundaries are defined by the FWHM along the optic axis. This value is used as an estimate for the minimum distance between two consecutive *x-y* images along the optic axis (minimum step size). Since the overlap between two consecutive *x-y* images is minimal, photo-induced damage is avoided without losing any information. However, a good signal-to-noise ratio in the detected fluorescence intensity may allow one to observe differences in the 90-100% region. Using appropriate signal-processing algorithms, fine structures may become visible. Techniques such as these are well known from video-enhanced Nomarski or video-enhanced fluorescence microscopy [Inoué, 1986]. Imposing a negative offset and a high gain on the input signal increases the contrast of

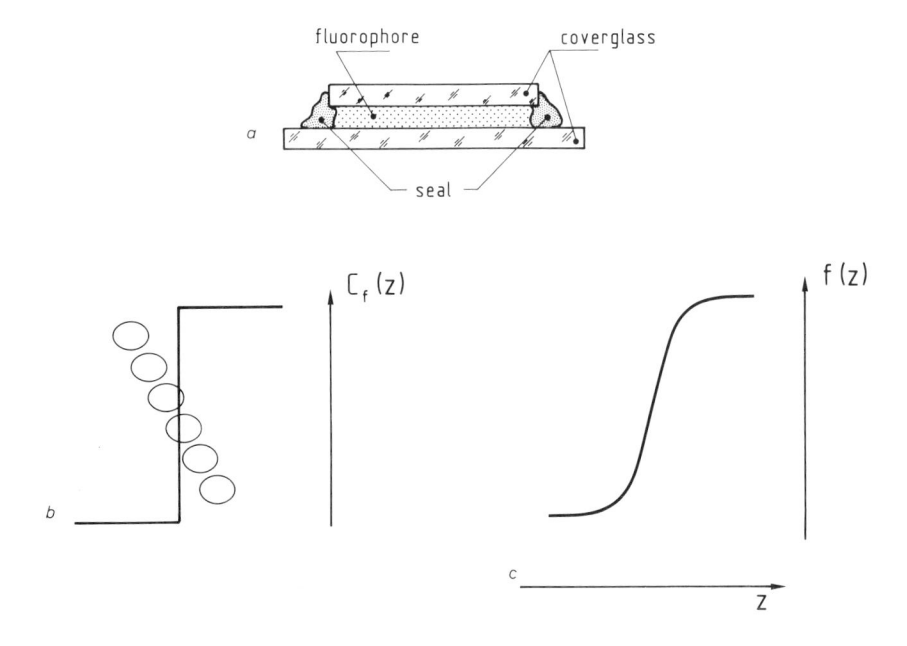

Figure 7.2. Measuring the depth discrimination properties of a confocal fluorescence microscope: (a) A layer of fluorophore (e.g. 10^{-5} M Rhodamine 6G) is mounted on a microscope slide. Spacers make sure that the layer has a thickness of at least 50μm. (b) The focal volume is moved into the fluorophore. The concentration of the fluorophore $C_f(z)$ as a function of z is a step function. (c) The response $f(z)$ of the confocal fluorescence microscope as a function of z. The slope can be interpreted in terms of depth discrimination.

objects [Bacallao and Stelzer, 1989]. The image *improves* and previously faint objects become visible on a video monitor. Results like these obviously lead to considering other criteria that describe the depth discrimination properties in a confocal fluorescence microscope. A definition for a term like "visibility" would, however, be missing the point. It is important to characterize the depth discrimination properties of a confocal fluorescence microscope and to express them in a clearly defined manner, since all other characteristics may be derived mathematically.

II.1. Generating x-z images

Conventional fluorescence microscopy generates images perpendicular to the optic axis in the x-y plane. A confocal laser fluorescence microscope is not, by itself, an imaging device, as only the signal from the focal volume is recorded. Consequently, an image is constructed

Microtomoscopy 203

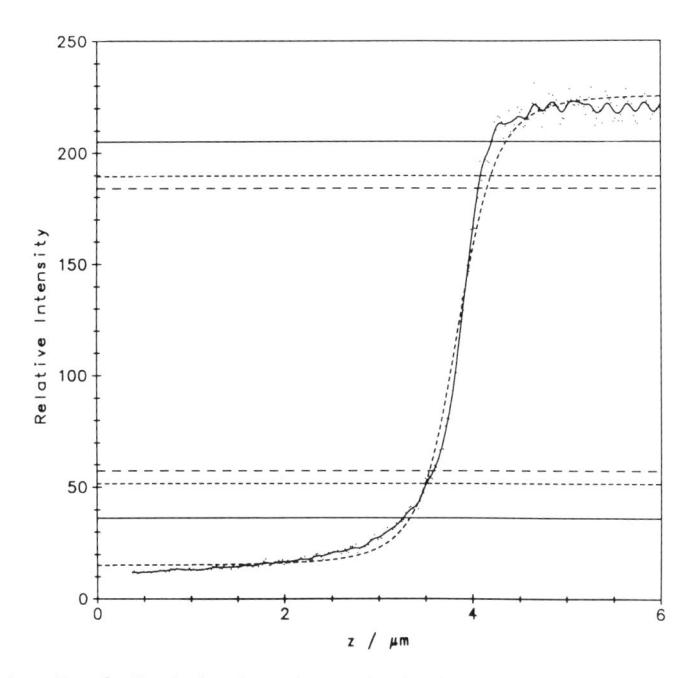

Figure 7.3. Depth discrimination of a confocal microscope. A result of an experiment as it is outlined in Figure 7.3. Displayed is the integrated intensity integrated as a function of z. The dots return the measured signal, the full line the filtered signal, and the broken line a theoretical least squares fit curve. The horizontal lines indicate the 10%-90%, 17.3%-82.7% and 20%-80% pairs that may be used to judge the quality of the signal. (Excitation 476nm, emission above 520nm, lens 100X/1.32.)

by scanning this focal volume through the sample and registering the signal as a function of spatial coordinates. Hence an image can be generated in any arbitrary plane or volume. One of these can be the classical x-y image, but full three-dimensional images can also be collected by scanning x-y planes as a function of z. Of course, it is also possible to generate directly x-z images [Wijnaendts et al., 1985] which can be compared to slicing or cutting the cell optically in comparison to the standard technique of actually slicing frozen sections of cells and observing the slices under a normal fluorescence microscope, Figure 7.4.

Various methods for fast z-scanning are possible. Similar to the scanning in x and y, the scanning along the optic axis can be achieved by scanning the object, the objective or the beam. Beam scanning is achieved by scanning the back focal plane of the microscope objective, usually by moving a mirror system. Object scanning is simply achieved by moving the object, and is directly comparable to focusing a classical microscope. For this work it was found that microscope objective scanning was the best technique, because there is minimal distortion when the object remains at rest. Shifting the back focal image is practical only for small changes in position.

As pointed out before, the minimum step size along the optic axis is larger than the lateral resolution. Oversampling along the optic axis will increase the photo-induced damage without improving the signal. If the lateral resolution is optimal (in the order of 150nm) and the aspect ratio (the ratio of the scales along the x- and the z-axis) is one, then three to four times as many lines have to be recorded per picture than pixels per line to fill an image. This is achieved by scanning with steps that are smaller than the minimum step size. An alternative is to generate in-between lines by appropriate algorithms after lines at optimal positions have been recorded. The thickness of directly scanned x-z images is one y-line, but several lines could be scanned and extended views or stereo pairs may be used to display the information. The requirements for scanning along the optic axis are in principle the same as those for scanning in an x-y plane: the minimum step size should be in the 100nm range and the repeatability with which the same plane is accessed should be around 50nm. It is important to note that stages for conventional microscopes are very often insufficient for three-dimensional microscopy. Temperature changes can cause the object to drift out of the sample volume during the time it takes to collect a set of images.

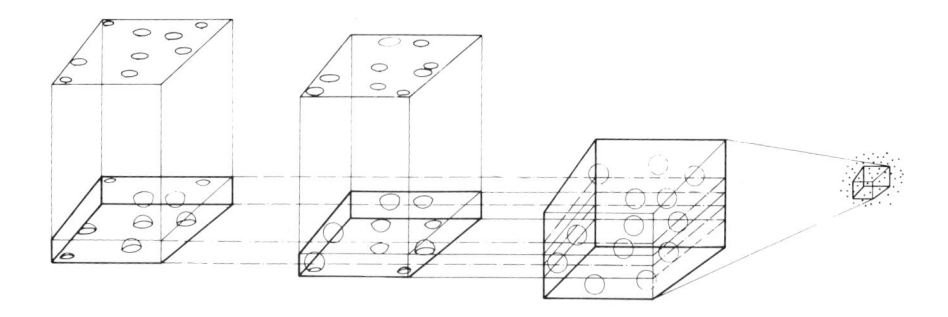

Figure 7.4. Serial sectioning in a confocal fluorescence microscope. The confocal fluorescence microscope is used to observe the three-dimensional distribution of fluorescent spheres in a cube. The observed images are projections through sections of the cube. The thickness of the sections is determined by the depth discrimination properties of the microscope. Even when applying the minimum step size, the information in two sequential x-y images will overlap. Appropriate 'neighbourhood criteria' allow the direct reconstruction of three-dimensional distributions from the raw data set.

III. Biological applications of confocal fluorescence microscopy

III.1. Consequences for biological applications

A number of important considerations must be kept in mind:

(1) Confocal fluorescence microscopy will may not dramatically improve an image recorded from thin (in the order of $1\mu m$) samples in comparison to video-enhanced fluorescence microscopy.

(2) The confocal fluorescence microscope should be operated with a high numerical aperture lens.

(3) The field size is limited by the numerical aperture of the lens and the number of pixels per line (and the number of lines per frame). A display size of 512 x 512 will limit the field to approximately $60\mu m$ x $60\mu m$ when using an objective lens with a numerical aperture of 1.3.

(4) Oversampling should be avoided to reduce the photo-induced damage. This means that the microscope should not be operated below the optimum field size and that the step size for series should not fall below the minimum step size (usually in the order of $0.6\mu m$).

(5) The sample preparation should be carefully reviewed, taking into account that quality criteria derived from conventional fluorescence microscopy are insufficient.

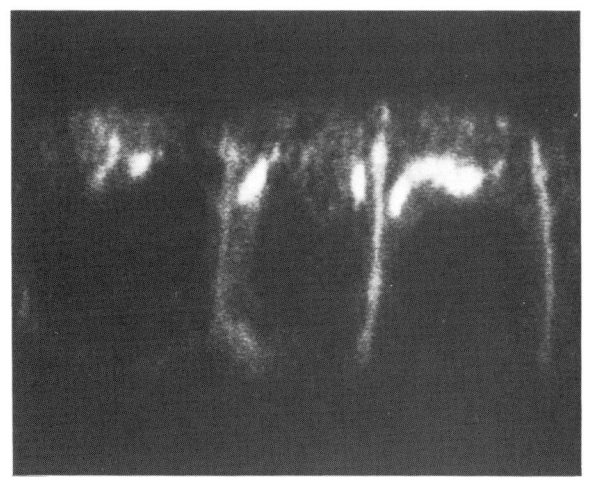

Figure 7.5. x-z image of MDCK cells that have incorporated C_6-NBD-ceramide. The dark areas in the middle are the nuclei of the cells. The bright object above the nuclei in each cell is the Golgi complex. The bright stripes on either side of the nuclei correspond to the lateral plasma membrane. Clearly visible is the C_6-NBD-ceramide in the cytoplasm. (The aspect ratio is one, the step size is 60nm and each line is averaged 8 times. It took 20 sec to record this image and the field size is $30\mu m$.)

III.2. Applying confocal fluorescence microscopes

Two important applications of the confocal fluorescence microscope that can give an idea of how the instruments [Marsman et al., 1983; Stelzer and Wijnaendts, 1986, 1987, 1989; Stelzer et al., 1988, 1989] can be used in biological cell research are: investigations of cells such as the Madin-Darby Canine Kidney (MDCK) cells [Simons and Fuller, 1985] which are line-cultivated on non-transparent filters that cannot be readily observed with conventional techniques, and colocalization experiments on PC-12 cells. Figures 7.5-7.9 show some typical results obtainable only with a confocal fluorescence microscope.

III.2.1. Metabolism of C6-NBD-ceramide in MDCK cells

To study the intracellular transport of newly synthesized lipids in MDCK cells, a fluorescence ceramide analog was used as the probe [Lipsky and Pagano, 1983, 1985; van Meer et al., 1987]. The ceramide analog was readily taken up by the cells from liposomes at a temperature of 0°C, and, having penetrated the cell, the fluorescent ceramide

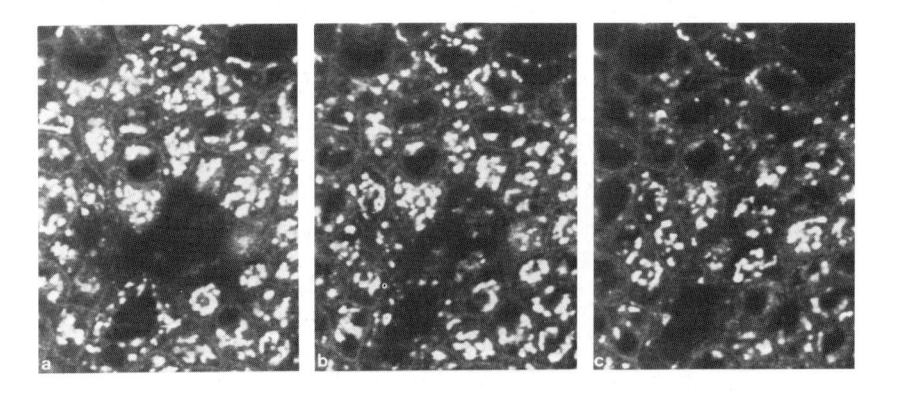

Figure 7.6. z-series of x-y images of MDCK cells that have incorporated C_6-NBD-ceramide. Three images recorded in three different planes in the cell monolayer are shown. The dark areas in the middle of the cells are the nuclei. These are surrounded by the Golgi complex and a faint staining in the cytoplasm. The cells are fully developed (120 hrs old) and the Golgi apparatus is therefore found above the nucleus. (The field size is 65μm and the step size is 1μm.)

analog accumulated in the Golgi area, Figure 7.5, as long as the temperature did not exceed 20°C. Previous biochemical characterization [Hansson *et al.*, 1986] had shown that fluorescent sphingomyelin and glucosylceramide are produced. The confocal fluorescence microscope can be used to study the three-dimensional intensity distribution as a function of the internalization state, Figures 7.6 and 7.7. An analysis of the fluorescence pattern after the cells had been kept at 20°C for one hour revealed that the fluorescent marker concentrated in the Golgi complex itself. Little fluorescence was observed at the plasma membrane or in the cytosol. After raising the temperature to 37°C for one hour, intense plasma membrane staining occurred and there was an overall loss of fluorescence from the Golgi complex. The addition of BSA to the medium in contact with the apical membrane cleared fluorescence in this membrane domain only, while the basal and lateral domains retained their fluorescence. The basolateral fluorescence could be depleted by adding BSA to the medium that was in contact with the basal membrane. These findings then led us to three conclusions [van Meer *et al.*, 1987]:

(1) The fluorescent ceramide products were delivered from the Golgi complex to the plasma membrane.

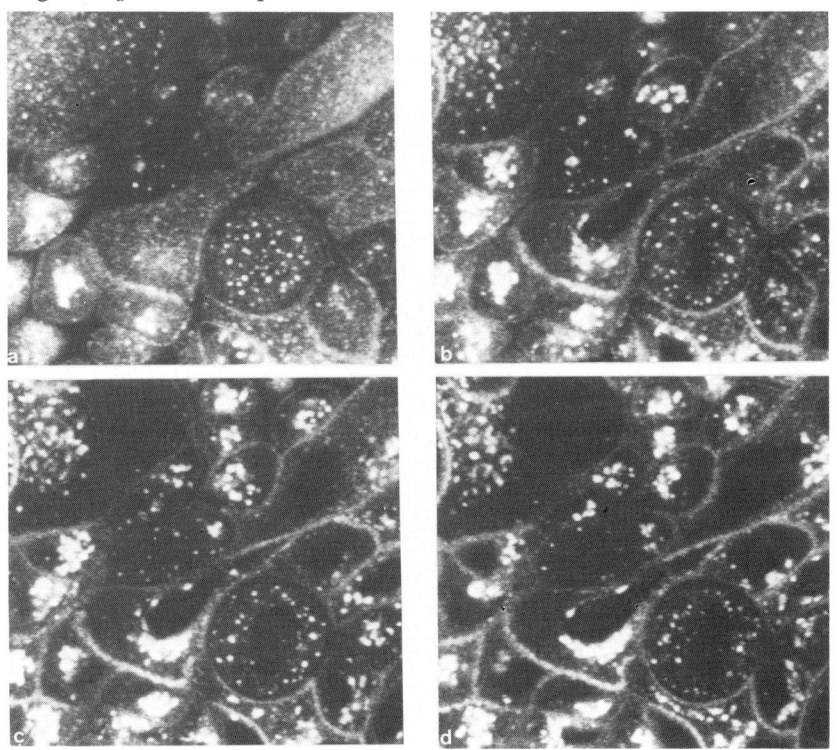

Figure 7.7 cont.

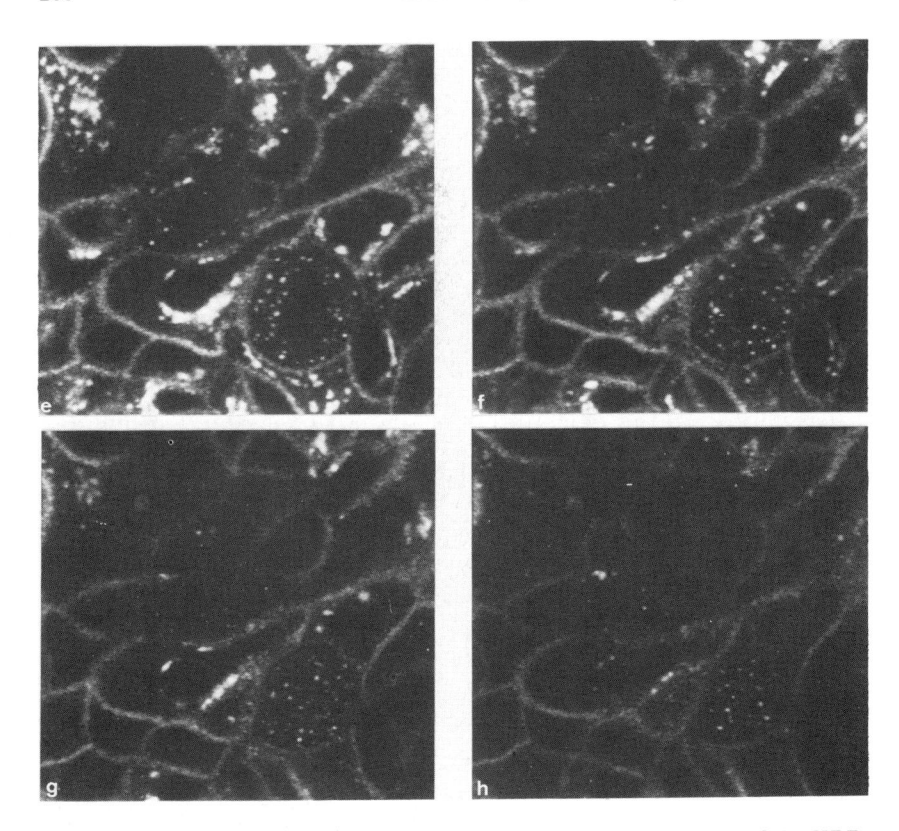

Figure 7.7. z-series of x-y images of MDCK cells that have incorporated C_6-NBD-ceramide. Eight images from a series of twenty-four have been selected. Images (a) and (b) show the sections recorded in the apical membrane, images (c) through (f) were recorded in planes of the nuclei, and (g) and (h) show the fluorophore distribution below the nucleus and near the basal plasma membrane. Clearly visible are the staining of the apical membrane, the Golgi complex, the nucleus, the lateral plasma membrane and the cytoplasm. The cell in the lower right quarter undergoes mitosis. The cell is rounded up, the Golgi apparatus is taken apart and the fine black ring indicates that no fluorescent product is transported to the plasma membrane. The Golgi apparatus is purposely over-exposed to show the labeling of the apical plasma membrane. (The field size is 75μm. The step size is 1.2μm in this figure, but was 0.4μm in the original series.)

(2) The products accumulated in the external leaflet of the membrane bilayer.

(3) The fatty acyl-labeled lipids were unable to pass the tight junction between the basolateral and the apical side.

Quantitation was performed biochemically and correlated with the data generated by the confocal fluorescence microscope. Together the data showed that the product concentration was higher on the apical than on the basolateral membrane. The investigation suggests that the fluorescent analog is sorted in MDCK cells in a similar way to their natural counterparts.

III.2.2. Colocalization of secretogranin I-containing vesicles and the Golgi complex in PC-12 cells

PC-12 cells cultivated on coverglasses grow fifteen or more micrometers high. A high primary product of these endocrine cells is secretogranin I. The pathway of secretion proteins from the endoplasmic reticulum to the plasma membrane is usually the subject of studies on these cells, since the secreted protein is available in large amounts. Although this allows simple biochemical studies, the high density tends to make microscopical observations extremely complicated. As a result, the confocal fluorescence microscope should be the first choice of instrument when investigating this system in PC-12 cells. In a double labeling experiment [Rosa et al., 1989], PC-12 cells were stained with an anti-Golgi antibody and a second antibody to secretogranin

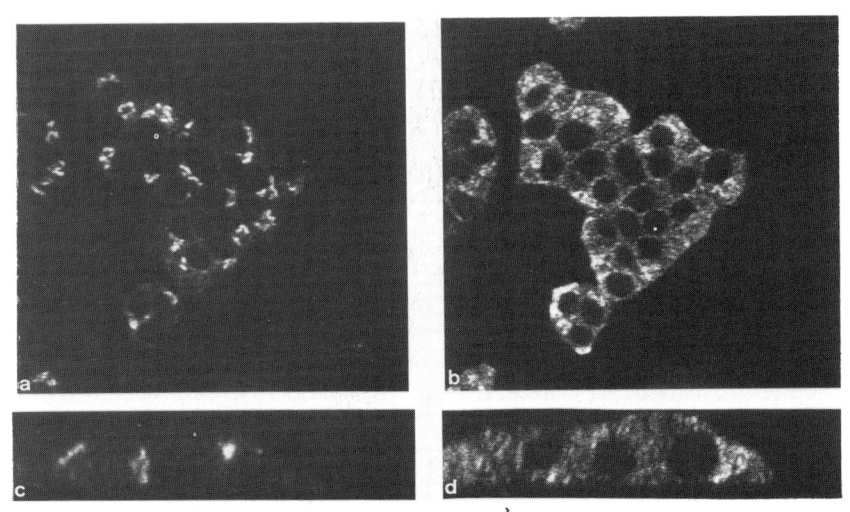

Figure 7.8. Sections through double-labeled PC-12 cells. Images (a) and (c) show a number of PC-12 cells that were labeled with a Rhodamine-conjugated antibody to make the Golgi complex visible. The bright area, corresponding to the Golgi complex, is found either beside or above the nucleus. Images (b) and (d) show the same cells labeled with an FITC-conjugated antibody to make secretogranin I-containing vesicles visible. The granules containing the secretogranin I are found above, below and beside the cell nucleus in the cytoplasm. Clearly visible are domains in which these vesicles are concentrated. Images (a) and (b) (the field size is 120μm) show only one x-y section through the cell monolayer, while images (c) and (d) (the field size is 60μm, the aspect ratio is one, the step size is 120nm) show one x-z section.

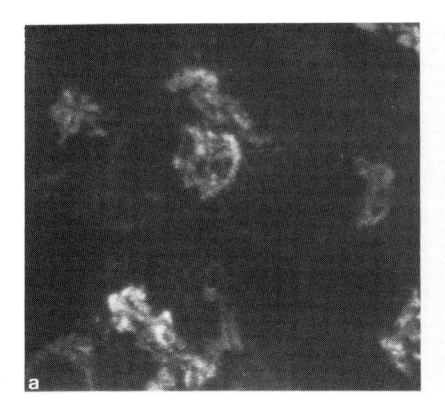

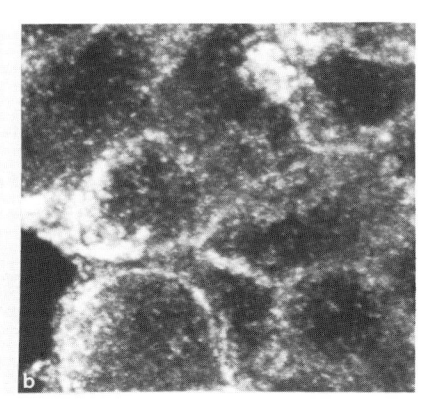

Figure 7.9. Extended views of double-labeled PC-12 cells. Image (a) shows the Golgi complex and image (b) the secretogranin I-containing vesicles. The cells are the same as those shown in the center in Figure 7.7. However, a series of images was recorded, including a stereo pair. The left (unshifted) images are shown. (The field size is $30\mu m$, the step size is $0.6\mu m$ and 15 images were used to record the series.)

I. The cells were then scanned twice: the fluorescein-labeled samples were observed with the 476nm line and the rhodamine with the 514nm line of an argon ion laser. Different filter sets ensured that no overlap from one dye became visible while observing the other. We found that one series returned the three-dimensional intensity distribution of the fluorophore that labeled the Golgi complex, and that the other series returned the three-dimensional intensity distribution of the fluorophore that labeled the secretogranin I-containing vesicles. The individual series have been used to calculate stereo images, Figures 7.8 and 7.9. Figure 7.8 also shows two x-z images of the same cells, but at half the magnification. The secretogranin I-containing vesicles are found throughout the cell, but seem to concentrate in the periphery of the cells. These areas of high concentration do not colocalize with the Golgi complex. In fact, the Golgi complex excludes a high concentration of secretogranin I, which was the expected result.

IV. Conclusions

When considering the usefulness of the confocal fluorescence microscope in cell biology, it should be kept in mind that as modern conventional fluorescence microscopes are excellent devices, confocal fluorescence microscopes should only supplement conventional devices that already exist. However, it can be expected that three-dimensional microscopy will become a technique as common as conventional microscopy, and hence will have an important influence on the biological systems that will be studied. When optically thick specimens are observed, there is no simple alternative to a confocal fluorescence microscope.

V. Acknowledgements

We thank G. van Meer, K. Simons, P. Rosa and W. Huttner for their collaboration in cell biology and C. Storz, R. Stricker and R. Pick for excellent technical assistance.

References

Bacallao, R., Stelzer, E .H. K., (1989). Preservation of biological specimens for observation in a confocal fluorescence microscope and operational principles of confocal fluorescence microscopy. In: *Methods in cell biology.* Tartakoff, A. M. (Editor). Academic Press, Orlando. In press.

Hansson, G. C., Simons, K. and van Meer, G., (1986). Two strains of the Madin-Darby Canine Kidney (MDCK) cell line have distinct glycolipid compositions. E.M.B.O. J., **5**, 483-489.

Inoué, S., (1986). *Video microscopy.* Plenum Press, New York.

Lipsky, N. G. and Pagano, R. E., (1983). Sphingolipid metabolism in cultured fibroblasts: microscopic and biochemical studies employing a fluorescent ceramide analogue. Proc. Natl. Acad. Sci. USA., **80**, 2608-2612.

Lipsky, N. G. and Pagano, R. E., (1985). Intracellular translocation of fluorescent sphingolipids in cultured fibroblasts: endogenously synthesized sphingomyelin and glucocerebroside analogues pass through the Golgi apparatus en route to the plasma membrane. J. Cell Biol., **100**, 27-34.

Marsman, H. J. B., Brakenhoff, G. J., Blom, P., Stricker, R. and Wijnaendts-van-Resandt, R. W., (1983). Mechanical scan system for microscopic applications. Rev. Sci. Instrum. 54, 1047-1052.

Rosa, P., Ursula, W., Pepperkok, R., Ansorge, W., Niehrs, C., Stelzer, E. H. K. and Huttner, W., (1989). An antibody against secretogranin I (chromogranin B) is packaged into secretory granules. J. Cell Biol., **109**, 17-34.

Simons, K. and Fuller, S., (1985). Cell surface polarity in epithelia. Annu. Rev. Cell Biol., **1**, 243-288.

Stelzer, E. H. K. and Wijnaendts-van-Resandt, R. W., (1986). Applications of fluorescence microscopy in three dimensions: microtomoscopy. S.P.I.E., **602**, 63-70.

Stelzer, E. H. K. and Wijnaendts-van-Resandt, R. W., (1987). Nondestructive sectioning of fixed and living specimens using a confocal scanning laser fluorescence microscope: microtomoscopy. S.P.I.E., **809**, 130-137.

Stelzer, E. H. K. and Wijnaendts-van-Resandt, R. W., (1989). Fluorescence microscopy in three dimensions: microtomoscopy. In: *Cell structure and function by microspectrofluorometry.* Cohen, E. (Editor), 131-143. Academic Press, San Diego.

Stelzer, E. H. K., Stricker, R., Pick, R., Storz, C. and Hänninen, P., (1989). Confocal fluorescence microscopes for biological research. S.P.I.E., **1028**, 146-151.

Stelzer, E. H. K., Stricker, R., Pick, R., Storz, C. and Wijnaendts-van-Resandt, R. W., (1988). Serial sectioning of cells in three dimensions with confocal scanning laser fluorescence microscopy: microtomoscopy. S.P.I.E., **909**, 312-318.

van Meer, G., Stelzer, E. H. K., Wijnaendts-van-Resandt, R. W. and Simons, K., (1987). Sorting of sphingolipids in epithelial (MDCK) cells. J. Cell Biol., **105**, 1623-1635.

Wijnaendts-van-Resandt, R. W., Marsman, H. J. B., Kaplan, R., Davoust, J., Stelzer, E. H. K. and Stricker, R., (1985). Optical fluorescence microscopy in three dimensions: microtomoscopy. J. Microsc., **138**, 29-34.

Wilson, T. and Sheppard, C. J. R., (1984). *Theory and practice of scanning optical microscopy.* Academic Press, London.

8. Confocal Brightfield Imaging Techniques Using an On-Axis Scanning Optical Microscope

C. J. COGSWELL AND C. J. R. SHEPPARD

I. Introduction

Confocal microscopes have become an established tool in a wide range of professions, mostly owing to rapid commercial development of a few microscope designs optimized to image a rather narrow range of object characteristics specific to each discipline. In the fields of metallurgy, electronic device manufacture and industrial inspection, the prevalence of subjects with highly reflective surface morphologies has led to the development of confocal reflection brightfield microscopes for viewing and measuring three-dimensional surface topographies. In the bio-medical fields, the great upsurge in the use of fluorophore-conjugated antibodies to label a vast range of cell constituents has stimulated the development of commercial laser-scanning instruments optimized for the reflected fluorescence mode. These instruments, if they provide confocal reflection brightfield capabilities at all, generally perform as if the optics were added as an afterthought, with little regard for the quality of performance or for understanding and evaluating the images produced. Another class of confocal microscopes includes the real-time or tandem-scanning designs, which use Nipkow discs and incoherent light sources. Although these have been used successfully to examine biological subjects such as teeth and bone in reflection, they are more suitable for looking at highly reflecting or scattering objects because they can detect only a small fraction of the incident illumination.

In view of the fact that few commercial instruments have reflection modes which can produce high quality confocal images of weakly scattering biological subjects, we believe that there is still a great need to develop new techniques so that the full advantages of confocal microscopy can be realised. Since biologists continue to employ as standard routine a wide range of non-fluorescence conventional microscope techniques such as brightfield, phase contrast and differential interference contrast, in both transmission and reflection, we believe that there is great potential for acquiring new information by carefully incorporating these traditional optical configurations into a confocal design. This chapter discusses some of the advantages as well as the pitfalls that may be encountered in designing and applying non-fluorescence confocal systems (henceforth referred to collectively as brightfield systems) and some of the factors to be considered in interpreting the resulting images.

II. Microscope design considerations

II.1. Choosing the optimum design configuration for brightfield optics

To start, we must consider what type of microscope design will be most appropriate to our goal of obtaining high-resolution images (in both the axial and transverse directions) in a variety of possible confocal brightfield configurations. Since many biological subjects contain only small variations in refractive index, reflection signals will often be weak, necessitating the use of some type of laser illumination. Confocal microscope designs which use laser sources fall into two categories:

(1) Off-axis beam scanning, in which some type of mirror arrangement or beam deflector is used to scan the illuminating beam through an objective and across a stationary subject [Carlsson et al., 1985].

(2) On-axis specimen or objective scanning, in which the illuminating beam remains stationary on the optical axis and either the subject (on the microscope stage) or the objective is scanned [Sheppard, 1977].

Although these two designs may at first appear to be very similar in their overall imaging properties (see Chapter 3 in this volume for a more detailed description) they are not similar with respect to their overall potential for producing maximum resolution, aberration-free reflection images [Cogswell et al., 1990].

A simple way of comparing the performance of the two types of instruments is to mount a front-surface mirror at the specimen plane and limit the transverse scanning pattern to a single point or line, while simultaneously changing the focus of the microscope over a precise distance (along the z-axis). Plotting the resulting signal intensity from the detector plane versus the axial position of the focus gives a method of measuring axial resolution (i.e. full width at half maximum (FWHM) measurements of the central peak) as well as a technique for detecting the presence of aberrations (i.e. deformities in the shape of the central peak and/or the presence of side-lobes). We refer to this type of measuring technique as a $V(z)$ plot. Figures 8.1 and 8.2(a) show a comparison of the performance of the same high numerical aperture (N.A.$=1.4$), highly corrected (planapo) objective on both a beam-scanning and an on-axis scanning microscope. Figure 8.1(a) demonstrates the appearance of a typical x-z image of the mirror sample using a beam-scanning instrument. The beam-scanning microscope used in this example was a Bio-rad MRC-500 used in the confocal reflection mode. In this type of imaging configuration the beam is made to scan repeatedly across a single horizontal line (i.e. the y-axis position of the beam is held constant) while the focus position is stepped through several planes in the vertical z dimension. Figures 8.1(b-d) are graphs of signal intensity, I, versus position in depth (z-axis) which were created by plotting the signal strength (brightness) along a vertical line drawn through the x-z images at some chosen value of x. Figure 8.1(b) is a graph from near the centre while Figures 8.1(c) and 8.1(d) are from opposite edges of the scanned field.

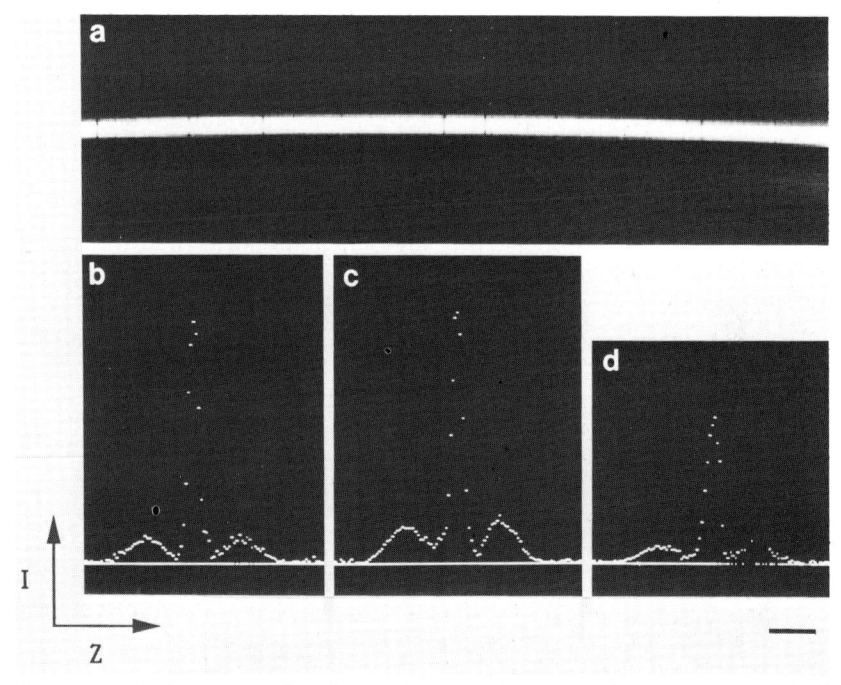

Figure 8.1. Performance of a planapochromat, N.A.=1.4 objective on an off-axis beam-scanning microscope. (a) A typical x-z image from a front-surface mirror placed at the specimen plane. In this example, $y = 255$ in vertical screen coordinates (centre of field). Dark vertical lines correspond to surface scratches on the mirror. (b) $V(z)$ plot of intensity, I, versus depth in z for $y = 255$, $x = 415$ in horizontal screen coordinates, (centre of field). (c) $V(z)$ plot for $y = 0$, $x = 735$, (top edge of field). (d) $V(z)$ plot for $y = 511$, $x = 735$, (bottom edge of field). This $V(z)$ method shows that there are variations in axial resolution (FWHM measurements of the central peak), signal intensity (height of central peak) and aberrations (shape and strength of side-lobes) as the beam scans across the field in x and y. Horizontal scale=2μm in z.

Variations in signal shape and intensity become apparent as the beam is scanned across the field in both the x and y directions. Although these variations are probably not much of a factor in obtaining usable fluorescence images (owing to the need to use larger detector pinholes and perhaps signal averaging for most biological applications) they most certainly must be considered to be an impairment to producing diffraction-limited images in brightfield reflection. One further drawback to the beam-scanning design is that to obtain medium or low magnification images a lower power (and consequently lower N.A.) objective must be utilised. This results in loss of axial (and lateral) resolution as well as increasing the likelihood that aberrations will be present.

Figure 8.2(a) is a plot of intensity versus depth in z using the same objective and mirror sample on a specimen-scanning microscope at the University of Oxford. The fairly narrow peak and small side-lobes represent good axial resolution with a minimum of aberrations, which of course will remain constant over the entire image field since the beam remains stationary on-axis. Interestingly, the peak can be made even narrower (i.e. axial resolution improved) and the side-lobes further reduced if a high numerical aperture (N.A.=1.3) fluorite (non-flat field) objective is inserted in place of the planapo lens, Figure 8.2(b). We interpret this to result from the fact that the planapochromat objective, because it is corrected for flat field, exhibits some residual aberrations on-axis. When used on the beam-scanning microscope, the fluorite objective also showed a slight improvement in resolution and a reduction

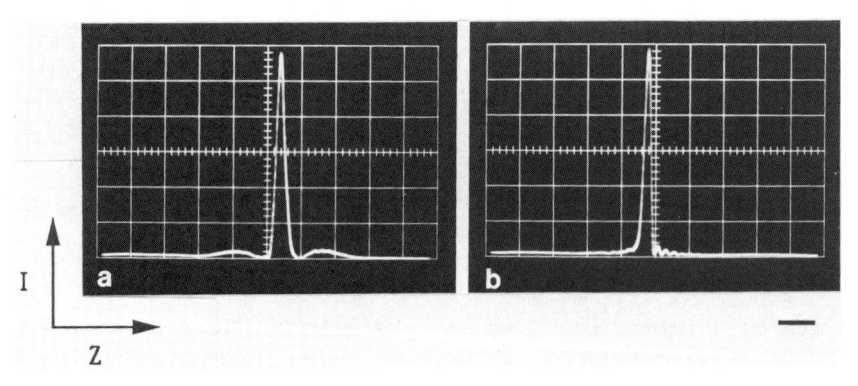

Figure 8.2. (a) $V(z)$ plot shows the performance of the same planapochromat, N.A.=1.4 objective used in Figure 8.1, but here it was tested on an on-axis specimen-scanning microscope. The narrow central peak and small side-lobes indicate good axial resolution and a minimum of aberrations which, of course, remain constant since the specimen is scanned rather than the beam. (b) $V(z)$ plot from another objective (fluorite, N.A.=1.3) used on the specimen-scanning microscope shows an improved axial response (narrower peak) and even smaller side-lobes. Horizontal scale=2μm in z.

of side-lobes when measured in the centre of the field, but this condition rapidly deteriorated as the beam scanned toward the periphery: the signal intensity became extremely variable and aberrations such as curvature of field and astigmatism were noticeably present.

Because of the difficulties inherent in the beam-scanning design, we have found it preferable to use an on-axis specimen-scanning configuration. This design eliminates the difficulty of correcting off-axis aberrations in the objective lens and has the further advantage of being able to use a single, high numerical aperture objective for the entire range of desired image magnifications; low-power imaging being achieved by simply increasing the specimen scanning distance. Still another advantage of the on-axis configuration is that a variety of brightfield techniques may be utilized which would be difficult or impossible to achieve with a beam-scanning system. Transmitted differential phase contrast, full-colour reflection, and differential interference contrast are but a few of the techniques described in this chapter which show great promise for biological imaging using a specimen-scanning confocal microscope.

II.2. Scanning speed

Before continuing this discussion, it is probably important to address what we believe to be an unfortunate misconception linked to the specimen-scanning confocal microscope — that it is too slow to be used for imaging most biological material [Inoué, 1989]. The scanning rate of our microscope makes it possible to obtain a 256 x 256 pixel image in 2.5 to 3 seconds which, interestingly, is within the range of several commercial beam-scanning instruments. As more theoretical and experimental work is published in the area of confocal fluorescence microscopy, it is becoming increasingly accepted that there is a finite limit to the speed in which a usable fluorescence image can be obtained, owing to the interplay between such factors as laser intensity versus fluorophore bleaching, detector pinhole size versus axial resolution, and detection of photons from the fluorescing sample versus background noise [Inoué, 1989]. For a full-frame image, this temporal limit may be anywhere from a very few to many tens of seconds, depending on whether the signal must be averaged over several frames. In this light, the slower maximum rate of operation of the specimen-scanning design should no longer be considered a major obstacle to most applications in biological imaging.

Another misconception linked to this microscope design is that moving the stage causes detectable vibrations in specimens composed of fresh or unfixed biological matter, making the system unusable. We have made many fresh preparations of a variety of plant cells mounted in water with a sealed coverglass and detected no intra- or extracellular motion using our specimen-scanning microscope (for example, see Figures 8.5, 8.11, 8.13 and 8.14). However, we do not pretend to have made exhaustive studies of the suitability of this system for examining all possible types of fresh material. Certainly, as in all types of biological microscopy, there will be limitations. However, a full appreciation of the advantages and limitations of the instrumentation to be used will influence the development of techniques for optimizing sample preparation.

III. Reflected brightfield

Because of its ability to produce clear images from narrow planes of focus while at the same time eliminating out-of-focus flare, the confocal microscope is extremely well-suited to reflected brightfield applications which are normally difficult or impossible to achieve with a conventional microscope. However, the simplicity with which images can be obtained in confocal reflection should not mislead the observer into making quick and frequently erroneous conclusions about the nature of the original specimen. Just as microscopists and biologists have had to consider carefully the characteristics of the sample as well as the instrument when developing the scanning and transmission electron microscopes and, more recently, when developing fluorescent probes for the confocal fluorescence microscope, so must we consider the many variables which contribute to acquiring and correctly interpreting confocal brightfield images. In this section, we will discuss some of the more important of these variables such as the refractive index of all regions surrounding the sample, the physical characteristics of the subject with respect to how it will affect the incident laser light (i.e. Rayleigh scattering, absorption or transmission of certain wavelengths) and the quality and alignment of the optical system.

III.1. Sample characteristics

In order to demonstrate the many applications of confocal reflected brightfield microscopy in biology, it is useful to consider first the physical and optical properties of the subjects being examined. For ease of discussion, we have chosen to divide these sample characteristics into four general categories:
  (1) Reflective surfaces.
  (2) Refractive index boundaries within samples.
  (3) Changes in absorption and reflection properties with wavelength (i.e. full-colour imaging).
  (4) Sub-wavelength scattering objects.

III.1.1. Reflective surfaces

From within the broad range of biological subjects suitable for confocal reflection techniques, probably those most readily imaged and easily understood are ones with strongly reflecting surfaces that possess some sort of regular surface morphology (topography). Examples of subjects that fit this category are insect exoskeletal components, vertebrate teeth, pollen grains, seeds and the waxy cuticle layer on plant surfaces. Specimens can be prepared very simply, i.e. mounted in air with no coverglass, and viewed with high N.A. reflection objectives. Interpreting the resulting images is equally straightforward, since most or all of the detected signal can be attributed to scattering from surfaces, due to the relative non-reflectivity of the material beneath. An illustration of this type of subject can be seen in Figure 8.3 which is a series of images of a portion of a compound eye from an adult wild-type fruit fly, *Drosophila melanogaster*. The eye was simply stuck to the surface of a slide and examined (in air) with a highly corrected reflection objective (0.85 N.A.). Figure 8.3(a) is a reflected brightfield image similar to that obtainable with a conventional microscope (i.e. no detector pinhole). Figures 8.3(b-d) show the same sample at three planes of focus, $2.5\mu m$ apart, but now with a $5\mu m$ pinhole in front of the detector to give confocal reflection images. Owing to the lack of signal from out-of-focus surfaces, these images are somewhat difficult to interpret when viewed individually. Therefore, in order to obtain a better understanding of the three-dimensional morphology, the confocal images can be combined in one of several ways. Figure 8.3(e) is an example of the auto-focus method of displaying depth information. It was created by comparing successive images from a through-focus series and storing the brightest value at each pixel position. This technique was also applied to an eye from a mutant strain of the same *Drosophila* species, illustrated in Figure 8.3(f), to show the severe irregularities in ommatidia development and the much less rounded and more ridge-like surface morphology of the entire compound eye.

Two other techniques for displaying three-dimensional information are probably worth illustrating at this point, since they are particularly applicable to surface topographic imaging. The first is the surface profile technique which encodes the position in depth of the brightest pixel from the through-focus series comparison. This depth

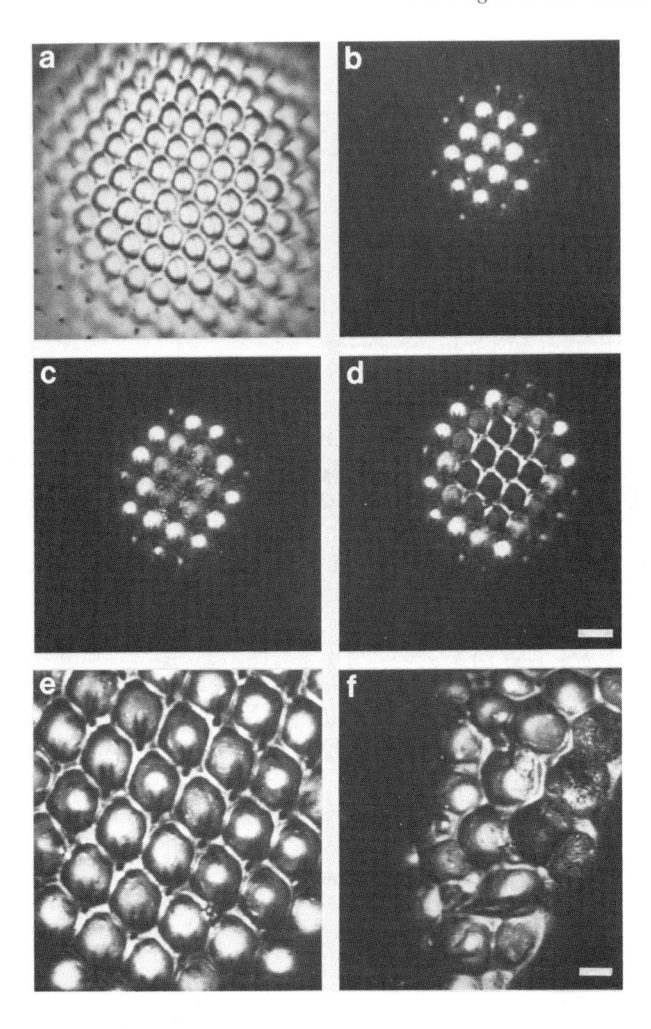

Figure 8.3. Reflection images of the compound eye of a fruit fly, *Drosophila melanogaster*, illustrate the advantages of using confocal brightfield techniques for examining biological subjects having highly reflective surface morphologies. (a) Reflected brightfield image with detector pinhole removed gives a poor axial response similar to a conventional microscope, i.e. out-of-focus planes are present in addition to the plane of focus. (b-d) A series of confocal reflection images produced by inserting a 5μm pinhole in front of the detector and changing the focus by 2.5μm intervals in depth. Now only reflections from within a very narrow plane of focus (optical section) are detected in each image. Scale=20μm. (e) Auto-focus image of a wild-type eye, created by comparing successive images from the through-focus confocal series and storing the brightest signal at each pixel position. (f) Auto-focus image of an eye from a mutant *Drosophila* illustrates an irregular, ridged morphology. Scale=10μm. Samples were stuck to the surface of a slide and examined with an N.A.=0.85 reflection objective, corrected for no coverglass.

information is displayed as a grey-level value in the resulting image, with darker grey values corresponding to signals from deeper planes of focus, Figure 8.4(b). The surface profile image can be further enhanced using pseudo-colouring routines. A second three-dimensional display method uses an auto-focus image, Figure 8.4(a), and the depth information encoded in the surface profile image to form an isometric projection, Figure 8.4(c). This technique illustrates the overall rounded surface typical of a wild-type eye as well as the regular array morphology of the ommatidia.

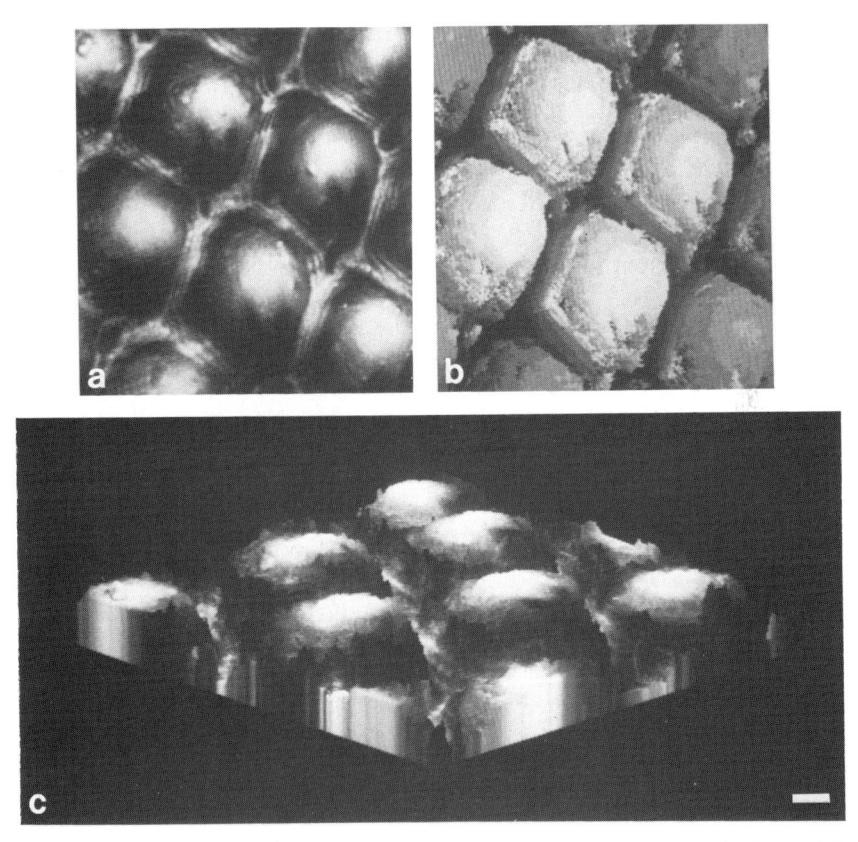

Figure 8.4. A higher magnification view of a wild-type eye of *Drosophila melanogaster* demonstrates additional techniques for displaying three-dimensional information. (a) An auto-focus image. (b) A surface profile image created at the same time as (a) but here the depth information from the focus series is represented by successively darker grey values. This creates a grey-level map of the surface morphology. (c) An isometric projection calculated from depth information in the surface profile combined with the auto-focus image. Scale=5μm.

III.1.2. Refractive index boundaries within samples

It is generally much more difficult to extract meaningful information from confocal reflection imaging techniques when the regions of interest are contained within a semi-transparent sample. This is true of a large number of biological preparations such as living whole mounts or cultures, fixed and embedded tissues, and cell or chromosome squashes. These subjects are usually placed on a microscope slide in some kind of fluid medium with an overlaying coverglass. Each of these more or less horizontal refractive index boundaries will reflect a portion of the incident light depending on such factors as angle of incidence, orientation (or tilt) of the boundary with respect to the optic axis, and the actual difference between the two refractive indices. Even when oil immersion objectives are used to match closely the refractive indices of the lens elements and coverglass, there is some loss of illumination from reflections before the initial plane of the specimen is reached.

An even more complex situation usually arises when focusing on the external boundary of the biological specimen. Strong reflections will occur at this interface if the refractive index of the object is different from the surrounding mounting or embedding medium. If the biological structure is a cell wall or membrane, then it probably will have a different refractive index from the underlying cytoplasm, and thus it will produce still another reflecting boundary (dependent on orientation as before) when this plane of focus is reached. The pattern of reflections from refractive index (phase) boundaries will continue throughout the entire focus series, with strong phase steps producing the brightest signals. Figures 8.5(a-d) illustrate this phenomenon using a fresh leaf peel from the monocot, *Rhoeo spathacea*. The tissue was mounted in water under a sealed coverglass and viewed with an oil immersion, N.A.=1.4, objective lens. The extremely bright (white) regions in Figure 8.5(a) are reflections from the waxy cuticle layer of the leaf surface, which forms a strong refractive index boundary with the underlying cells. The signal from these regions is saturated because the overall gain of the system has been increased in order to image the much weaker phase boundaries within the underlying stomatal structure. Focusing down into the specimen at 2μm intervals, Figures 8.5(a-d), shows that stomatal guard cells containing chloroplasts and accessory cells containing pigment granules have enough variations in refractive index to produce weak but easily detectable signals in confocal reflected brightfield, provided they are far enough below the cuticle layer boundary so as not to be overpowered by reflections from its plane of focus. This masking by highly refractive regions is, of course, the reason why it is extremely difficult to use reflection brightfield techniques to view biological specimens with a conventional microscope.

When imaging biological subjects of this type, further complicating factors must also be taken into account, including refraction or attenuation of the illuminating rays by inclusions in the sample such as lipid globules or opaque bodies, and interference fringes due to narrow layers of differing refractive index. These sorts of effects frequently make it extremely difficult to interpret images from complex biological structures obtained from planes of focus more than a few microns in depth below the top of the subject.

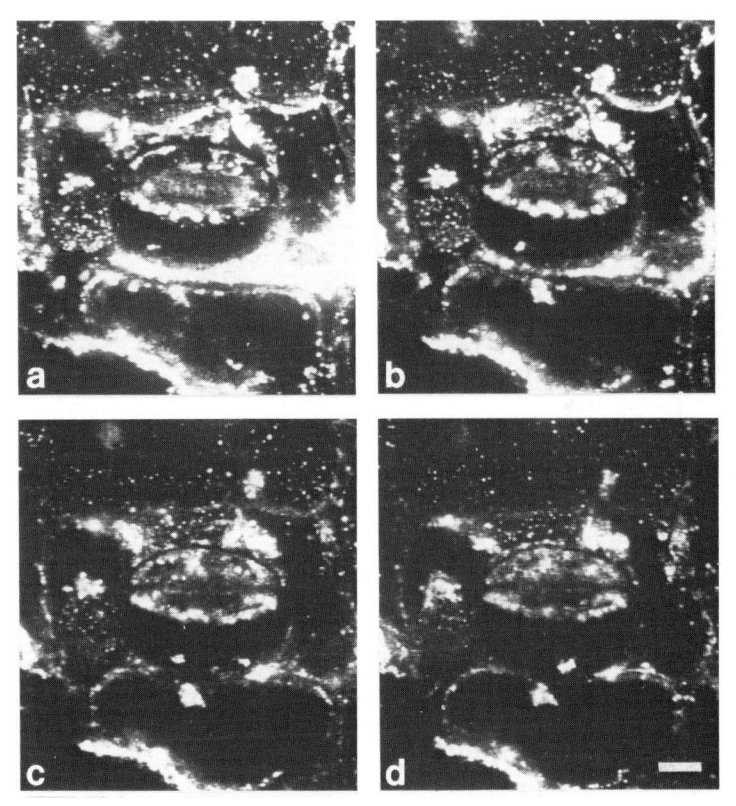

Figure 8.5. Advantages of using confocal reflected brightfield techniques for imaging refractive index boundaries within biological samples. (a-d) A series of images from a fresh leaf peel of *Rhoeo spathacea* at 2μm focus intervals. A stoma with its two half-moon shaped guard cells is seen in the centre of the field. The gain of the system was increased to allow viewing of the weakly reflecting stomatal structures. Thus, the signal saturated when focusing on the plane of the highly reflective waxy cuticle layer boundary, visible as a large white patch in (a). The optical sectioning capability of the confocal microscope makes it possible to selectively adjust the sensitivity of the system in order to detect weakly reflecting structures (provided they are not so close to a strong phase boundary as to be masked by its bright signal). Scale=15μm.

III.1.3. Changes in absorption and reflection properties with wavelength: full-colour reflected brightfield imaging

In addition to producing reflections at phase boundaries, biological structures may contain natural pigments or artificial stains which cause these reflections to vary in intensity depending on the wavelength of the illuminating source. For more than a century, coloured biological stains have been used in conventional light microscopy to enhance visibility and aid in identifying tissue sub-structure and cell constituents. We believe the ability to image colour information in biological microscopy to be such an important tool that it should be made available in contemporary confocal brightfield microscope systems.

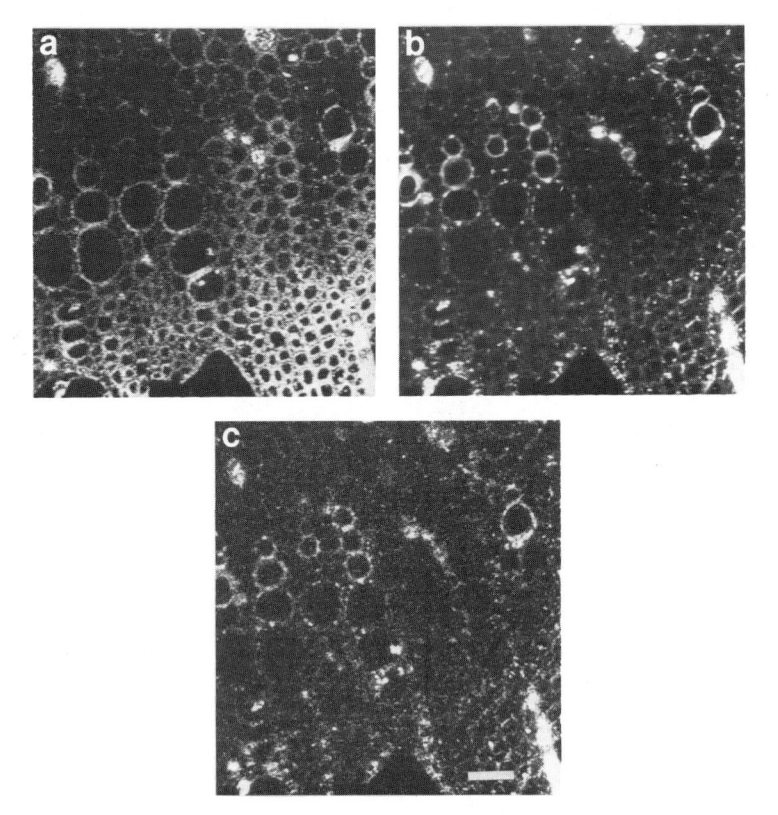

Figure 8.6. Confocal reflected brightfield techniques using red, green and blue lasers to detect differentially-stained tissue. Subject is a cross-section from a woody plant stem, *Chodanthus puberulus*, stained with alcian blue and safranin. The three images simultaneously produced with the (a) HeNe (red) laser, (b) frequency-doubled YAG (green) laser, and (c) HeCd (blue) laser can be combined to form a full-colour image. Scale=20μm.

To obtain full-colour reflected confocal images with our on-axis design, we have incorporated three lasers into our system (HeNe: 633nm; frequency-doubled YAG: 532nm; HeCd: 442nm) and simultaneously detected their signals using three separate photomultipliers. This technique can be performed more easily using apochromatic or fluorite objectives to reduce chromatic aberration effects. It also requires careful axial and transverse positioning of the three pairs of incident and detector pinholes (one confocal pair for each laser) to ensure precise registration of the resulting images. This requirement is much more easily attained with an on-axis scanning microscope, since off-axis lens aberrations, which will usually vary for different wavelengths, need not be considered. Figure 8.6 shows three reflected brightfield images of a cross-section from a woody plant stem, *Chodanthus puberulus*, fixed, embedded and subsequently stained with alcian blue and safranin. Figure 8.6(a) was made using the red HeNe laser, Figure 8.6(b) with the frequency-doubled YAG (green) laser and Figure 8.6(c) with the HeCd (blue) laser. All three images were produced at the same time and were also displayed in the red, green and blue channels of a colour monitor to produce a full-colour representation. This technique shows clearly the variations in reflectivity versus wavelength of the differently stained cell walls.

III.1.4. Sub-wavelength scattering objects

Biological samples may contain sub-structures which fall into the general category of isolated, sub-wavelength diameter particles which act as purely scattering objects. Prime examples of this type of subject are colloidal gold particles of only a few nanometers diameter, which are commonly used to label specific antibody probes for immunocytochemical experiments. The confocal reflected brightfield microscope is especially well-suited for detecting and locating (in three dimensions) these immunogold probes. However, in order to draw accurate conclusions from confocal reflection images of immunogold-labelled preparations, it is imperative to provide non-labelled control samples for comparison imaging. An example of the difficulties that can occur while imaging these gold probes is a comparison study of the plaques associated with Alzheimer's disease in human brain sections, which was recently performed in our laboratory (unpublished results). In this study we observed three serial sections: one which had been labelled with an appropriate plaque antibody linked to 1nm gold, a second which was the same as the first except that, in addition, the gold had been silver-enhanced to produce particles greater than 50nm in diameter, and a third which had been treated with 1nm gold but no antibody. Even though the label in the section containing silver-enhanced gold was clearly visible in transmission using a conventional brightfield microscope while the other two were not, when viewed with the confocal reflection brightfield microscope all three appeared the same (i.e. plaque-like regions contained brightly reflecting particles in all three sections)! Clearly, small phase objects occurring naturally in the tissue or resulting from preparation techniques were producing reflection signals that were so bright as to overpower any signal that might have been present from the gold-labelled probe.

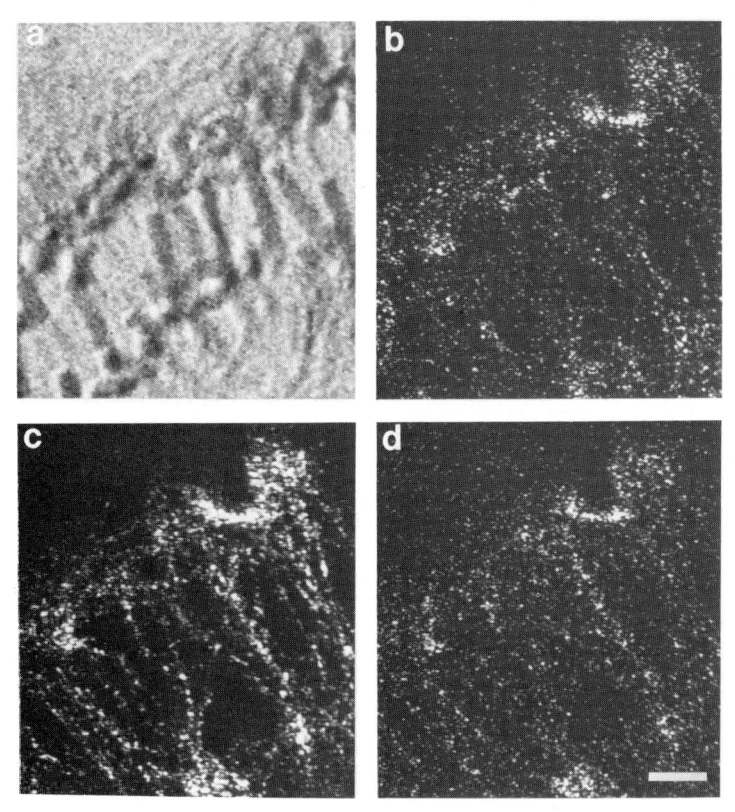

Figure 8.7. Response of sub-wavelength gold particles to confocal reflected brightfield techniques using red, green and blue laser light. Subject is an anaphase from the endosperm of an African blood lily, *Haemanthus katherinae*, which has microtubule bundles labelled with 15nm immunogold. (a) Transmitted differential phase contrast image (non-confocal), included here for reference, to show the location of the chromosomes within the microtubule bundles which comprise the mitotic spindle. (b-d) Confocal reflected brightfield images of the microtubule bundles in the same sample region using (b) HeNe (red), (c) YAG (green), and (d) HeCd (blue) lasers. The image produced with the green laser (c) shows the brightest signal which is characteristic of Rayleigh scattering by sub-wavelength gold particles. Scale=5μm. (*Haemanthus* preparations courtesy of Dr. J. Molé-Bajer and Dr. A.S. Bajer, University of Oregon.)

Sub-wavelength diameter gold particles exhibit another optical feature which is potentially useful when analysing images for the presence of immunogold labelling. Unlike silver particles, the resonance properties of gold molecules cause them to scatter green wavelengths more strongly than red or blue [Born and Wolf, 1980]. (This should not be confused with the optical behaviour of thin gold films in reflection which appear pink to red in colour.) Figures 8.7(b-d) show three images simultaneously produced using red, green and blue light, of a mitotic endosperm nucleus in the African blood lily, *Haemanthus*, in which microtubules were labelled with a tubulin antibody linked to 15nm gold. The microtubules in the image made with the green laser, Figure 8.7(c), are the brightest and most clearly delineated. This green peak response is to be expected from the Rayleigh scattering properties of sub-wavelength gold particles. (In this experiment a control specimen containing no tubulin antibody was examined and no microtubules were visible.)

These types of subjects are also well-suited to three-dimensional reconstruction and display techniques such as computer-generated stereo pairs.

III.2. Improving visual quality and resolution

In addition to carefully evaluating the physical and optical properties of the biological specimens to be examined with the confocal brightfield microscope, it is equally important to align and manipulate the instrument itself in order to obtain images with high visual quality and maximum resolution. For example, inserting additional optical elements or reconfiguring existing components in the microscope may lead to considerable improvements in the final image. In particular, we will consider three areas where careful adjustments to the on-axis design could prove beneficial:

(1) Correcting lens aberrations.

(2) Adjusting the detector pinhole size to obtain maximum lateral resolution with minimum noise.

(3) Using detector arrays to further increase resolution.

III.2.1. Correcting lens aberrations

Assuming that the user can identify by conventional means certain features of the preparation to be viewed (such as coverglass thickness, mounting/embedding medium refractive index and approximate sample thickness), it is then possible to minimize optical aberrations in the microscope if one has an understanding of how the objectives and optical elements in the system behave under controlled test conditions. For example, the $V(z)$ technique described earlier in this chapter can be used to test the performance of a specific objective when the mirror reflector used as the test object has been modified with various thickness coverglasses and mounting media. For each of these known test samples, weak spherical or cylindrical correction lenses (< 1 dioptre) can be added to the beam path (before the rear pupil of the objective) to correct for any residual effects which commonly occur, such as spherical aberration and astigmatism. Figures 8.8(a-d) illustrate the effects that varying the coverglass thickness can have on the axial resolution of an objective and also the improvements that can be realized when correction lenses are inserted in the beam path. Figure 8.8(a) is a $V(z)$ plot of the signal from a mirror sample that has had a number 1.5 coverglass (0.17mm thick) mounted to its surface with a thin layer of immersion oil. The objective used in

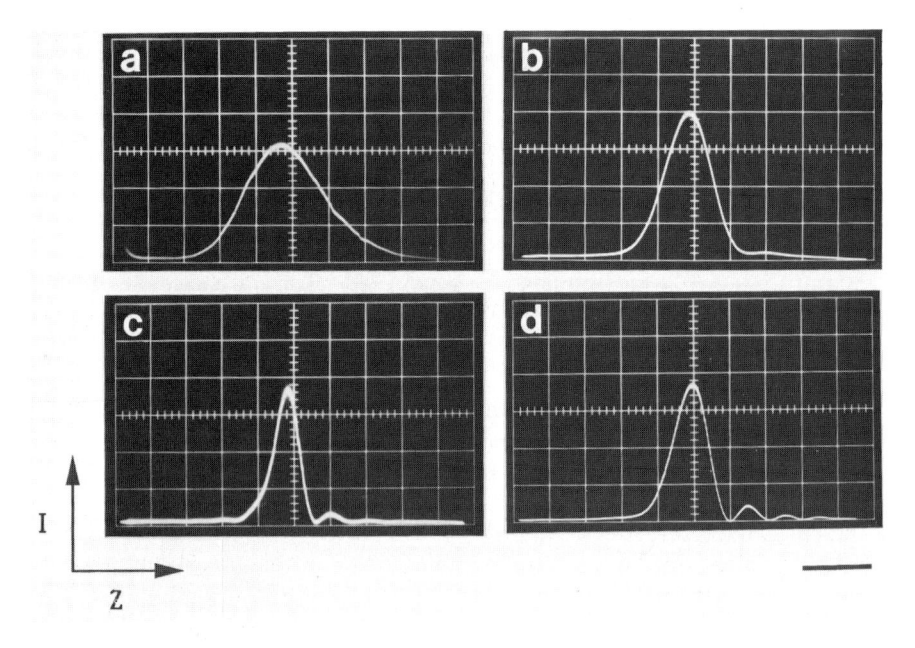

Figure 8.8. Changes in axial resolution produced by varying coverglass thickness or by inserting weakly correcting spherical lenses into the beam path. (a-d) $V(z)$ plots using an infinity-corrected, planapochromat, N.A.=1.4 objective and a mirror sample. (a) $V(z)$ plot resulting from mounting a number 1.5 coverglass (0.17mm thick) to the mirror surface with a thin layer of immersion oil. (b) $V(z)$ plot showing improvements in signal strength and axial resolution, using the same sample as in (a), produced by inserting a -0.25 dioptre spherical lens into the beam path to correct for aberrations. (c) $V(z)$ plot from a number 1 coverglass (0.14mm thick) immersed to the mirror and corrected with the same weak spherical lens. (d) $V(z)$ plot using the same correcting lens but no coverglass on the mirror. Scale=14μm in z.

this figure was an infinity-corrected planapochromat, N.A.=1.4. Figure 8.8(b) illustrates how the amplitude of the signal can be increased and the full width at half maximum (FWHM) dimension of the peak intensity improved (narrowed) when a weak spherical correction lens (-0.25 dioptre) is inserted into the beam path. Figures 8.8(c) and 8.8(d) show the best response we were able to obtain using correction lenses when the mirror had a number 1 (0.14mm thick) coverglass and no coverglass respectively. Clearly, for this particular objective and mounting medium, the best axial resolution and signal strength is obtainable with a number 1 coverglass appropriately corrected.

In order to minimize aberrations when looking at an actual biological specimen, it is of course necessary to use an objective of the correct tube length and, ideally, one that can be internally corrected for coverglass thickness as well. Even then, the fact that the system is focused through different types and thicknesses of mounting medium needs to be considered. It may ultimately prove desirable to mount some kind of reflective foil or evaporated metal film onto the microscope slide adjacent to the biological specimen so that a $V(z)$ test might first be performed on a known reflecting object which has, as nearly as possible, the same external optical properties (coverglass, mounting medium, etc.) as the biological subject. While conducting such a test, further correcting lenses could easily be added to the optical system.

III.2.2. Adjusting pinhole size to obtain maximum lateral resolution

An often overlooked feature of scanning optical microscope systems which utilize coherent light is their ability to improve lateral resolution (in the x-y plane) as well as resolution in depth [Sheppard and Choudhury, 1977]. This is because the object is effectively imaged using both incident and collected light, and may provide as much as 1.4 times the lateral resolution of a conventional microscope. Besides being adversely affected by aberrations, resolution is dependent on pinhole size and critically dependent on alignment of the pinhole on the optic axis, both in the axial (along the z-axis) and transverse directions.

To illustrate factors that improve lateral resolution in our reflected confocal brightfield design, a region out of the centre of a late mitotic anaphase nucleus in endosperm tissue of the African globe (blood) lily, *Haemanthus katherinae*, was imaged at 9,000 diameters magnification. The microtubule bundles have been labelled with 15nm immunogold. Figure 8.9(a) shows a few of these microtubule bundles imaged with a fluorite 100X, 1.3 N.A., oil immersion lens and a

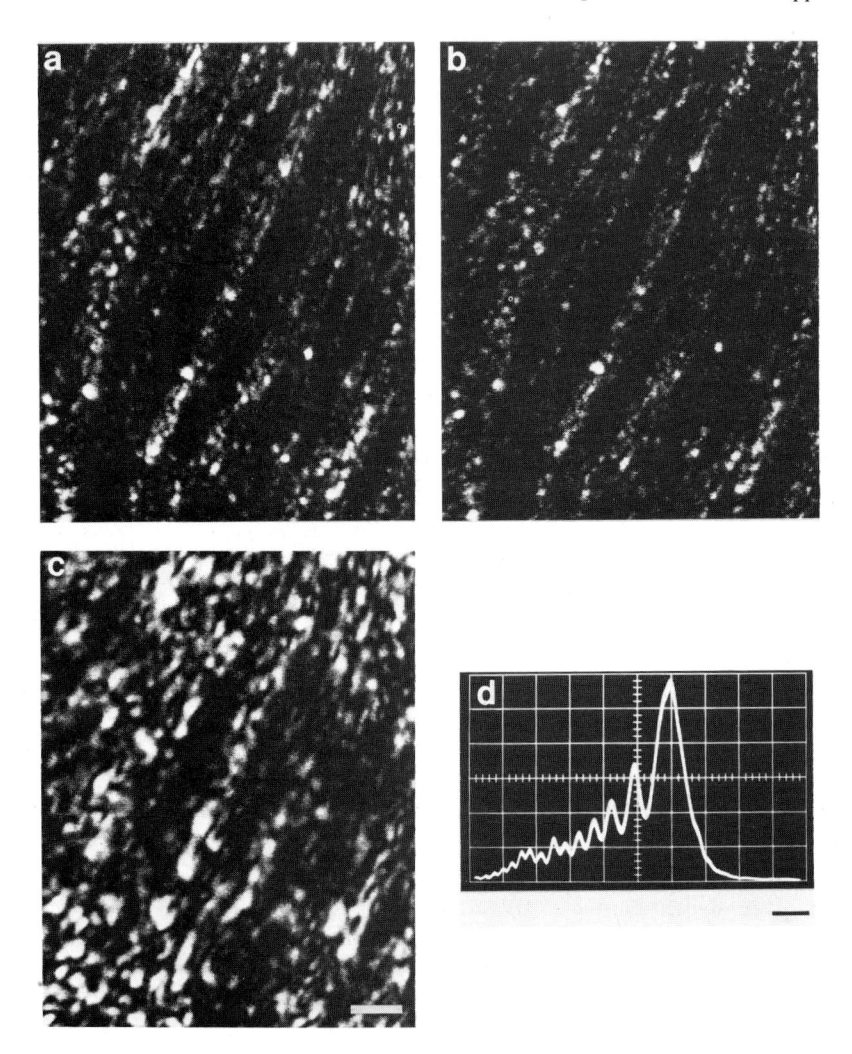

Figure 8.9. Changes in transverse resolution produced by varying the detector pinhole diameter or by introducing aberrations into the system. (a-c) Enlarged view of *Haemanthus* 15nm immunogold-labelled microtubule bundles using confocal reflected brightfield optics and a fluorite, N.A.=1.3 objective. (a) A 10 micron pinhole was placed in front of the detector. (b) Changing to a 2.5 micron detector pinhole shows an improvement in lateral resolution. Meaningful comparisons were made possible by measuring the background density and holding it constant during all steps in the photographic process. (c) Image of the same region using the same objective with its tube length correcting lens removed, i.e. the image plane (detector) is much farther away than the 160mm optimum tube length. Aberrations present in this configuration greatly reduce resolution. (d) $V(z)$ plot of this incorrect tube length condition in which side-lobes on the left side of the central peak result from spherical aberration effects of high numerical aperture. (a-c) Scale=2μm. (d) Horizontal scale=2μm in z.

10μm diameter pinhole in front of the detector. Changing to a 2.5μm detector pinhole, Figure 8.9(b), gives an improvement in lateral resolution which is in keeping with predictions based on theory [Wilson and Carlini, 1987]. On the other hand, Figure 8.9(c) shows the effects of using an objective that has not been precisely corrected for tube length. The microtubule bundles in the resulting image are virtually impossible to distinguish owing to the presence of spherical aberration, which is confirmed by the presence of multiple side peaks in the $V(z)$ plot in Figure 8.9(d).

It should be mentioned that all of these factors affect axial resolution as well, and in particular we have found that it is very sensitive to even small amounts of aberration.

III.2.3. Using detector arrays to further increase resolution

It is possible to further increase both the axial and transverse resolution of the confocal microscope by using a detector consisting of an array of rings [Sheppard and Cogswell, 1990]. One method for implementing this type of array detector configuration is to subtract an image formed using a somewhat larger detector pinhole from the image formed by a small pinhole (i.e. one that provides maximum resolution, as discussed in the previous section).

To illustrate this technique, experiments were conducted using our specimen-scanning confocal microscope and two photomultiplier detectors. A pinhole of 5μm diameter was placed in front of one detector and a 150μm diameter pinhole was positioned in front of the other. These were selected as being optimal for the 1.3 N.A. objective lens used in the system. Initially, $V(z)$ measurements were recorded for each detector using a front surface mirror as the sample. Following careful alignment of the two detector signals both in the axial as well as transverse directions, the amplitude of the signal from the detector having the larger pinhole was adjusted to be half that of the smaller. Subtracting this half-amplitude signal from that of the smaller pinhole produced a narrower central peak (improved axial resolution) in the resulting $V(z)$. This effect is illustrated at the top of Figure 8.10 where (A) was made using the 5μm detector pinhole, (B) was with the 150μm detector pinhole and (A-50%B) represents the first signal minus the second at half amplitude. To test for improvements in lateral resolution, the same system was used to observe 15nm immunogold-labelled microtubule bundles in a mitotic anaphase nucleus of *Haemanthus katherinae* endosperm (illustrated at the bottom of Figure 8.10). As one would expect, the resolution of fine structure is noticeably better in (A) as compared to (B). The image produced by using the difference signal (A-50%B) shows an even greater improvement in transverse resolution, although at the expense of introducing some background noise. The axial resolution has also improved to the extent that it is difficult to see continuity along the strands of microtubule bundles, their resolution in depth being limited to a narrow plane.

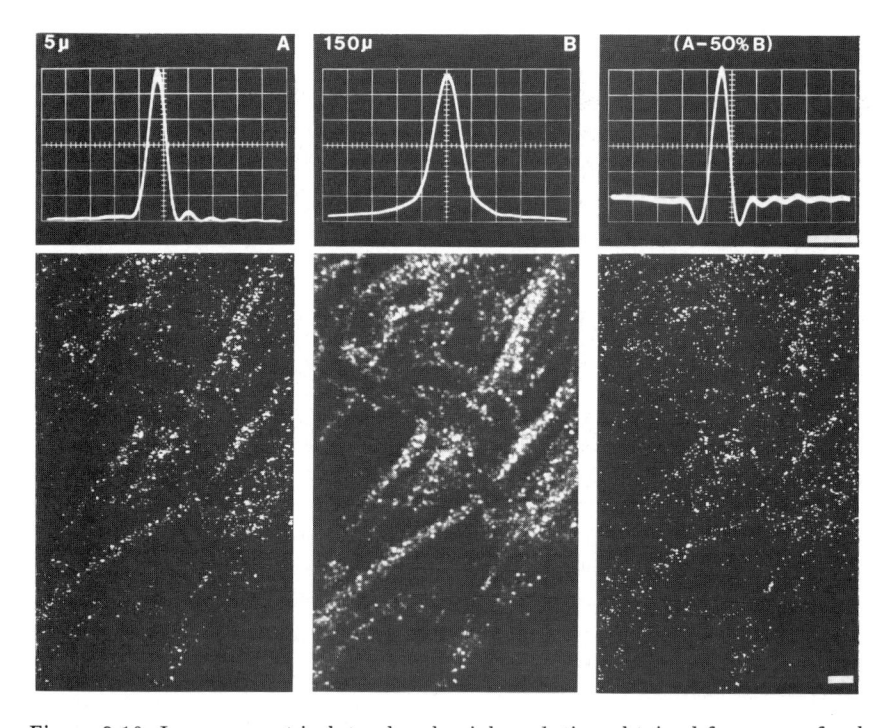

Figure 8.10. Improvement in lateral and axial resolution obtained from a confocal reflected brightfield configuration using two detectors. At the top of the figure are $V(z)$ plots showing the axial resolution response while at the bottom are actual images of an enlarged region of a mitotic metaphase in endosperm tissue of *Haemanthus katherinae*. Microtubule bundles have been labelled with 15nm immunogold. Objective was a fluorite 100X, N.A.=1.3. (A) Image formed with a 5μm diameter pinhole in front of one photomultiplier detector. (B) Image formed with a 150μm diameter pinhole in front of a second detector shows reduced resolution. (A-50%B) Image formed by subtracting a portion (50%) of the 150μm pinhole detector signal from the 5μm pinhole detector signal. This shows an increase in lateral resolution and such an improvement in axial resolution (optical sectioning) that it is now difficult, with just a single image, to see continuity along the microtubule strands. Horizontal scale=1μm. *Haemanthus* scale=2μm.

III.3. Improving visibility of image features by electronic differentiation

In the past decade, the development of digital image-processing and analysis techniques has had a strong impact in conventional and confocal microscopy. Although our microscope system commonly employs standard digital image-processing routines to enhance visibility of various specimen features, a description of these routines falls outside the scope of this chapter. However, it is probably worthwhile to

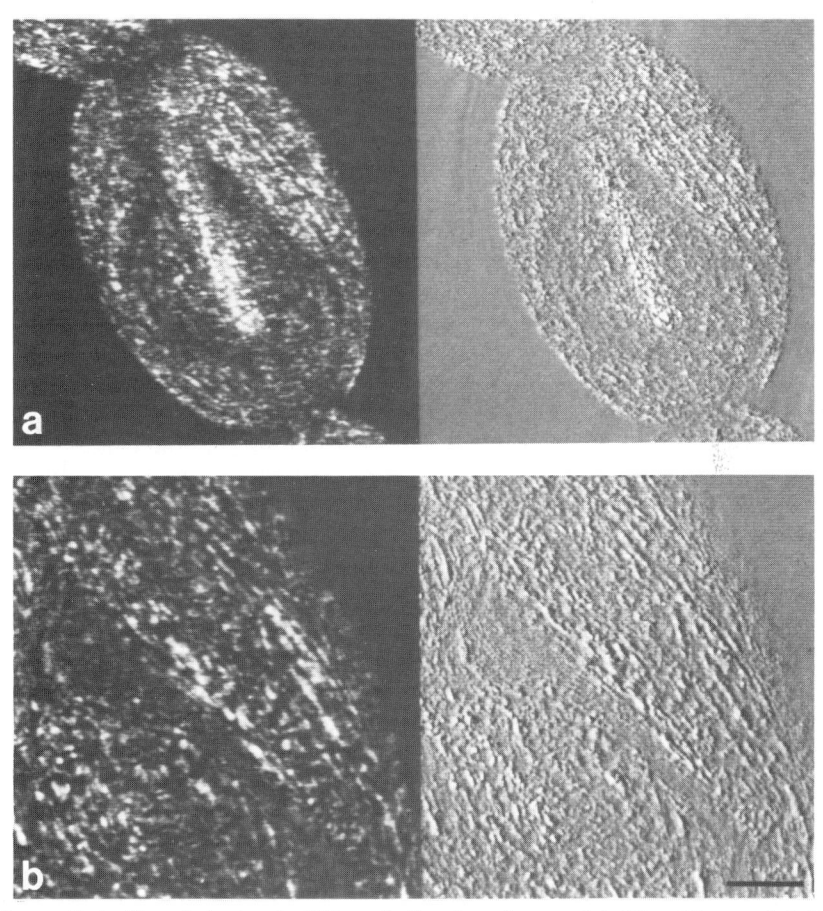

Figure 6.11. Use of electronic differentiation to enhance visibility of image detail. Subject is a living *Tradescantia* stamen hair cell mounted in water. (a) Shows an entire cell while (b) is an enlarged view of the upper right region, focused at the same plane (near the top surface). Left images are confocal reflected brightfield while right images have been electronically differentiated to reduce the distracting speckle appearance of the standard confocal configuration, thus making cell surface striations more visible to the observer. Scale=10μm.

briefly describe at this point a technique that we have used to improve the visual appearance of certain features in an image. This process electronically differentiates the analogue signal generated by the photomultiplier detector prior to its being digitized and displayed by the computer. Figures 8.11(a-b) show a low and a higher magnification view of a living *Tradescantia* stamen hair cell, with the left images made using the standard reflected confocal configuration and the right images produced by electronically differentiating the confocal analogue signal. This edge-enhancement technique was employed to assist the human observer in distinguishing meaningful cell surface structure (striations) from the speckle pattern that is commonly present in high magnification images produced using coherent (laser) illumination.

IV. Differential phase contrast

In a scanning microscope it is possible to include another extremely useful optical configuration which can produce non-confocal (increased depth of field) images in transmission at the same time as the confocal reflection images. This technique, called differential phase contrast (dpc), provides the opportunity for retrieval of information regarding slight changes in refractive index within a biological subject or between the subject and its surrounding medium. The system is able to detect refraction effects at edges (refractive index boundaries) which results in phase information being imaged in addition to the amplitude information (i.e. absorption) characteristic of conventional brightfield microscopes [Hamilton and Sheppard, 1984]. These refraction effects are dependent on the angle of incidence of the illuminating beam and the orientation of the edge (refractive index boundary) in the sample. Two images, each produced by an opposing half-aperture detector, will appear different wherever a component of an edge boundary parallel to the half-aperture division has produced a tilt in the resulting transmitted beams. When these two images are subtracted (and an offset midtone-grey signal added to improve visibility) the resulting image shows alternate highlights and shadows (in bas-relief) along refractive index boundaries.

In practice, we have found that simultaneously producing and displaying a differential phase contrast image of our biological subject while acquiring confocal reflection images gives an extremely useful frame of reference for locating otherwise indistinguishable internal features (such as nuclei, chromosomes, vacuoles, axons and cell bodies to name a few examples). Figure 8.7(a) shows an example of a transmitted dpc image of a *Haemanthus* anaphase which was produced at the same time as the confocal reflection images. The dpc technique clearly shows the location of the chromosomes within the mitotic spindle which is composed of microtubule bundles.

The substage split detector used in transmitted differential phase contrast imaging can also be readily adapted to other useful configurations. For example, simply adding the two images from each side of the split detector gives a conventional transmitted brightfield image, Figure 8.12(c), as compared to the dpc image in Figure 8.12(d). Displaying the images from each half of the detector side-by-side gives a stereo pair brightfield image, Figures 8.12(a) and 8.12(b). This is similar to stereo pair-producing techniques used in conventional microscopy. The subject in this figure is a *Haemanthus* anaphase labelled with 5nm immunogold.

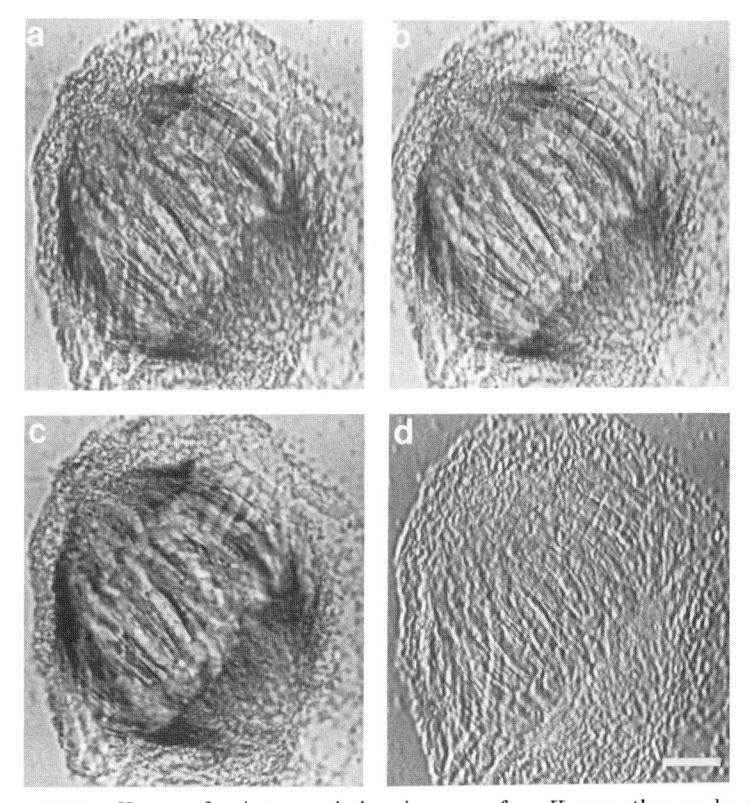

Figure 8.12. Non-confocal transmission images of a *Haemanthus* endosperm anaphase produced using a split detector in a variety of configurations. (a) and (b) are images from the left and right half of the split detector respectively which, when viewed together, form a stereo pair. (c) Transmitted brightfield image created by adding the signals from each half of the split detector. (d) Differential phase contrast image produced by subtracting the signal in (b) from that in (a) and adding an offset grey. Scale=5μm.

It is even possible to produce usable stereo pairs from a through-focus series in differential phase contrast if a high numerical aperture condenser (collector) lens is used between the specimen and the split detector to produce images of reasonably high axial resolution (ca. 1μm). This technique was employed to produce a stereo pair image of a portion of a living *Tradescantia* stamen hair cell, mounted in water between two sealed coverglasses, Figure 8.13. Even though the six dpc images used to create the stereo pair are not confocal, the axial resolution is good enough to produce a clear indication of the location of the cell nucleus and surface striations in depth.

Finally, it is possible to enhance or subdue specific spatial frequencies in differential phase contrast images by using a variety of detector aperture masks [Hamilton et al., 1984].

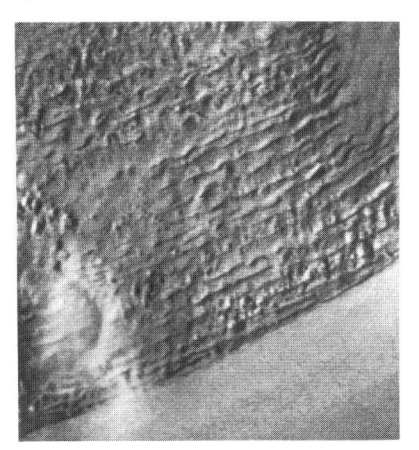

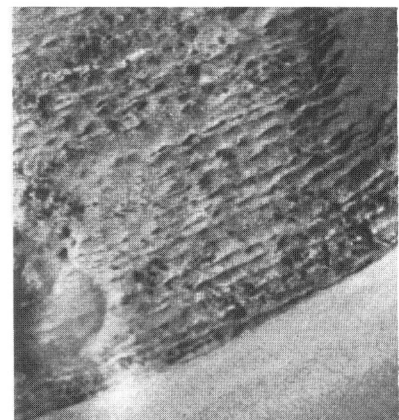

Figure 8.13. Stereo pair produced from a series of 6 *non-confocal* transmitted differential phase contrast images at 0.4μm focus intervals. Subject is a portion of a living *Tradescantia* stamen hair cell, mounted in water between two coverglasses. High numerical aperture immersion objectives (N.A.=1.3 and 1.25) were used above and below the sample to improve the resolution of the transmitted dpc images. Scale=5μm.

V. Confocal transmission

Interestingly, confocal microscopy in the transmission mode, although available for some time [Brakenhoff et al., 1979; Sheppard and Wilson, 1979], has received little attention in recent years, perhaps because of difficulties in aligning the optical system and maintaining its alignment during scanning. Confocal transmission microscopy does not exhibit as strong an optical sectioning property as in the reflection mode [Sheppard, 1986] but, combined with the facility for electronic contrast enhancement, it can result in useful information. The alignment problem stems from the fact that refractive index inhomogeneities cause the transmitted beam to move in the plane of the pinhole. Although it could be argued that the resultant variations in signal provide meaningful information, nevertheless it is difficult to maintain an adequate signal and be sure that the system is correctly aligned. This problem can be alleviated by reflecting the transmitted light back through the object with a phase conjugate mirror [Cathey et al., 1986]; however, some improvement can be obtained simply by reflecting the light with an ordinary mirror. The light transmitted through the object for the first time is focused onto a mirror and refocussed back onto the object. In this case, lateral movements of the transmitted light spot are cancelled on the return path, but not axial movements. This double-pass method was described by Sheppard and Wilson (1980) and is commercially available on the Lasertec microscope. We have used this technique with a pair of high aperture oil immersion objectives and a sample mounted between coverslips to avoid the introduction of unwanted spherical aberration. Figure 8.14 shows three stereo pair images of a *Tradescantia* stamen hair cell. Figure 8.14(a) shows a confocal reflected brightfield image which clearly delineates the surface striations in the cell wall. (This image is actually a projection of several planes of focus produced by using the auto-focus technique in which the maximum signal is retained during scanning.) The double-pass confocal transmission image of Figure 8.14(b) shows the cell nucleus and is again a projection, but this time generated by retaining the minimum signal. Figure 8.14(c) shows a double-pass confocal transmission differential phase contrast image, formed by using two detectors with a half-plane mask in the lens pupil. This method [Benschop, 1987] gives differential phase contrast in reflection from surfaces while also exhibiting confocal optical sectioning. We have found that analogous images in reflection from semi-transparent objects are difficult to interpret, as confocal reflection always gives differential contrast from axial variations in refractive index, and this results in a complicated behaviour when combined with half-plane apertures. However, these problems are overcome in confocal transmission and phase information is imaged.

VI. Confocal differential interference contrast (DIC)

The Nomarski DIC method [Nomarski, 1955] can be used in reflection in a confocal microscope and is especially suited to imaging subjects having strongly reflecting surface topographies with underlying opaque regions similar to the types of specimens described earlier in this chapter. The DIC optics provide surface texture information in

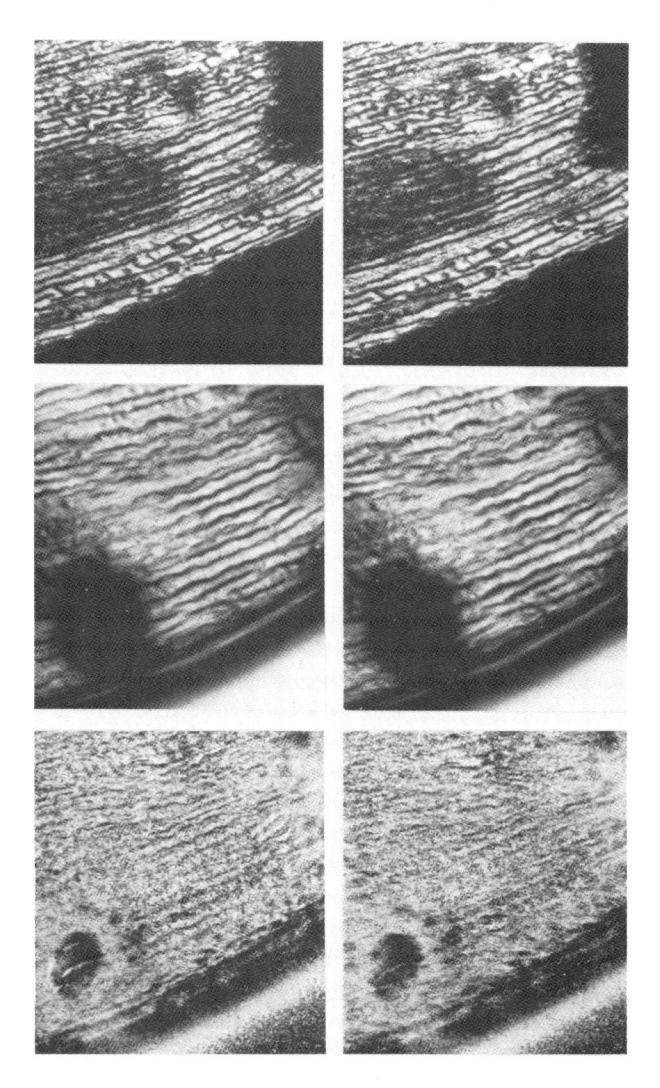

Figure 8.14. Stereo pairs produced using *confocal* double-pass transmission techniques both for brightfield and differential phase contrast imaging. Subject and objectives are the same as in Figure 8.13. (a) Confocal reflected brightfield (included here for comparison) shows surface striations in the *Tradescantia* stamen hair cell wall from 4 planes of focus at 0.2μm steps. (b) Double-pass confocal transmission from 4 planes of focus at 0.4μm steps shows the location of the cell nucleus in the lower left region, in addition to surface structure. (c) Double-pass transmission with two detectors and half-plane masks in the lens pupil gives confocal differential phase contrast (created from 8 planes at 0.2μm steps). This technique more clearly delineates the nucleus and shows surface striations in bas-relief. Scale same as Figure 8.13.

Confocal Brightfield Imaging Techniques 239

bas-relief, which is superimposed on the already strong optical sectioning property of confocal reflection. Figures 8.15(a) and 8.15(b) show a comparison of the brightfield and DIC techniques using a commercially manufactured integrated circuit as a subject. The images were created using the auto-focus technique over 5 planes of focus at 0.4μm steps. To obtain DIC, a Nomarski-modified Wollaston prism was inserted above the rear pupil of the objective and an analyser placed in the reflected beam before the detector converging lens. As always, the on-axis design of our system allows the use of full-aperture, high N.A. optics at any magnification. This provides high axial resolution and thereby limits the bas-relief information characteristic of DIC to being within the depth of the currently resolved optical section.

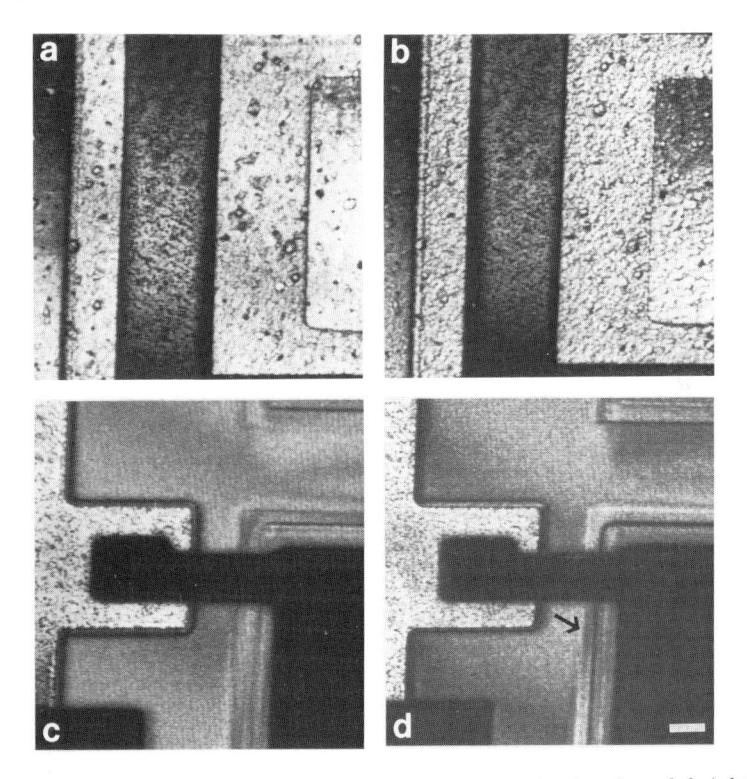

Figure 8.15. Comparison of images created using confocal reflected brightfield versus confocal reflected Nomarski DIC techniques. Subject is a portion of a commercially manufactured integrated circuit. Objective is oil immersion, N.A.=1.3. (a-b) Auto-focus images from 5 planes, created using brightfield (a) and Nomarski optics (b), show the increased delineation of surface roughness due to the bas-relief effects of Nomarski. (c-d) Brightfield versus Nomarski auto-focus images of a different region of the sample show an improvement in the ability of the DIC technique (d) to accurately delineate phase edges (i.e. small changes in surface height denoted by arrow). Scale=10μm.

Figure 8.16. Confocal reflected Nomarski DIC of an integrated circuit using an N.A.=1.3, oil immersion objective. (Top) Stereo pair (for normal viewing) created by storing the brightest signal from 12 successive planes in a through-focus series at 0.5μm focus steps. This technique shows fine structure of surfaces in bas-relief at successive planes of focus. (Middle) Same stereo pair as above but arranged for viewing with crossed eyes. (Bottom) An isometric projection of the same 12 images showing surface details and height information. Scale=10μm.

In addition to enhancing surface texture, we have found that the use of DIC avoids problems in interpretation of images from phase objects which may be caused, for example, by small surface-height changes. Figures 8.15(c) and 8.15(d) show another region of the same integrated circuit focused on a shallow-ridged structure just above the base substrate. The reflected Nomarski image clearly delineates the ridge-groove morphology which is distorted in the brightfield image.

In order to demonstrate the potential of DIC for producing three-dimensional images with well-delineated phase edges and surface fine-structure, a series of 12 images at $0.5\mu m$ focus steps was used to form an auto-focus image starting at the top of the same region of the sample, Figure 8.16 (top left). There is no problem in utilizing the auto-focus method since DIC gives a combination of differential phase and amplitude information. Therefore, it is not necessary to add a midtone-grey offset to the displayed image, as in the differential phase method with half-plane apertures described earlier. In Figure 8.16 (top right) the same imaging technique was used but with images from successive planes of focus displaced laterally by one pixel to produce a stereo pair. Figure 8.16 (bottom) shows a computer reconstruction of an isometric projection (looking across the sample from the upper left corner) which was formed using the differential interference contrast images.

Although the confocal reflected Nomarski technique has been illustrated here using a highly reflective metallic subject, one can conceive of many potential applications in the biological sciences: for example, imaging surfaces of seeds, spores and pollen grains.

VII. Conclusion

In this chapter we have attempted to show that, with an on-axis scanning microscope, a range of brightfield configurations is possible with all the advantages inherent in a confocal design. These have the potential for providing new and extremely useful information especially in applications using living or freshly prepared tissue, immunogold and silver probes, naturally pigmented or differentially stained material, or subjects with highly reflective surface topography. The microscopist must not assume, however, that conventional methods of biological sample preparation or even high quality conventional microscope objectives will perform in predictable ways on a confocal instrument — careful consideration of how the image is formed and what is being imaged is paramount in correct usage of the system and accurate interpretation of the results. Therefore, for these techniques to find widespread usage in the biological sciences, continued improvements in confocal optical instrumentation must proceed in tandem with new developments in sample preparation in order for clearly understandable images to be generated.

Acknowledgements

This work was performed while the authors were at the Department of Engineering Science, University of Oxford. We wish to thank D.K. Hamilton, University of Oxford, for frequent input and assistance with the experiments and instrumentation. We thank the Plant Anatomy and the Cytology Departments, Royal Botanic Gardens, Kew, for the loan of plant specimens. We also wish to express our appreciation to A.S. and J.M. Bajer and J. Karpilow for providing biological samples, H.M. Howard and S.A. Poston for photographic assistance and D.E. Wimber, University of Oregon for editorial comments.

References

Benschop, J. P. H., (1987). Confocal differential phase contrast in scanning optical microscopy. Proc. S.P.I.E., **809**, 90-96.

Born, M. and Wolf, E., (1980). *Principles of optics.* pp. 661. Pergamon Press, Oxford.

Brakenhoff, G. J., Blom, P. and Barends, P., (1979). Confocal scanning light microscopy with high aperture immersion lenses. J. Microsc., **117**, 219-232.

Carlsson, K., Danielsson, P. E., Lenz, R., Liljeborg, A., Majlöf, L. and Åslund, N., (1985). Three-dimensional microscopy using a confocal laser scanning microscope. Opt. Lett., **10**, 53-55.

Cathey, W. T., Johnson, K. M. and Cormack, R. H., (1986). Phase conjugate resonator for high-resolution scanning microscopy. J. Opt. Soc. Am., **A3**, P24.

Cogswell, C. J., Sheppard, C. J. R., Moss, M. C. and Howard, C. V., (1990). A method for evaluating microscope objectives to optimize performance of confocal systems. J. Microsc. To be published.

Hamilton, D. K. and Sheppard, C. J. R., (1984). Differential phase contrast in scanning optical microscopy. J. Microsc., **133**, 27-39.

Hamilton, D. K., Sheppard, C. J. R. and Wilson, T., (1984). Improved imaging of phase gradients in scanning optical microscopy. J. Microsc., **135**, 275-286.

Inoué, S., (1989). Foundations of confocal scanned imaging in light microscopy. In: *The handbook of biological confocal microscopy.* Pawley, J. (Editor), 1-13. I.M.R. Press, Madison, Wisconsin.

Nomarski, G., (1955). Microinterferometrie differential a ondes polarisées. J. Phys. Radium, **16**, 9-135.

Sheppard, C. J. R., (1977). The scanning optical microscope. I.E.E.E. J. Quantum Elec., **QE-13**, 100D.

Sheppard, C. J. R., (1986). The spatial frequency cut-off in three-dimensional imaging II. Optik, 74, 128-129.

Sheppard, C. J. R. and Choudhury, A., (1977). Image formation in the scanning microscope. Opt. Acta, 24, 1051-1073.

Sheppard, C. J. R. and Cogswell, C. J., (1990). Confocal microscopy with detector arrays. J. Modern Opt. To be published.

Sheppard, C. J. R. and Wilson, T., (1979). Effect of spherical aberration on the imaging properties of scanning optical microscopes. Appl. Opt., 18, 1058-1063.

Sheppard, C. J. R. and Wilson, T., (1980). Multiple traversing of the object in the scanning microscope. Opt. Acta, 27, 611-624.

Wilson, T. and Carlini, A. R., (1987). Size of the detector in confocal imaging systems. Opt. Lett., 12, 227-229.

9. Direct View Confocal Microscopy

M. PETRÁŇ, A. BOYDE AND M. HADRAVSKÝ

I. Introduction

I.1. Historical background

Despite the constantly growing importance of the light microscope as a tool for science, the development of the very core of the microscope was stagnant for a quarter of a millennium: its development was steady but slow, and without a deeper theory until the work of Abbé and Rayleigh a hundred years ago. The first substantial novelty improving the visibility of fine structures in unstained thin biological objects was Zernike's phase contrast, introduced fifty years ago. However, the classical resolution limit was impaired in phase contrast devices.

Microscopy of whole animal organs remained impossible because of the overlapping of images from many different levels of the object, only one being sharply focussed on, but buried in a mass of blurred images of object levels lying above and below. This obstacle was surmounted as a rule by the conventional approach of fixation, embedding, cutting or slicing, staining, mounting on a slide and covering with a coverslip. Many objects were difficult: among biological ones especially bone, teeth, and cartilage, but also the eye lens, insect organs, crustaceans, many plant organs and others. Slicing or grinding was mostly performed on dead, denatured objects, but even in these, such operations could not be carried out on unique museum specimens.

The first type I (simple) scanning microscope was built by Roberts and Young in the late 1940s [Young and Roberts, 1951]. The main advantages were the ability of several observers to see the image simultaneously on a TV display and the improvement in contrast through electronic manipulation, thus slightly improving upon the classical Rayleigh or Abbé limits which were defined around the sensitivity of the human retina. The same advantage also accrues in using scanning only on the detection side, which is effectively what happens when a video camera is attached to the normal light microscope.

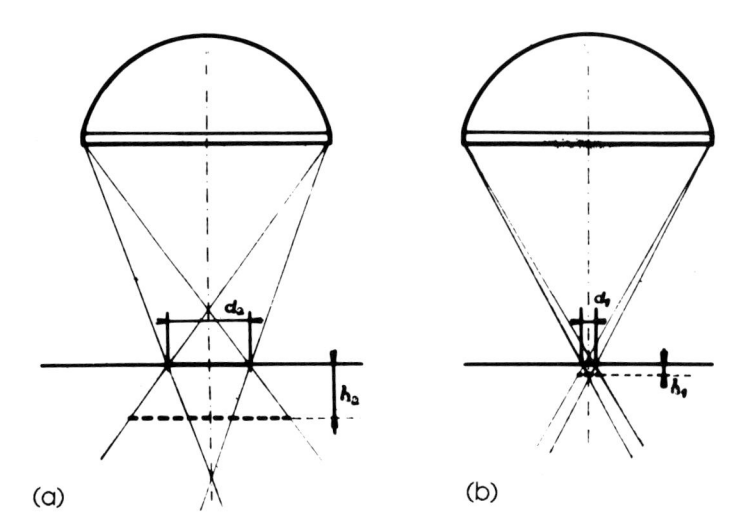

Figure 9.1. Diagram showing the illumination of a translucent specimen with a conventional objective in incident light: in (a) conventional illumination with the full cone of light, and (b) when the field of view is extremely narrow as in a confocal microscope. In both cases, the central double-cone volume is evenly and most strongly illuminated, and, in this volume, the instantaneous intensity of illumination per unit volume is identical in both cases if the same light source is used. In a confocal microscope, we also use exactly the same diagram to consider light collection from the samples: i.e., in (b) we collect light only from the very small (and short!) central double cone. This diagram shows the origins of the problem of trying to observe translucent specimens in reflected light. If at the same time we illuminate and image a semi-transparent object from above, we illuminate and image not only the very thin focussed plane, but also all the layers above and below the plane of our interest (from which we wish to make the image) with nearly the same intensity. In (a), the out-of-focus planes reflect large amounts of light back into the objective, and this scattered light, which does not carry any information about the plane of our interest, usually completely whites out, and wipes out, the useful image. (b) predicts the improvement expected for every single image point in a "double" (tandem, confocal) scanning microscope.

I.2. Double scanning: tandem scanning

Double scanning, meaning scanning in both illumination and detection, seems to have first been achieved in the device patented by Minsky (1961). This microscope (which is described in full in Chapters 1 and 3) imaged a single aperture in the intermediate image plane into the object and in turn imaged that illuminated volume onto a matching aperture in front of the detector. The scanning was achieved by moving the specimen, and frame rates of one frame per 10 seconds could be achieved. Minsky's invention was still unknown when Petráň and Hadravský were confronted with the problem of observing animal nervous cells intravitally, at that time impossible in whole brains as well as in most other organs in vertebrates. It soon became clear why experience and tradition had exactly determined the situations in which it was possible to observe, intravitally, tissues such as epithelia, capillary blood vessels, glands, nervous fibres, muscle fibres, etc. in such classical objects as the cuticular fold of the nail, the mesentery, the web of the frog's foot, the gills of fish, etc. Similar tissues in other organs and other animals usually gave much worse images in classical microscopes. To become independent of such choices and get images of better quality of any living tissue, it was decided to try scanning. Here it was soon found that it was necessary to scan twice: first in illumination and then in the image, to prevent scattered light forming a halo around the image of the actually illuminated object "point," Figure 9.1.

I.3. Disc scanning

Petráň and Hadravský (1966a) opted for a multiple-aperture (multibeam) scanning disc approach, which has the merit that it reduces the frame time necessary if the field is scanned by a single point. Any degradation of the image when single-point scanning is replaced by multiple-point scanning may be much less significant than the degradation caused by the inevitable coarseness accompanying any single-point scanning so far implemented. The theoretical spatial resolution of the "tandem scanning microscope" (TSM) may be the same as that of any other Type II confocal microscope, as will be shown below. The temporal resolution in tandem scanning can be a few orders of magnitude better than in single-point scanning, as is self-evident.

The first TSMs (or "direct view," "type II" or "confocal" microscopes, as they were christened many years later) came into being in the fall of 1965. To test the principle without having to build a complicated light source as well, the cheapest sufficiently bright light source was used — the sun. The development of the TSM began only after this trial, which first demonstrated the feasibility of real-time scanning — which we take to mean scanning at least as fast as normal cinematographic or TV rate, although, as we shall see later, much faster scanning speeds can be obtained.

The improvement in resolution both in the x-y plane and in the z (optic axis) direction was recognised in the Petráň and Hadravský patents (1966a, 1966b, 1967) and shown both theoretically and experimentally by Petráň et al. (1968). The first experiments with live animals were conducted at Yale University in 1966. The first public recognition of the merit of the TSM was given at the Brno Exhibition of Inventions in 1969, followed by the Vienna Fair of Inventions in

1970, at which gold medals were awarded.

In the first prototype TSMs, standard optical components were used in the construction: only in later designs were special components fabricated and an attempt made to minimise the number of optical surfaces and, in particular, to abandon all plane optical surfaces perpendicular to the optic axis. Thus, Petráň et al. (1968) describe the use of prisms in the beam inversion system, already abandoned in the microscope fabricated in Plzen in 1970. The first TSMs contained eleven plane reflecting surfaces. In later versions, four mirrors were used in the head, transforming the central symmetry of the disc into the identity of its opposite areas. Two additional mirrors were used to orient the objective vertically and the illumination bench horizontally. The many reflecting surfaces perpendicular to the optic axis in the initial designs required the use of polarised light and a circularly polarising filter placed between the objective and the specimen in order to reduce the significance of reflections at plane optical surfaces within the microscope [Francon, 1961].

II. Choice of aperture disc for the rapid scanning method

A rotating scanning disc with holes is inherently a much more precise, steady and durable device than any vibrating device can be. It is the most simple single contrivance allowing isotropic scanning with many spots (holes) simultaneously.

Using such a device, a stationary specimen is scanned with a set of many hundreds or a few thousands of isolated equidistant small light spots together covering approximately one percent of the object field at one instant. These spots travel over the field, covering it completely in a small fraction of a second. These spots are images, as formed by the microscope objective in its object plane, of holes in an opaque sheet. Images of an instantaneous position of the set of holes illuminate this object plane in isolated "points," which are actually spots with dimensions close to the central portions of diffraction patterns. The light which has to form these spots may be reflected at many different places on its journey, but because of the high objective aperture, the strongest reflections take place in the most strongly illuminated object plane (which is conjugate to the plane of the perforated sheet: the image plane). Any object point scattering (or fluorescing under) the illuminating light thus becomes a (secondary) light source, the luminance of which depends strongly on the local properties of the object and on the position of the point. This reflected (or fluorescent) light is collected by the microscope objective to form the image of the reflecting "point." This image is formed in the vicinity of the perforated sheet, and only the light reflected exactly in the object plane of

Direct View Confocal Microscopy

the objective is collected exactly in the plane of the sheet, in a "point" congruent with the hole through which the light has been directed to the objective. Here it is "filtered" through a hole so that only the light passing through the geometrical image of the transmitting hole, i.e., passing through an equal hole on the opposite side of the disc, is allowed to proceed farther and to take part in forming the image of the illuminated "point" (describing the optical properties of the matter to be imaged). This is made possible because the place where the image of the illuminating point will be formed is exactly known in advance: only the amount of light has to be determined experimentally and exactly reproduced in the image. To be exact, the centre of the image at the point is "measured," and this is assured by the dimensions of the hole in the rotating disc.

III. Some ways in which disc scanning can be implemented

III.1. Reflection or transmission?

The reflection case is more important for several reasons. First, it is an advantage to be able to examine a specimen without physically sectioning it, and the reflection case is the only one possible for bulk samples. Second, there may be an overwhelming proportion of light which has not interacted with the sample and which does not carry information in the transillumination case. Third, transillumination in a TSM requires two exactly matched objectives, an ideal which can hardly be achieved.

III.2. Reflection: one or two discs?

For every point, one hole can be used to illuminate it and another different hole to function as the gate which allows the light to pass into the image. This can be accomplished in two ways: either each hole is dedicated, either for the illumination scan or the image scan, or there must be two different discs, one for each scan, rigidly bound together. This latter approach may have advantages: the holes could have different diameters for illumination and for image filtration, and the mirror system for the separation of image-forming and illuminating rays could be simpler. Further, the holes in the disc on the image side may be provided with optical power [Lau, 1960] to fill out the whole exit pupil of the eyepiece and also the pupil of the photographic or TV camera. It therefore does not make sense to employ a photographic objective with an extremely large aperture (small f-number).

III.3. Reflection: one disc with central symmetry

However, there can be a simpler arrangement in which any hole serves at one instant for illumination and at another moment (after the disc has turned through one-half a revolution) for image filtration, the cycle repeating continuously. Thus it is also possible to use only one aperture disc for both scans provided that the pattern of hole distribution in the aperture disc has central symmetry, so that there are always two identical patterns in both fields (in the illumination field and in the image field). Such an arrangement is much simpler kinematically and much more compact, but it imposes higher demands on the geometry of disc design, Figure 9.3, and on the precision of fabrication, and prevents the use of different hole shapes for the two scans. Nevertheless, we consider the latter solution preferable because it enables us to build a device superficially resembling a conventional microscope, Figures 9.2, 9.4, 9.5 and 9.6.

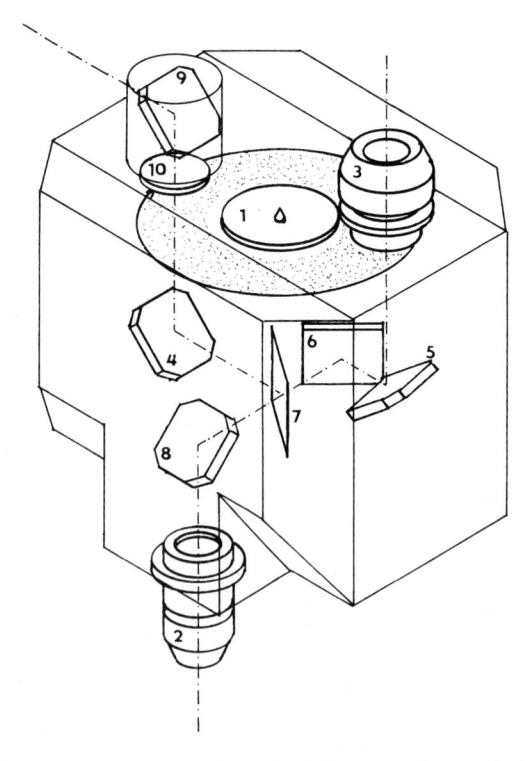

Figure 9.2. Diagram showing the layout of the essential optical components in the "Confocal 2002" TSM from Pilsen (Plzen). 1. Rotating Nipkow disc. 2. Objective. 3. Eyepiece. 4.,5.,6.,7. Internal mirrors. 7. Thin semi-transparent mirror. 8. Output mirror. 9. Input mirror. 10. Field lens.

Direct View Confocal Microscopy 251

Figure 9.3. Pattern of holes in a 1985 vintage Plzen disc.

Figure 9.4. Stereoscopic view of the head of the UCL TSM built in Pilsen and delivered to UCL in December 1983.

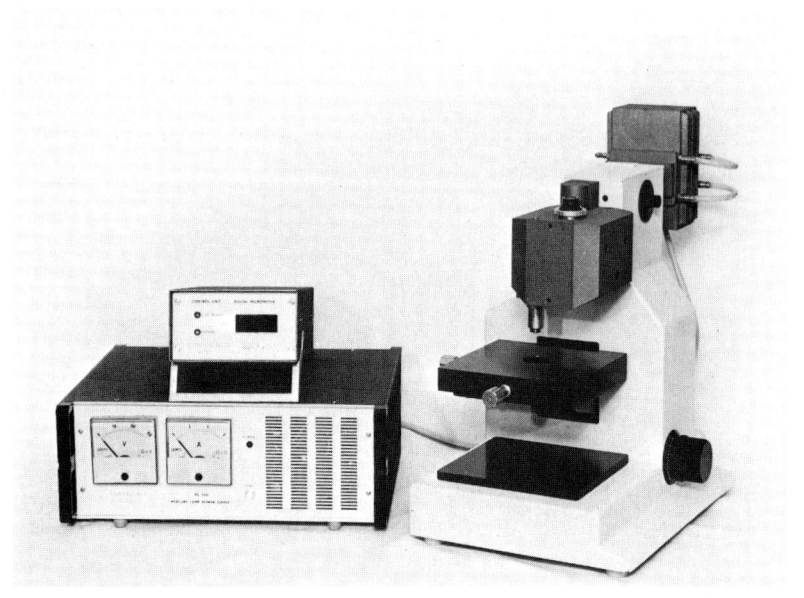

Figure 9.5

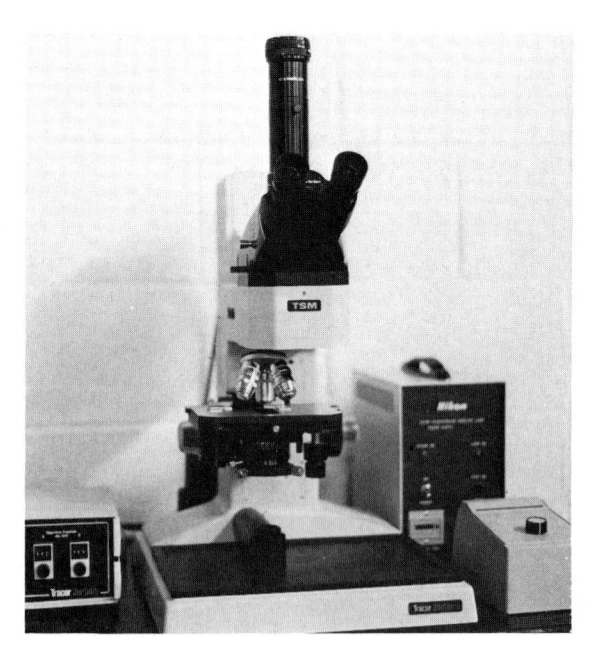

Figure 9.6

Figures 9.5 and 9.6. Two modern production TSMs. Figure 9.5 shows the "Confocal 2002" made by JZD Komorno, Plzen CSSR. Figure 9.6 shows the TSM made by Tracor Northern, Madison, Wisconsin, USA.

For bulk semi-transparent objects, this is the most practical approach. Both sets of holes are optically separated from each other by a semi-reflecting (semi-transparent) mirror. The simplest way to accomplish the outlined principle is to perform both scans with only one set of scanning holes located on a single disc, a derivative of the invention of Nipkow (1884).

III.4. Reflection: the autocollimation case

However, it is also possible to use each hole twice at the same instant for both scans, i.e., for both illumination and detection simultaneously. We may call such a device an auto-collimation TSM or a one-sided TSM: OTSM. Our own decision to use a two-sided disc was made after seriously considering the merits and demerits of the one-sided disc.

In the OTSM case, the light path must be divided by a beamsplitter "above" the disc, so that the full strength of the illuminating light impinges on the side of the disc toward the eye, where a non-negligible portion is reflected into the "eyepiece." The light passing back through the disc and carrying information from the specimen is very much weaker. Thus the clear disadvantage of the OTSM arrangement is that (depending upon the reflective characteristics of the top surface of the disc) up to 99% of the illuminating light must be excluded from the sum of this overwhelming DC component with the extremely small fraction of light which may be reflected in the desired object plane. To exclude the light reflected from the top of the disc, one may use crossed-polars and a quarter-wave plate [Francon, 1961; Egger and Petráň, 1967; Petráň et al., 1968; Kino, 1989], thereby losing between 75 and 90% of the available light and also adding eight more optical surfaces perpendicular to the light path. The disc must be made of a perfectly polished and clean material to reflect the 99% of unwanted light at a central stop [Xiao and Kino, 1987] or the disc may be tilted to reflect the light at an eccentric stop [Kino, 1989]. (Both arrangements give rise to a great deal of general unwanted scatter close to the relay optics.) In total, therefore, the OTSM is more complicated: with a glass disc, there are 12 more dioptric surfaces perpendicular to the optical path. The advantage of using the one side of the disc is that there are no mirrors below the disc to be aligned.

In considering further the relative advantages of the original and autocollimation variants, it should be noted that in the former case, it is easy to stop down the illumination using a field diaphragm — and thus to improve the available contrast further — without imaging the reflection of the diaphragm. In the autocollimation device, the field diaphragm would reflect in the same way as the aperture disc if it were not relayed by some more complicated optical system.

III.5. Transmission TSM: one disc with central symmetry

A design for a transmitted light TSM using opposite sides of a complex aperture disc was published in the second of the original Petráň and Hadravský patents (1966b). A principal difficulty in constructing such a TSM will be to find exactly matched, flat field objectives, and another will be to prepare the appropriate specimen mounted between two identical coverglasses. Focussing would also be a complex operation involving three separate focussing translations.

III.6. Transmission TSM: two discs

Sheppard and Wilson (1981) considered a transmission TSM design employing two identical discs ganged to rotate at the same speed by being attached to the same axle, but this empirical design was by way of illustrating a discussion of the basic operating principles and has not been realised in practice. (Transmission CSLM will be much more easily accomplished in a unitary-beam, object-scanning microscope.)

IV. Illumination

The ideal illumination source for a TSM would be broad, flat and incoherent. Unfortunately, we have to compromise when using real devices. Probably the best approach to the above requirement is provided by tungsten ribbon lamps, but they are not as bright as mercury or mercury-xenon arcs, which are equally suited to visible, ultraviolet and infrared illumination. Arc lamps, however, have the disadvantage of non-uniformity in the intensity of illumination but this may be overcome with Ellis's bent light guide method (1985) [also see Inoué, 1986].

It has recently been proposed that it may be desirable to use scrambled laser light, such as may be achieved by passing laser light through a bent quartz fibre, using that as a point source which is then appropriately expanded [Ellis, 1979]. This might prove to be useful in applications where the advantage of using polychromatic illumination could be dispensed with. It could even be a considerable advantage in defined fluorescence applications.

A possible alternative source is hafnium carbide: a disc of this material is heated by radio frequency radiation [Peek, 1957]. The brightness may be three times greater than the brightest incandescent lamps, with the advantage of a broad uniform source.

The advantage of using broad, flat and structureless illumination sources is that critical illumination may be employed. With sources such as mercury arc lamps, coherent illumination, such as Köhler, must be used. To achieve uniform illumination over the whole field, the scanning aperture disc must be uniformly illuminated over the entire field which is used in the illumination of the object, along with an aperture angle which should not exceed the real aperture of the objective (in order to avoid unnecessary scattered light).

V. The aperture disc

V.1. Disc size: why not Nipkow's original design?

As considered elsewhere in this volume, scanning can be accomplished in various ways and with various devices. However, if we wish to deal with more than one point simultaneously, so that all the scanning light "points" and all the scanning "gates" must be exactly spatially and temporally bound, there is only one device that seems to be practical: something based on the scanning disc as conceived by Nipkow (1884), although substantially modified.

The modifications will include the dimensions, number, and arrangement of holes. The disc described by Nipkow (1884) and used by Baird in early television contained one spiral set of holes at regular angular distances, arranged so that at one instant only one hole was present in the image field. As soon as this hole left the field the next one entered it. One such disc scanned the image to be transmitted, and a second disc reproduced the image on the receiver end of the line covering a broad, uniformly lit plane. The illumination intensity was changed so as to represent the brightness of the point in the image field, which was determined by the instantaneous position of the hole and so corresponded with the illumination in the hole on the transmitter end. We cannot use Nipkow's original design in a TSM, as we will now show.

Assuming that the field in the eyepiece focal plane is a square of side 16mm, and that a 100X oil immersion objective is used, this corresponds to $160\mu m$ ($25,600\mu m^2$) at the object. Such an objective has a resolution (according to Rayleigh's formula) of $0.24\mu m$, i.e., 667 lines per field. To scan with one hole at a time, the holes would have to be at tangential distances of 16mm at the disc, and the disc would therefore have to be $16 \times 667/\pi = 3400$mm diameter. This is obviously both ridiculous and unnecessary. It is sufficient to have the holes much nearer to each other and hence to have more than one in the field at one time. On the other hand, their dimensions should correspond to the resolution of the microscope objective, which we assume to be of the order of $0.24\mu m$ in the object space and $24\mu m$ in the image space. To have some reasonable factor for the suppression of "cross talk" (transmission of light through holes other than the corresponding holes on the image side) we arrange that the fraction of the image plane occupied by holes lies in the range 1/100 to 1/1000, whereupon we find that the distances between the holes will be 10 to 30 times their diameter, i.e., 240 to 720μm respectively. The exact value may perhaps differ according to the object being examined and the demands of the study.

The sizes of all practical discs made to date have been in the range of 75 to (mostly) 100mm diameter. This is because of the fixed distance at which the intermediate image is located in the common (160mm tubelength, RMS standard) objectives, Figure 9.2.

V.2. Pinhole layout

The geometry of the packing of the holes in the disc is determined by the requirement of scanning the field in the object uniformly. The factors which have to be considered for photometrical reasons (to get an evenly bright field of view) are as follows:

(1) The holes, or strictly their images in the object field, should give an even brightness of field when a structureless object is observed. This condition separates into two others:

(a) The radial dimensions of the holes should be equal, and chosen so that either there would be no overlap of traces of holes lying on neighbouring radii, or else the overlap would be regular. In either case the dimension should be optimised so as not to give rise to (optical) ripple.

(b) It is simpler to design a layout in which the tangential dimension of the hole will be proportional to the radius vector on which it is situated.

These requirements are achieved through the use of Archimedean spirals as the lines on which the holes are to be located. The pattern of hole distribution (to secure regularity in field illumination) may be either quasi-rectangular (roughly square packing at the middle of the aperture zone) or quasi-hexagonal (approximately regularly hexagonal at the same median circle.) The hexagonal arrangement has the advantage of greater distances between holes. If all the holes have to be the same size because of limitations of manufacturing technique, for example, then it is necessary to design a more sophisticated pattern of hole distribution. Another possibility is the use of a specially designed neutral density filter to compensate for any radius-dependent non-uniformity of image brightness. Further constraints are applied in the case of the double-ganged disc by the need to arrange for hole pairs to lie on the same diameter of the disc and at the same radius. This constraint is removed in the one-sided disc. Although it has not yet been experimentally examined, there could be an advantage in orienting rectangular aperture holes so that the direction in which the diffracted light is preferentially propagated is that in which there is the greatest distance to images of neighbouring holes.

In practice, the Plzen discs, Figure 9.3, have been made with holes which are broader toward the periphery of the disc in order to maintain a constant transmittance, making it possible to use a regular geometric pattern in the layout. The Tracor discs have been made with uniform square holes, which introduces the constraint in the disc layout just mentioned. Kino and colleagues (personal communication, 1989) have designed one-sided discs in which the packing pattern is not uniform, varying between extremes of pseudo-square and pseudo-hexagonal packing over different portions of the disc. These discs also have an unequal distance between holes in the tangential (scanning) and radial directions.

V.3. Pinhole size in the disc and spot size at the object

The optimal pinhole size is the result of several compromises.

(1) The distances between the holes must not be too small, because "crosstalk" can occur more easily (and the spurious image can be much brighter) when the holes are too close together.

(2) The holes must be broader than the thickness of the disc material, or they must be conical.

(3) The number of holes must be adequate to the resolving power and the "photometrical ripple" in the image.

(4) The dimension must be adequate for the resolving power of the microscope, i.e., smaller than the central spot in the Airy disc on the eyepiece side of the objective, where it is about $30\mu m$ for a 100X/1.3 objective at a wavelength of 550nm.

(5) The total area of holes must be "negligible," for example, much less than 1% of the active area of the Nipkow disc.

The dimensions of the illuminating spot in confocal microscopy have been discussed theoretically in several recent contributions from Sheppard (1988, 1989), Wilson and Carlini (1989), Brakenhoff et al. (1989) and Kino et al. (1989). All of these considerations are relevant to the TSM case, but it is also necessary to consider the results of practical experience, which have shown that the aperture size is better when it is a little larger than the values predicted by theory (and the distance between the holes greater than the smallest limit given by theory).

In achieving the illumination spot size in a TSM, we are limited by the magnification of available objective lenses, which determines the demagnification of the illuminating pinholes. In the best possible case we may have a 200X demagnification, but this unnecessarily limits the available field of view. The maximum usual magnification with high aperture immersion lenses is 100X. To achieve a diffraction limited spot, we would therefore choose holes in the disc of much less then 25μm in diameter.

In the case of the double-ganged disc, limiting the aperture size increases the difficulty of aligning the images of the conjugate apertures on opposite sides of the discs.

In practice, in discs with real holes (copper foil or silicon), the usual hole dimensions are in the range of 20μm to 80μm. Even this larger hole diameter gives rise to a very dramatic improvement in z resolution compared to the non-confocal case.

V.4. Disc materials in practice

The ideal disc material will be such as to permit the fabrication of holes a few microns in diameter and with no tunnelling, light guiding or shadowing effects due to the thickness of the disc material. Three basic approaches have so far been adopted.

In the microscopes made in Plzen, Figures 9.2, 9.3, 9.4 and 9.5, a very thin (10 or 20μm) copper or nickel foil is etched by standard photolithographic procedures, reproducing a master pattern. Because such a disc will not support its own weight, it has to be supported at both the central unused area and at its periphery. It should be prestretched on the peripheral support before etching to avoid geometrical deformations.

In the discs manufactured by Tracor Northern Inc., Figure 9.6, the properties of single crystal silicon are exploited to produce uniform etch pits which just penetrate a wafer of uniform thickness. All the pits and holes are square, and they have the same orientation over the entire disc, which gives rise to disadvantages in the design of the ideal space-filling pattern. Difficulties might also arise in using this material with infrared wavelengths beyond the absorption edge of Si, when a metal coating should be applied to limit the transparency of the disc material.

The discs may be fabricated by standard semiconductor industry technology from chrome on glass. As with silicon discs, such devices are stable and self-supporting, requiring no peripheral ring as copper discs do. Precise hole patterns can be generated. However, there are disadvantages to be anticipated from the reflections from the top and bottom of the glass support: examples of the worst case are reflections from surfaces perpendicular to the optic axis (or nearly perpendicular if the disc is tilted). The transmittance characteristics of the glass may create problems for violet and ultraviolet work.

V.5. Practical design information on discs to date

Most of the Plzen discs to date have used a quasi-hexagonal arrangement of about 20,000 holes on 10,000 different radii, the hole pairs at the same radius lying opposite each other on the same diameter (i.e., at an angular distance of 180°). The mean hole diameter is about 30μm, Figure 9.3. Then the (geometrical) image of the hole in the object plane of a 100X immersion objective will be 0.3μm in diameter, close to the classical resolution limit of such an objective. Scanning with such a disc is equivalent to a closed circuit television with 10,000 lines per picture, 20 times the number of lines and 400 times the number of "pixels" in ordinary television. These 10^8 pixels are more than the ratio between the area of the field of view and the area inside the first zero of the (object-side) Airy disc of the highest power objective available, which is 1×10^5 (for an eyepiece field of 16mm diameter). Hence the area for the core of an Airy disc contains about 400 pixels (which may overlap partially) in the case of a 100X/1.4 objective. In smaller power objectives this figure will be lower, but it will hardly fall below 30. We may conclude that the practical resolving power will not be impaired in this case by the finite number of scanning lines in a TSM.

Tracor have packaged their discs in a removable cartridge, thus simplifying the exchange of discs if a new design becomes available.

V.6. Slits

In the recent past, interest has been reawakened in the use of slits rather than discrete round, square, or hexagonal apertures in confocal scanning microscopy [Sheppard and Mao, 1988]. The first slit TSM was described by Egger et al. (1969). Slits have the disadvantage of being "less confocal" and anisotropic in their confocality, but the positive advantages of transmitting more light and being easier to align.

V.7. Driving the disc

The disc has to be centred with the greatest possible precision on the shaft which is to spin it, irrespective of whether one or two sides of the disc are employed. This is a practical manufacturing problem which has been overcome wherever the TSM has been constructed.

The speed at which the disc may be rotated may, in the end, be limited by the strength of the disc material as well as by the introduction of vibrations at high rotating speeds if the disc is not properly balanced. These considerations have not caused problems in practice.

It is already possible to spin a disc at such a speed that problems might be encountered in imaging certain long decay-time fluorescence or photoluminescence processes.

The bearings of the motor (or disc assembly if this is separate) must obviously be good enough not to introduce vibrations into the head of the microscope which will move the specimen in relation to the objective lens. The more important aspect of maintaining disc centration lies in ensuring the uniformity of illumination over microscopic areas of the field of view, i.e., to make sure that the disc functions as designed.

It is our practical experience that well-designed disc mountings and bearings will last in unattended and reliable function for many years, in fact for far longer periods than one would expect trouble-free operation from electronic devices.

VI. Other components

VI.1. Mirrors and beam-splitters

The beam-splitter should be extremely thin ($\leq 5\mu$m) if it is not to double the image or introduce astigmatism. The beam-splitter may be provided with appropriate interference coatings to give desired transmission and reflection in the ultraviolet, visible and near-infrared range.

All other components (with the exception of the optics in the illuminating source) are front-surface, plane mirrors, Figure 9.2. All could be given appropriate coatings to improve performance in violet and near-ultraviolet regions to enhance the fluorescence performance.

VI.2. Filters

A variety of well-chosen interference filters will be required to optimise the utilisation of the TSM in fluorescence microscopy. Cheaper broadband glass filters will be usable in many applications.

Heat-rejecting filters will be required at the beginning of the light path to protect other filters as well as sensitive samples. On the other hand, TSMs work into the infrared, and when working with a TV camera it would be desirable to use long pass filters to reject visible wavelengths.

VI.3. Alignment aids

The practical design of the TSM head should incorporate arrangements to permit the alignment of the mirrors with respect to the optic axis, the centring of the beam splitter to ensure equal pathlengths on both sides of the two-sided design, and the centring of the aperture disc axle within the framework of the optical layout. Such arrangements should in theory need to be used only once, at the stage of manufacture of the instrument. Practical experience has taught us that the final alignment should be conducted on site, and that it will only be necessary to re-align the mirrors in a TSM head following abuse of the instrument.

Alignment is performed by inspecting both sides of the disc simultaneously, viewing backwards into the TSM head via the beamsplitter. Depending on the design of the head, this may be done by removing the cover plate for the side of the head, or by using a low-power microscope to look through the objective mount (through a low-power objective). The illumination and detection sides of the microscope head are uniformly illuminated by separate red and green light sources, so that the two sets of holes on the opposite sides of the disc can be recognised. Adjustments are made to bring the images of the holes into perfect overlap across the entire field of view (when they will appear single and yellow).

VI.4. Objective lenses

The "mechanical tubelength" of both the Plzen and the Tracor TSMs is 160mm. Objectives designed for 160mm or 170mm tubelength can be used. For longer tubelength objectives, an appropriate prolonging nose piece must be used. Infinity-corrected objectives are not preferred, because their use also requires the employment of an additional lens between the objective and the beam-splitter, with at least two more almost flat reflecting surfaces perpendicular to the optic axis.

A manufacturer's cited tubelength of 160mm should be verified: older Leitz objectives have a tube length of 170mm. This is the length between the thread which accepts the objective and the upper end of the tube: the eyepiece is recessed into the tube. In TSMs, the distance between the objective thread and the disc should be shorter by the amount of the length (also standardised) between the upper end of the tube and the position of the intermediate image formed by the objective (in the Zeiss standard it is -10mm).

Microscope objectives were developed, and are highly specialised, for observation of either polished metals and rocks, or bacteriological and haematological smears, or thin and ultrathin slices of prepared biological objects. Only exceptionally are microscopes used for other objects.

It is clear from theory, as well as from experience, that only an insignificant fraction of the enormous number of the types of objectives which have been produced are well suited for scanning microscopes, and none of them is suitable for all objects, all purposes and all devices. The time is ripe to encourage progressive manufacturers of microscope optics to market special objectives for confocal microscopy, bearing in mind that such objectives will also benefit other branches of LM. The demands to which such objectives should respond will differ only slightly with respect to different TSMs and CSLMs and with respect to different objects. We can divide these demands into several groups with respect to their generality and urgency.

The general demands are as follows:

(1) The corrections of aberrations must not be distributed between objective and eyepiece: the objective must be compensated within itself.

(2) The chromatic correction must be much better than was the case in classical microscopes. Because fluorescence microscopy will be in great demand, the corrections should also be extended into the near ultraviolet and far red.

(3) Because Rayleigh scattering is inversely proportional to the fourth power of the wavelength, it can be anticipated that the near infrared could be of interest, especially when better penetration in depth is attempted.

(4) It is possible that mirror objectives will receive more favour because of the need for better and broader chromatic corrections and the lower influence (in TSM and CSLM) of central obscuration on the image quality. It may even be desirable to obscure the central portion of the objective aperture for the sake of improvement of lateral resolution and of image contrast. The longer free working distance of mirror objectives can also be of interest, particularly as regards micromanipulation.

(5) TSMs create the need for the highest aperture objectives with very low magnifications. In terms of resolution in conventional microscopes, high apertures for low-power objectives are useless, and hence their production was never attempted. In confocal microscopy, the high aperture is necessary for axial resolution (i.e., to ensure a rapid decrease in image brightness with axial distance from a focussed-on point) as well as for lateral resolution and image brightness. Thus high apertures are highly desirable even in the lowest-power objectives.

(6) In confocal microscopy, immersion serves not only to raise the numerical aperture, but is important for the suppression of surface reflections. Hence different types of water immersions will be necessary for microscopy of living matter, and variable index immersions will be needed for microscopy of rocks, fossils, ceramics, plastics and industrial and food materials.

(7) If specimens must be covered with coverslips with a refractive index other than that of the object or the medium before the objective (often air), thick coverslips can be more useful than the standard 0.17mm. It could be advantageous to correct some objectives for thick glass (e.g., the glass used in chambers for counting blood cells) and to standardise this greater thickness.

(8) As in phase contrast microscopy, monochromatic light (isolated mercury bands with mean wavelengths 254, 365, 405, 436, 578, 586nm, etc.) can be used for the most critical work. One can therefore envisage special objectives corrected for one wavelength and given antireflective coatings for that wavelength.

There is also a need for special purpose objectives for use in biology and medicine. The most important objective features are water immersion, high aperture, and long free working distance. Small residues of field curvature can usually be tolerated. Very good suppression of internal reflections is important. Hence it is necessary to reduce the number of component lenses, to apply the best antireflection layers, and perhaps to apply antireflecting layers even on some cemented surfaces as well as on the first (immersed) surface.

The special requirements for work in areas such as paleontology, geology, ceramics and the glass industry include variable index immersion and extremely long free working distances. This could be a further strong impulse for the return of mirror objectives of different and rather strange types.

In the semiconductor industry, it will be important to employ the highest aperture dry plan objectives. The objectives should be corrected for the visible and near infrared band, some of them even for farther infrared to observe "physiological" and "pathological" hot spots. The objectives should be very well plan-corrected for wide fields, and should be strain-free to permit the best use of polarised light. Zoom objectives (transfocators) would enable the easy switch from broadfield to highest magnification.

Hard objects can damage the front surface of the objective. There would be good grounds for using harder material for the objective front lens component, e.g., spinel or quartz glass.

VI.5. Eyepieces and relay optics

Relay optics have to transfer the image in the plane of the disc to the final image plane, which can be infinity in the case of the Ramsden-type eyepiece. Any more complicated relay optics, e.g., Figure 9.6, should be designed with the minimum number of antireflection-coated optical surfaces to minimise additional light losses.

Because the TSM image (with the mirrors in their normal, simplest configuration, Figure 9.2) is side-inverted — the image of the left hand looks like the image of the right hand — a viewer incorporating a single plane mirror may be used to side-correct the image, also making it easier to interpret and permitting the use of any type of eyepiece.

VI.6. Cameras, TV cameras and image intensifiers

A considerable advantage of the TSM is that there is an observable image which can be recorded on normal photographic materials. It is well known that the density of information which can be recorded on a photographic emulsion greatly exceeds the realistic possibilities of TV technology. The TSM image is a real colour image and can be recorded by any suitable positive or negative colour film material.

The TSM image may be photographed with any optical camera. Most often we have used standard cameras with standard lenses focussed at infinity and placed close above the eye-piece. Any TV camera may be employed. For low light-level work, there will be the requirement for an image intensifier and camera combination or SIT, ISIT, or CCD cameras. We stress again that, provided that the entrance pupil of the camera lens is not smaller than the exit pupil of the microscope, it is of no benefit to hanker after (very expensive) very low f-number camera lenses. The result may be perceptibly worse than with cheaper lenses, because of the thicker glass and the increased number of components and surfaces. Greater sensitivity can only be achieved with a brighter source, a better beam-splitter and a more sensitive detector.

VI.7. Stands: rigidity versus flexibility

The requirements for rigidity of the stand and specimen carriage in a TSM are no different from those of any other microscope in which it is expected to record a high resolution image. Thus it should be a realistic requirement to limit mechanical vibration to less than 0.05μm. In the TSM scenario, the most important self-sources of vibration are the disc motor and bearings, as well as any motors used in cooling fans for cooled CCD cameras and for temperature-controlled systems such as may be used in vital microscopy.

The flexibility offered by the TSM approach to CSLM is perhaps exemplified by the version first built in Plzen for UCL [Petráň et al., 1985a, 1985b] in which the entire "real" microscope, Figure 9.4, was mounted, together with its light sources, on the arm of a crane fixed to a large flat base. Thus the TSM could be offered to really massive specimens and rapidly brought to focus, completely bypassing the need for mechanical specimen preparation. Alternatively, the head may be incorporated into an instrument no bigger than any standard research LM, and may be accompanied by perfectly conventional mechanical specimen stage arrangements, Figures 9.5 and 9.6.

VI.8. Stage xyz automation

One of the great advantages of the TSM is that it may be used to record extended-focus images, otherwise known as range images, by through-focussing whilst photographing. It is therefore likely that practical manufactured TSMs will incorporate an automated feature for precision mechanical through-focussing. For the highest precision work, such movement will be of the objective lens relative to the specimen, and will probably be performed by piezo-electric devices. For lower-resolution work, or where a great z range is to be covered, it will probably be more convenient to resort to the use of motorised stage automation.

Figure 9.7. TSM image of enamel prism cross-sections in the superficial enamel of a molar tooth of Paranthropus (Australopithecus) Boisei, 2 million years BP fossil hominid from East Africa. A ribbon-like reflection (at left) corresponds to the surface of the tooth, to the right of which is a relatively featureless layer (the true surface zone prism-free enamel): to the right again, the enamel prism boundaries are seen as reflective features. 100X/1.30 Oil. Fieldwidth 76μm.

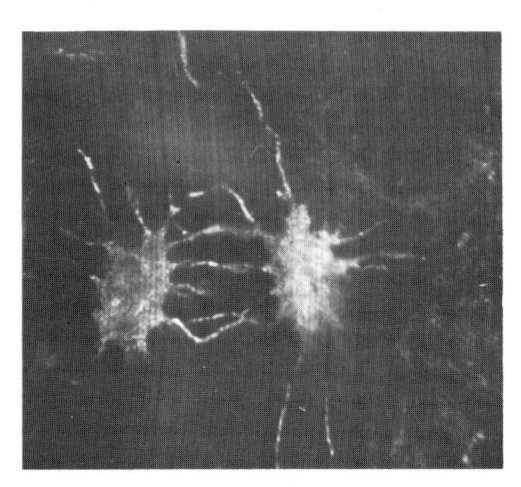

Figure 9.8. Osteocyte lacunae in human skull bone imaged in the Plzen TSM at UCL in 1985. The detail of the collagen fibre bundles can be recognised in the walls of the osteocyte lacunae. Focussing up and down at the TSM, one can recognise osteocyte lacunae at different depths within the bone. 100X/1.3 Oil. Fieldwidth 72μm.

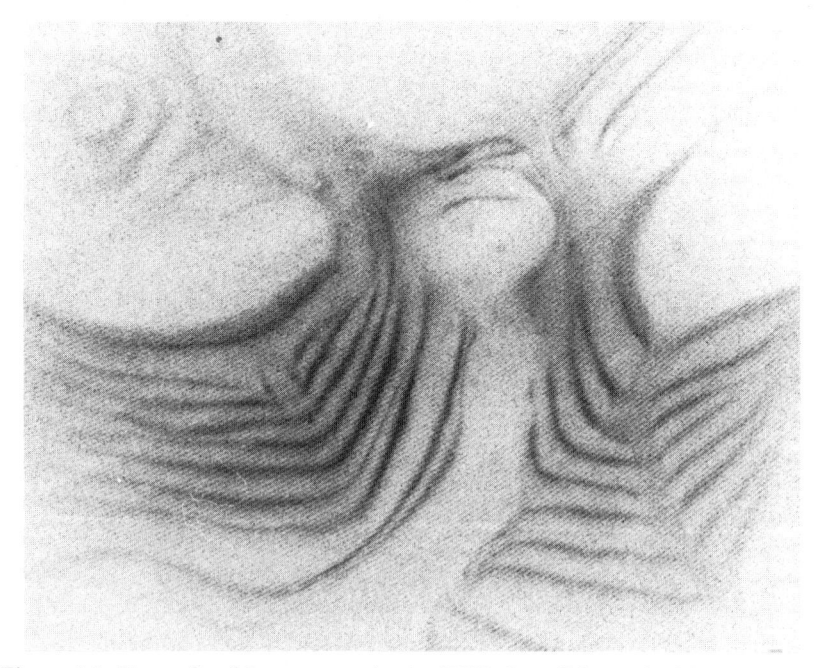

Figure 9.9. Example of fluorescence in the TSM. Rat tibia cortical bone: tetracycline was administered by injection at weekly intervals with one two-week gap. Print from Kodachrome transparency showing yellow tetracycline fluorescence lines (black in this negative image) under blue-violet excitation. 20X/0.8 Glycerine. Fieldwidth 224μm.

Direct View Confocal Microscopy 267

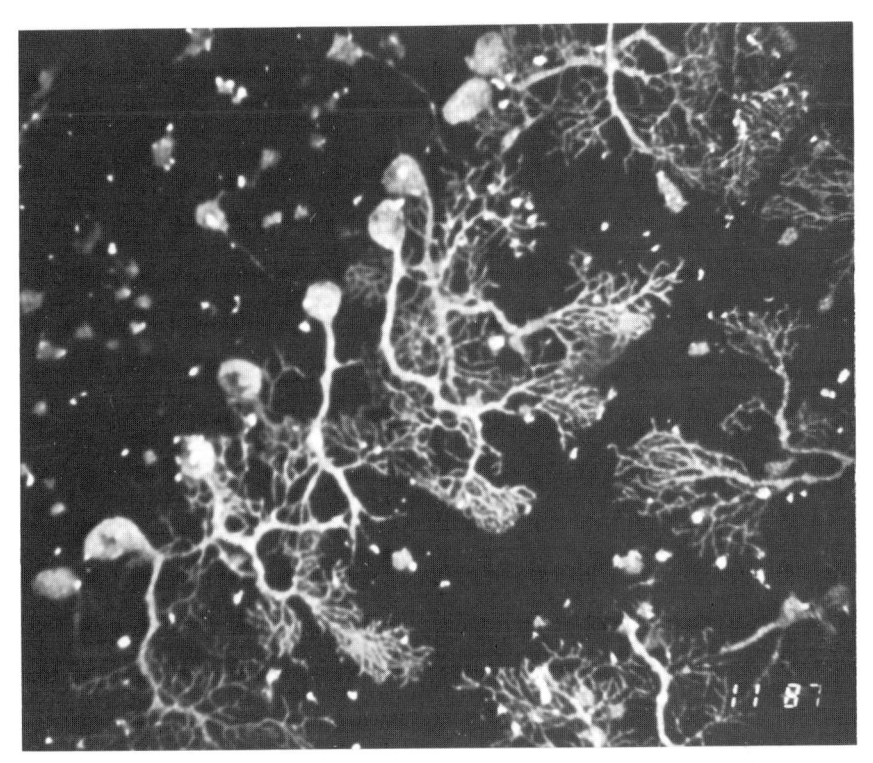

Figure 9.10. Extended-focus image of Purkyne cells in mouse cerebellum, 40μm range. This image was recorded by successively refocussing the (1983 Plzen) microscope whilst recording a single photographic image. 25X/0.65 Oil. Field height = 260μm.

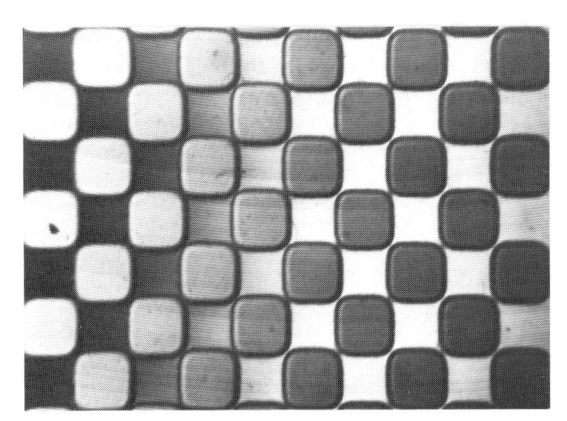

Figure 9.11. Silicon calibration standard consisting of 10μm pitch, 1.5μm deep squares etched in a chequerboard pattern. The sample is slightly tilted (by <> 1°) such that squares which are white at one end appear black at the other end of the tilted field of view. (1983 vintage UCL Plzen TSM). 100X/1.3 Oil.

Stereoscopic images in a TSM may also be recorded directly by through-focussing along two slightly inclined vertical axes [Boyde, 1985b, 1987, 1989a]. At high resolution, the precision requirement for the combination of a small x with a large z movement will be better achieved through the use of piezo-electric devices. Thus, although the TSM can function alone without any modern electronic encumbrances, Figures 9.7-9.11, we foresee that a microcomputer would be a standard accessory merely for controlling through-focus image recording — and, once incorporated, can be used for more sophisticated image processing.

VI.9. Simultaneous illumination for non-confocal modes

The reflection TSM can be combined with parallel non-confocal modes in several different ways. The simplest, and perhaps the most often employed, is the transillumination of a relatively thin sample being examined in confocal mode in the TSM. This provides a simultaneous non-confocal image (in almost any LM mode), the disc functioning only as a neutral density filter on the detection image recording side of the microscope [Boyde, 1989b]. It is not absolutely necessary to provide a separate source of illumination to operate in this mode, since the waste light transmitted through the specimen can be reradiated through the sample using a mirror. In the more general case, however, a normal sub-stage illumination system, such as is found on any standard LM, will be provided.

It is also possible to provide an additional source for non-confocal epi-illumination (e.g., via an extra beam-splitter below the objective mount or via fibre optic guides). This second source may be coloured to distinguish confocal and non-confocal information. Non-confocal epi-illumination modes can also be achieved by removing or bypassing the disc. No doubt manufacturers will improve the arrangements for removal and realignment of the disc for this purpose: here may lie a real advantage of the one-sided disc configuration.

VII. Image processing and computers

Feeding the TSM image via a TV camera to an image-analysing computer may have several uses. A few years ago, we used background subtraction to remove the (then prevalent) undesirable curved lines which originated from non-uniformity of illumination due to inadequate disc design and manufacture. An image was acquired with the microscope defocussed, usually above the top plane of the object, and this was subtracted from the desired image.

Subtracting one image from another acquired from a nearby focus plane shows features which lie only in the one plane as black, and those which lie only in the other as white. Such images simplify the means of making a valid three-dimensional count [Howard et al., 1986].

As with any other confocal LM image, more elaborate computer image processing can also have important advantages. In our own experience, two-dimensional FFTs are used to determine spatial frequencies in the image. Using the appropriately masked reverse transform, unwanted spatial frequencies can be removed and desired frequencies enhanced. Similar manipulations can, of course, also be made with analogue techniques of photo-processing.

Other uses of computer image processing include the acquisition of x maps series and the redisplay of such information in isometric projections, Figure 9.12.

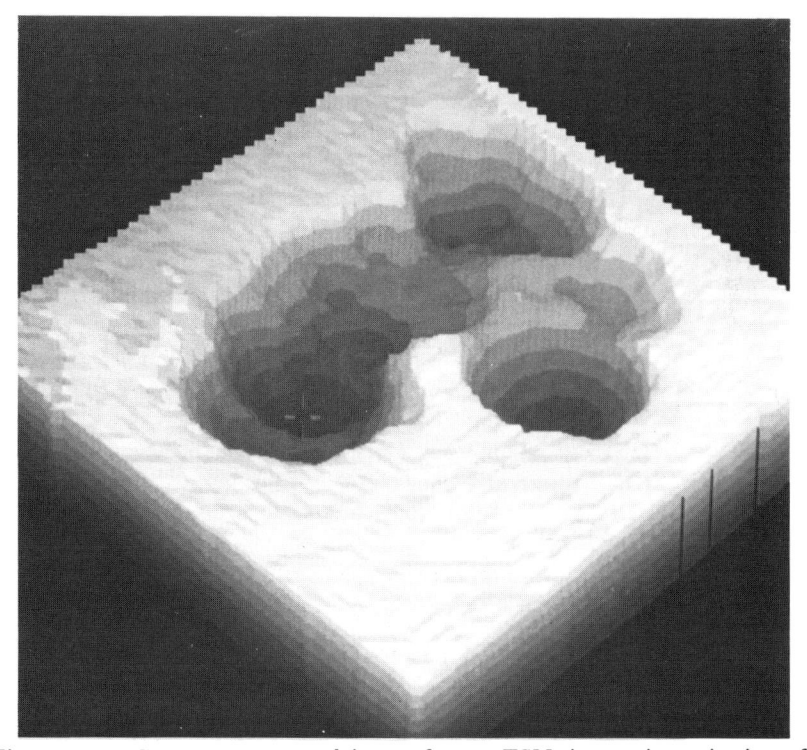

Figure 9.12. Computer-generated image from a TSM: isometric projection of a z map (image recording the focal depth at which maximum signal was returned during automatic through-focus sequence) of SEM specimen of resorption lacuna complex made by single chick osteoclast in sperm whale dentine (this osteoclast differentiated from marrow cells in 19 day culture). Tracor TSM and 8500 image-analysing computer. 60X/1.4 Oil. Maximum depth in pit is 25μm. Fieldwidth = 166μm.

VIII. Applications, specimen preparation and imaging modes

VIII.1. The main areas of application

The main areas of application of the TSM are in studying samples which are difficult to manage by other means, or in which the depth of information would be too great, thus reducing contrast. We summarize them as zoological and botanical micromorphology and microphysiology, cell and developmental biology, medicine and pathology (autopsies and biopsies), "difficult" organs such as eyes, bones, Figure 9.8, and teeth, Figure 9.7, prosthetic materials including dental filling materials and bone implant materials, archaeology, palaeontology, geology, mineralogy, materials science, semiconductors and integrated circuits, ceramics, industrial materials, food products and all kinds of forensic investigation.

We provide a bibliography of applications, and examples which are explained in the captions to Figures 9.7-9.12.

VIII.2. For which kinds of specimens is TSM most suited?

Specimens which are most suited to study by TSM are those which require the resolution of three-dimensional fine structure near a relatively flat overall specimen surface. The origin of the specimen contrasts should lie in reflection (refractive index differences) or fluorescence (chemistry). The need for the specimen to be nearly flat results from the fact that the high aperture lenses which we will use have a very small free working distance. If it is necessary to extract subsurface information, we will nearly always consider the use of immersion lenses. If it is desirable to measure the distance into the sample at which a particular light-scattering event occurs, it will be necessary to match the refractive index of the immersion medium to the mean index of the sample. Of course, this will also be necessary to reduce aberrations.

As regards living animal tissues, it should be noted that there are quite sufficient reflections in live cells and tissues to enable the image to be read without any other means of improving the contrast [Petráň et al., 1986]. Tissue fixation frequently removes the fine refractive index differences present in live tissue. This is frequently forgotten, and disappointing results may be obtained in reflection microscopy of preserved specimens.

VIII.3. Specimen preparation

For this reason, it may be necessary to employ reflective or fluorescence stains with preserved zoological specimens [Boyde, 1985a]. There is obviously a great deal of work to be done in developing these aspects of specimen technology, as well as in borrowing from established work in other areas of light microscopy.

Hard samples will frequently need to be cut and polished before examination in the TSM, principally for the reason that the sample would otherwise scratch the (expensive) objective lens. Apart from this, we would note that the principal reason for using a TSM might be to avoid all requirements for real specimen preparation.

VIII.4. Fluorescence

The utility of the TSM in fluorescence microscopy was not originally envisaged. However, it has been shown in practice, Figure 9.9, that this may be one of its major areas of application: we note that the greatest rate of increase in interest in using the CSLMs followed upon the realisation that they made fluorescence immunocytochemistry so simple.

The special difficulties in fluorescence TSM are:

(1) The problem given by some objective lenses of longitudinal chromatic aberration.

(2) The problem of very low light-level fluorescence.

The latter is overcome by finding the location in the specimen in the reflection (or other non-confocal imaging) mode; the former by employing a lens without such aberration.

VIII.5. Reflective stains

The usefulness of reflective stains has been shown in many TSM studies to date. Particularly the reactions which deposit metallic silver or gold or insoluble metal salts in, or on, the tissue have the most dramatic effect, e.g., the Golgi impregnations used by neuroanatomists, Figure 9.10, and muscle cell cytologists, and the immunogold and silver-enhanced immunogold reactions developed in the first instance for EM immunocytochemistry. However, many organic dyes are also reflective, and we can expect a considerable advance in this area of investigation.

VIII.6. Reflective coatings

TSMs give a high contrast image of the surface layer of the sample, obviously highest in the case of a dry sample viewed in air. In order to reduce the relative significance of this surface reflection, we use immersion lenses. However, immersion lenses may also be used in imaging rough surfaces of biological hard tissue and geological samples, in order to be able to record detail from steeper topographic slopes than can be captured with a dry lens: the reduction in the surface reflection may then be a problem, particularly if there is a good index match between the immersion fluid and the material in question. The image may then be improved by employing a metallic reflective coating such as is used on SEM samples, Figure 9.12. Both information and beauty may be enhanced by using metals with contrasting colours to code for slopes facing in contrasting directions, exploiting, for example, the silvery colour of aluminium, the green of gold and the reddish colour of copper.

VIII.7. Chromatic aberration colour-coding for depth

If the TSM is illuminated with white light, and if the objective lens has a significant longitudinal chromatic aberration, it will be confocal for different colours at slightly different focal positions. A plane reflector will change its colour as it is moved through focus. A facetted reflector will show contrasting colours which differ as a function of the distance of the local facet from the lens. Thus this chromatic aberration effect gives rise to an image which is colour-coded for depth. The range is set by the characteristics of the objective and the spectrum of the lamp: tungsten has a more useful spectrum than the line-dominated spectrum from a mercury arc. With the appropriate choice of conditions, it should be possible to map surface height variations quantitatively for nearly-plane reflectors from a single focus position. The height co-ordinates would be taken from the position of each pixel in the chromaticity diagram, i.e., the relative output in red, green and blue of a colour TV camera.

VIII.8. Stereoscopic image recording

Through-focussing during photography, repeated twice on inclined axes, provides the simplest direct means of obtaining stereoscopic views at the limit of resolution in light microscopy [Boyde, 1985b, 1989b]. The operator selects the specimen volume to be reconstructed by direct view in the TSM. An image corresponding to one direction of view through this selected volume is recorded whilst changing the mechanical focus of the microscope. The image is therefore an extended-focus view in that direction. The same volume is then photographed through the same vertical interval, but along an inclined vertical axis, which corresponds to a different direction of view. The two photographs are viewed as a stereoscopic pair to give a three-dimensional impression. Being able to record the stereo pair in the TSM as an integral part of the procedure of photographing the field of view is an important advantage for the practising microscopist.

Any means for achieving the simultaneous combination of vertical movement and horizontal movement may be employed. The simplest would be to employ two mechanical movement devices, both with their axes of movement nearly parallel to the usual (z) focussing direction, but each tilted slightly to correspond to the direction of view through the "stacked" image. For routine application to high-resolution stereo imaging, the through-focussing needs to be done under precise remote control [Boyde, 1989b]. The most precise systems are those operated by piezo-electric devices, which are also relatively simple to control by computer. The xz movements can be applied either to the specimen or to the objective lens: the latter makes it simpler to use the microscope to examine specimens which are either bulky or which may move as a consequence of an experimental manipulation, e.g., in a loading test.

In practice, objective magnifications as low as 1.5X and as high as 200X have been used with this direct stereo method. As regards the speed of acquisition of the stereo images, we can presently (1989) acquire stereopairs within a two-second acquisition period (one second per pass), either photographically or via CCD or TV camera and framestore.

The stereo-pair images may be overlapped at the top or the bottom of the field of view, or outside the field altogether. The overlap plane is the level at which the two members of the pair have the same x shift. Any contribution(s) to the image(s) which do not derive from the real three-dimensional distribution of features within the object will be perceived as lying in this plane in the stereo view: they are the same in both members of the pair and arise, for example, from dirt on lenses and lines resulting from imperfect discs. To put this undesirable contribution in the background (at the deepest level of the reconstructed volume), the sequence for recording the images is to start focussed at the top of the desired volume and to record the first photo whilst focussing down into the specimen with the required ratio of simultaneous movement in x. The second photo is taken whilst focussing upwards with the same direction of movement in x. The common plane in the two images is, then, that at the end of the first photo and at the start of the second — both with the focus at its deepest level into the sample. If required, an additional increment of movement in the same sense in x applied at the bottom focus level will cause the overlap to lie beyond the deepest level in the reconstructed volume. Unwanted contributions to the image then lie completely outside the three-dimensional optical model.

When recording with short exposure times, it is better to begin each pass at the same level to compensate for any delay in the commencement of movement, and to arrange that the oversampled layers are the same in both images (best at the top in this case).

Given the availability of computer image frame-store capacity, stereo-pairs can be generated from the set of images acquired in a single, vertical through-focus pass just as is done in LSCMs [Cox and Sheppard, 1983; Brakenhoff et al., 1985; Carlsson et al., 1985; Wijnaendts-van-Resandt et al., 1985]. Whether photography or computer image-processing are employed, stereopairs generated by the single pass method show more blur than those acquired by the double pass procedure.

Confocal microscope stereo images have the interesting property of parallel projection geometry: perspective is eliminated, because every imaged feature is recorded when it is in the same plane in the optical system and hence "measured" with the same "scale". This is an advantage for three-dimensional measurement because the theory for parallel projection is much simpler.

VIII.9. Colour-coding without a computer

The depth of any TSM (or CSLM) range image can be far greater than the depth of the field, giving rise to colour-coding owing to chromatic aberration of the objective in the TSM. However, the range of coding for depth can be extended (by pseudo-colour) by changing the colour of the illuminating or image-recording light with a sequence of filters, so that the reflective features lying within different depth bands are imaged in contrasting colours [Boyde, 1987].

Combining methods of conveying the depth information can be useful. A mono-coloured mono-image of a thick volume (e.g., one half of a stereo pair) taken by through-focussing whilst recording does not give much information as to the real arrangement of structural features. A single, colour-depth-coded image could be understood, given the information about the vertical separation corresponding to particular colours. A mono-coloured stereo image would show the real distribution. However, the effects are cumulative: the colour-depth-coded stereo images are easily interpreted even by those with limited experience in the interpretation of stereo images. However, artificial colour-coding does away with one of the prime advantages of the TSM: that it works in real colour in both reflection and fluorescence.

Using different colours to image different depths within the sample has implications for both lateral and depth resolution. Red might demonstrate lower resolution than blue according to the ratio of wavelengths, but red light suffers less scattering and would therefore be expected to produce less hazy images at greater depths.

VIII.10. Depth limitation

The limit to the depth to which one can image within a given object is set by the free working distance of the objective lens and by the light-scattering properties of translucent objects. Another limit is set in range (and stereo-) imaging of thick translucent objects, owing to feature overlap in the stacked image.

VIII.11. Particle counting and stereology

Unbiased estimation of particle density can be conducted in the TSM by counting the number of particles that lie within a counting "brick" (or three-dimensional counting frame), or which intersect one set of three of its six sides [Howard et al., 1986]. Such counting is conveniently done on-line in a TSM because of its real-time mode of operation. However, TSM range images, or better still the stereo images, are just such "bricks", and they could be used for retrospective particle counting.

VIII.12. Reflection interference contrast

The TSM gives excellent contrast for the interface zone of cells attached to flat substrates such as glass and plastic [Boyde 1989a; Paddock, 1989]. The contrasts are interpreted as due to interference in thin layers: either the thin gap between the cell and the substrate, or thin cytoplasm (lamelloplasm). It is thus especially suited to the dynamic observation of cell migration phenomena.

IX. Resolution and contrast

IX.1. Resolution from the standpoint of information theory

In the 1960s, much theoretical work was done in information theory showing that, in a defined information-processing machine, the (quantitative) flow of information (considered as a manifoldness) is constant [Lukosz and Marchand, 1963; Lukosz, 1966, 1967]. We can illustrate the meaning of this by considering the improvement in sensitivity and precision in measuring vertical co-ordinates using a multiple-beam interferometer microscope (according to Tolansky, 1948), when we simultaneously diminish, *ex principio*, the precision in measuring the transverse dimensions. In scanning microscopy, this means that we have to pay for the introduction of scanning (depending on the number and length of scanning lines) by losses in time resolution. From this standpoint, the TSM is considered a flexible connection between the classical (parallel) microscope and the Type II (unitary-beam) confocal scanning microscope. The flexibility is accomplished partly by the construction (we can design it with various numbers and dimensions of scanning spots), and partly by changes of scanning speed, scanned field dimensions and form — and hence the number of scanning lines and points — to fit the instantaneous demands of the particular object and aim.

IX.2. The equivalence of the TSM and other CSLMs

The equivalence of the multiple-beam scanning TSMs and the unitary-beam CSLMs can be shown by assuming that a sufficiently small object detail is visualized so that it is illuminated, and its image "filtered," by only one hole pair at any one instant. This can be achieved by limiting the field, appropriately by an iris diaphragm in the illumination. Obviously, we then get a true confocal microscope possessing a field area much smaller than in the conventional microscope, but still a hundred times greater than its (two-dimensional) resolution. This field is scanned only by one "point," with the same dimensions as one originating from a laser. If we now stop down the (illumination) field diaphragm, we find that the only change in the appearance of finely structured details (in specimens ranging from living tissues to replicas of fine gratings not resolvable by the same objective in a conventional microscope) is a gradual rise in image contrast accompanying the restriction of field. This can be used in practice for difficult features: beginning with a field as wide as the correction state of the objective permits, we locate the field centre on the detail of interest and then limit the field to improve the contrast, and hence to take full advantage of the resolution improvement (whilst conserving the time resolution).

IX.3. Methods which compete with overlapping advantages

(1) Nomarski DIC: A biologist considering the acquisition of a TSM for vital microscopy of bulky samples might like to try using conventional Nomarski differential interference contrast microscopy first. If the sample can be rendered as a slice less than 200μm thick, it is often possible to provide optical sectioning to the required xyz resolution. This is the only normal method of light microscopy which may provide one, and only one, of the advantages possessed by the existing TSMs, i.e., of improved z resolution.

(2) Video-enhanced contrast: Simply processing any (low-contrast) LM signal via a TV camera and monitor display allows for electronic contrast enhancements, which render smaller, low-contrast features visible. Oblique illumination and the use of polarised light improve matters even further, and the most beautiful demonstrations of the movement of previously "submicroscopic" features in living cells have been shown by Inoué (1986), Allen et al. (1985) and numerous others. The contrast in the TSM image may be similarly improved.

(3) Image deconvolution: Fay and colleagues (1989) show how the axial resolution in a conventional fluorescence microscope can be improved through image deconvolution.

X. Summary and conclusions

X.1. Present standing of the TSM

The most important features of TSMs which distinguish them from other CSLMs are compactness, sturdiness, durability, reliability, almost instantaneous response (i.e., shorter time constant), independence from computer hardware and software, polychromatic illumination, contrast dependent on field size used, and the lowest possible noise and hum (because no transformations of light into other forms of energy take place).

The TSMs are the only contemporary Type II or confocal microscopes able to work in real time, without electronic image-processing, together with electrophysiological and similar equipment.

X.2. Future directions

Further development of the TSM will probably include the upgrading of the precision of the aperture disc pattern, raised revolution speed, and improved high speed static and dynamic image recording. As regards its essential accessories, we await the introduction of new specially designed and dedicated objectives, the most important features of which will be better chromatic correction, low internal reflections (attained by design and by coating), the highest possible numerical aperture even in the lowest power objectives, and the highest power and highest aperture objectives possessing longer free working distance than is sufficient for classical mounted slices. We expect better immersion objectives for special cases — water immersion for living cells, glycerine and oil immersions for hard tissues, fossils and minerals — special shapes of objective fronts, and special hard materials for the front component of the lens.

Different kinds of apodised and/or annular objective apertures must be made both to improve their resolution and enhance image contrast. The revival of mirror objectives is also probable, perhaps in the form of a single (cemented) block of glass as in the objectives developed by Bouwers et al. (1951). Mirror objectives will be especially desirable for ultraviolet, infrared and above all for fluorescence microscopy, with greater gaps between the exciting and excited radiations. Objectives will probably also be developed in which there is a geometrical separation of image-forming and illuminating rays.

As can be seen, and as could be anticipated, this new method for morphological research is able to help in several branches of science, but poses more questions and problems for its own further development than it has solved. TSM allowed the first experimental demonstration that the microscopic space-time structure of complex matter — e.g., living matter — could be studied beyond the limits stated by conventional experience and beyond the theoretical limits based on classical optics. However, only the core of the new approach has been developed so far, and it has posed new, practical questions about how these fundamental results can be further improved and diversified according to the objects and questions studied.

XI. Acknowledgements

We thank Jiri Benes, Pavel Vesely, Frantisek Franc, and Miroslav Maly for their support in various ways. No thanks are due to Leonid Brezhnev et al., whose activities in 1968 caused substantial loss of acceleration, in this field as in others. In London, we acknowledge the financial support of the MRC and the SERC which were instrumental in putting the TSM back on the road. We thank Tim Watson for his close collaboration and Roy Radcliffe, Elaine Maconnachie and Maureen Arora for technical assistance.

XII. Applications bibliography

Baddeley, A. J., Howard, C. V., Boyde, A. and Reid, S. A., (1987). Three-dimensional analysis of the spatial distribution of particles using the tandem scanning reflected light microscope. Acta Stereologica, **6**, Suppl. 2, 87-100.

Benn, D. K. and Watson, T. F., (1989). Correlation between film position, bitewing shadows, clinical pitfalls and the histological size of approximal carious lesions. Quintessence, **20**, 131-141.

Boyde, A., (1985). Anatomical considerations relating to tooth preparation. In: *Posterior composite resin dental restorative materials.* van Herle, G. and Smith, D. C. (Editors), 377-403. 3M Co. St. Paul, MN ISBN-088159-601-9.

Boyde, A., (1987). Applications of tandem scanning reflected light mi-

croscopy and 3-dimensional imaging. Ann. New York Acad. Sci., **483**, 428-439.
Boyde, A., Ali, N. N. and Jones, S. J., (1985). Optical and scanning electron microscopy in the single osteoclast resorption assay. Scanning Electron Microscopy, (SEM Inc. - AMF O'Hare), **III**, 1259-1271.
Boyde, A., Hadravský, M., Petráň, M., Jones, S. J., Martin, L. B., Watson, T. F. and Reid, S. A., (1986). TSRLM: How it works and applications. Proc. 44th Ann. Meeting Electron Microscopy Soc. Amer., Bailey, G. W. (Editor), 84-87. San Francisco Press, San Francisco.
Boyde, A., Jones, S. J. and Dillon, C., (1989). Automated 3-D characterization of osteoclastic resorption lacunae. Microscopy and Analysis, **23**, Jan. 1989.
Boyde, A., Jones, S. J., Taylor, L., Wolfe, L. and Watson, T. F., (1989). Fluorescence in the tandem scanning microscope. J. Microsc. In press.
Boyde, A., Maconnachie, E., Reid, S. A., Delling G. and Mundy, G. R., (1986). SEM in bone pathology: review of methods, potential and applications. Scanning Electron Microscopy, **IV**, 1537-1554.
Boyde, A. and Martin, L., (1984). A non-destructive survey of prism packing patterns in primate enamels. In: *Tooth enamel IV*. Fearnhead, R. W. and Suga, S. (Editors), 417-421. Elsevier, Amsterdam.
Boyde, A. and Martin, L., (1987). Tandem scanning reflected light microscopy of primate enamel. Scanning Microsc., **1**, 1935-1948.
Boyde, A., Petráň, M. and Hadravský, M., (1983). Tandem scanning reflected light microscopy of internal features in whole bone and tooth samples. J. Microsc., **132**, 1-7.
Boyde, A. and Reid, S. A., (1986). 3-D analysis of tetracycline fluorescence in bone by tandem scanning reflected microscopy. Bone, **7**, 148-149.
Boyde, A. and Watson, T. F., (1989). Fluorescence mode in the tandem scanning microscope. Proc. Roy. Microsc. Soc., **24**, 7.
Boyde, A. and Wolfe, L. A., (1986). Block face microscopy for the interfacial region between bone and implant materials. J. Dent. Res., **66**, 859. (Abst. 223.)
Fortelius, M., (1985). Ungulate cheek teeth: developmental, functional and evolutionary interrelations. Acta Zool. Fennica, **180**, 1-76.
Freire, M. and Boyde, A., (1989). Study of Golgi-impregnated material using the confocal tandem scanning reflected light microscope. J. Microsc. Submitted.
Jester, J. V., Cavanaugh, H. D. and Lemp, M. R., (1988). In vivo confocal imaging of the eye using tandem scanning confocal microscopy (TSCM). Proc. Electr. Microsc. Soc. Amer., **46**, 56-57.
Jones, S. J. and Boyde, A., (1987). Scanning microscopic observations on dental caries. Scanning Microsc., **1**, 1991-2002.
Jones, S. J., Boyde, A., Ali, N. N. and Maconnachie, E., (1985). A review of bone cell and substratum interactions. Scanning, **7**, 5-24.
Lemp, M. A., Dilly, P. N. and Boyde, A., (1985). Tandem-scanning (confocal) microscopy of the full-thickness cornea. Cornea, **4**, 205-209.

Lester, K. S., Boyde, A. and Gilkeson, C., (1986). Marsupial and monotreme enamel. Scanning Microsc., 1, 401-420.

Petráň, M. and Hadravský, M., (1974). Employment of tandem scanning microscope in morphological research of living sensory systems. Activitas Nervosa Superior (Praha), 16 (4), 289.

Petráň, M., Hadravský, M., Benes, J. and Boyde, A., (1987). In vivo microscopy using tandem scanning microscope. Ann. New York Acad. Sci., 483, 440-448.

Petráň, M. and Sallam-Sattar, M., (1974). Microscopical observations of the living (unprepared and unstained) retina. Physiologia bohemoslov, 23, 369.

Reid, S. A., Smith, R. and Boyde, A., (1985). Some scanning microscopies of fibrogenesis imperfecta ossium. Bone, 6, 275-276.

Sallam-Sattar, M., (1974). New technique for in vivo correlation of morphological and electrophysiological events in the central nervous system. Ph.D. Thesis, Inst. of Biophysics Czechoslovak Academy of Sciences, Brno, Kralovopolska. Vol. 1 text, Vol. 2 illustrations.

Sallam-Sattar, M. and Petráň, M., (1974). Dynamic alterations accompanying spreading depression in chick retina. Physiologia bohemoslov, 23, 373.

Watson, T. F., (1989). A confocal optical microscope study of the morphology of the tooth/restoration interface using Scotchbond 2 dentin adhesive. J. Dent. Res., 68, 1124-1131.

Watson, T. F., (1989). Real-time confocal microscopy of high speed dental bur/tooth cutting interactions. J. Microsc. In press.

Watson, T. F., (1989). Tandem scanning reflected light microscopy for rapid histological evaluation of carious lesions labelled with fluorescent markers. In: *Evolution in dental care*. Elderton, R. J. (Editor). Proc. of the World Dental Conf., Bristol, 1988. In press.

Watson, T. F. and Boyde, A., (1984). The tandem scanning reflected light microscope (TSRLM) in conservative dentistry. J. Dent. Res., 62, 512.

Watson, T. F. and Boyde, A., (1985). Tandem scanning reflected light microscopy of fluorescent labelled composite interfaces. J. Dent. Res., 64, 664.

Watson, T. F. and Boyde, A., (1986). In-vitro dentine penetration of a GLUMA bonding agent. J. Dent. Res., 66, 835. (Abst. 8.)

Watson, T. F. and Boyde, A., (1987a). Tandem scanning reflected light microscopy: a new method for in vitro assessment of dental operative procedures and restorations. Clinical Materials, 2, 33-43.

Watson, T. F. and Boyde, A., (1987b). The use of fluorescent markers for studying the distribution of a dentine bonding agent between a composite restoration and tooth. Clinical Materials, 2, 45-53.

Watson, T. F. and Boyde, A., (1987c). Tandem scanning reflected light microscopy: applications in clinical dental research. Scanning Microsc., 1, 1971-1981.

Watson, T. F. and Boyde, A., (1989). An in-vitro study, using a confocal

Hauptgeschäft: **Frauenstraße 42**

Rechtswissenschaften · Steuerrecht
Medizin · Psychologie · EDV
Biologie · Geologie · Geographie
Mathematik · Physik · Chemie

Zweiggeschäft: **Bäckergasse 2 am H 1**

Volkswirtschaft

Betriebswirtschaft

juristische Studienliteratur

Zweiggeschäft: **Hüfferstraße 75**

im Klinikenviertel

Medizin

Wir freuen uns auf Ihren nächsten Besuch.

Krüper GmbH
UNIVERSITÄTSBUCHHANDLUNG

---

*Haubold*ESCHWEGE · 37269 Eschwege · Tel. 0 56 51 - 309-0

Krüper GmbH

Universitätsbuchhandlung

48143 Münster · Frauenstraße 42
☎ 02 51 / 4 17 65-0 · Bäckergasse 2
Fax 4 17 65-70 · Hüfferstraße 75

Anz.	Artikel	DM
1	Wilson, Confocal Microscopy	118,70
1	f. Anzahlung	20,—
		98,70

Verk. 3   incl. ___ % MWSt.

05.11.04

Rückgabe und Umtausch nur gegen Vorlage des Kassenzettels

*98.70

optical microscope, of the marginal adaptation of vitrabond TM light cured glass ionomer. J. Dent. Res., **68**, 577.

Wolfe, L. A., (1988). Ph.D. Thesis, University of London. The incorporation of titanium and hydroxyapataite reinforced polyethylene implants into rabbit bone.

References

Allen, R. D., Weiss, J. H., Hayden, J. H., Brown, D. T., Fujiwake, H. and Simpson, M., (1985). Gliding movement of and bidirectional transport along single native microtubules from squid axoplasm: evidence for an active role of microtubules in cytoplasmic transport. J. Cell Biol., **100**, 1736-1752.

Bouwers, A., Blaisse, B. S. and Bulthuis, H. W., (1951). Katadioptrisches System mit Hohlspiegel und Kittflaeche. Oesterreichisches Patent Nr. 177,937. Netherlands priority Mar. 22, 1951. United States Patent No. 2,704,417.

Boyde, A., (1985a). The tandem scanning reflected light microscope. Part 2 - Pre-MICRO 84 applications at UCL. Proc. Roy. Microsc. Soc., **20**, 130-139.

Boyde, A., (1985b). Stereoscopic images in confocal (tandem scanning) microscopy. Science, **230**, 1270-1272.

Boyde, A., (1987). Colour-coded stereo images from the tandem scanning reflected light microscope. J Microsc., **146**, 137-142.

Boyde, A., (1989a). Combining confocal and conventional modes in tandem scanning reflected light microscopy. Scanning, **11**, 147-152.

Boyde, A., (1989b). Direct recording of stereo-pairs from disc-scanning confocal light microscopes. In: *The handbook of confocal microscopy.* Pawley, J. B. (Editor), 147-151. IMR Press, Madison WI.

Brakenhoff, G. J., van der Voort, H. T. M, van Spronsen, E. A., Linnemans, W. A. M. and Nanninga, N., (1985). Three-dimensional chromatin distribution in neuroblastoma nuclei shown by confocal scanning laser microscopy. Nature, **317**, 748-749.

Brakenhoff, G. J., van der Voort, H. T. M, van Spronsen, E. A. and Nanninga, N., (1989). Three-dimensional imaging in fluorescence by confocal scanning microscopy. J. Microsc., **153**, 151-159.

Carlsson, K., Danielsson, P. E., Lenz, R., Liljeborg, A., Majlöf, L. and Åslund, N., (1985). Three-dimensional microscopy using a confocal laser scanning microscope. Opt. Lett., **10**, 53-55.

Cox, I. J. and Sheppard, C. J. R., (1983). Digital image processing of confocal images. Image and Vision Computing, **1**, 52-56.

Egger, M. D., Gezari, W., Davidovits, P., Hadravský, M. and Petráň, M., (1969). Observation of nerve fibers in incident light. Experientia (Basel), **25**, 1225-1226.

Egger, M. D. and Petráň, M., (1967). New reflected-light microscope for viewing unstained brain and ganglion cells. Science, **157**, 305-307.

Ellis, G. W., (1979). A fiber optic phase randomizer for microscope illumination by laser. J. Cell Biol., **83**, 303 abstract.
Ellis, G. W., (1985). Microscope illuminator with fiber optic source integrator. J. Cell Biol., **101**, 83 abstract.
Fay, F. S., Carrington, W. and Fogarty, K. E., (1989). Three-dimensional molecular distribution in single cells analysed using the digital imaging microscope. J. Microsc., **153**, 133-149.
Francon, M., (1961). *Progress in microscopy.* pp.161. Row Peterson and Co., Evanston IL.
Howard, V., Reid, S. A., Baddeley, A. and Boyde, A., (1985). Unbiased estimation of particle density in the tandem scanning reflected light microscope. J. Microsc., **138**, 203-212.
Inoué, S., (1986). *Video microscopy.* pp. 584. Plenum, New York.
Kino, G. S., (1989). Efficiency in Nipkow disc microscopes. In: *Handbook of biological confocal microscopy.* Pawley, J. (Editor), 93-97. IMR Press, Madison WI.
Kino, G. S., Chou, C-H. and Xiao, G. Q., (1989). Imaging theory for the scanning optical microscope. In: *Scanning imaging.* Wilson, T. (Editor), Proc. S.P.I.E., **1028**, 104-113.
Lau, E., (1960). Doppelmikroskop. In: *ABC der Optik.* Brockhaus, Leipzig.
Lukosz, W., (1966). Optical systems with resolving powers exceeding the classical limit I. J. Opt. Soc. Am., **56**, 1463-1472.
Lukosz, W., (1967). Optical systems with resolving powers exceeding the classical limit II. J. Opt. Soc. Am., **57**, 932-941.
Lukosz, W. and Marchand, M., (1963). Optischen Abbildung unter Ueberschreitung der Beugungsbedingten Aufloesungsgrenze. Opt. Acta, **10**, 241-255.
Minsky, M., (1961). Microscopy Apparatus. United States Patent No. 3,013,467. Filed Nov. 7, 1957; granted Dec. 19, 1961.
Nipkow, P., (1884). Elektrisches Teleskop. Kaiserliches Patentamt, Patentschrift Nr. 30,105. Applied Jan. 6, 1884; granted Jan. 6, 1884; published Jan. 15, 1885.
Paddock, S. W., (1989). Tandem scanning reflected light microscopy of cell-substrate adhesions and stress fibres in Swiss 3T3 cells. J. Cell Sci., **94**, 143-146.
Peek, S. C., (1957). Illumin. Eng., **52**, 96.
Petráň, M. and Hadravský, M., (1966a). Zpusob a zarizeni pro omezeni rozptylu svetla v mikroskopu pro osvetleni shora. Czechoslovak Patent No. 128,936. Applied Jul. 5, 1966; granted Feb. 15, 1968; published Sept. 15, 1968.
Petráň, M. and Hadravský, M., (1966b). Zpusob a zarizeni pro zlepseni rozlisovaci schopnosti a kontrastu optickeho mikroskopu. Czechoslovak Patent No. 128,937. Applied Jul. 5, 1966; granted Feb. 15, 1968; published Sept. 15, 1968.
Petráň, M. and Hadravský, M., (1967). Method and arrangement for im-

proving the resolving power and contrast. United States Patent No. 3,517,980. Filed Dec. 4, 1967; granted Jun. 30, 1970.

Petráň, M., Hadravský, M., Benes, J., Kucera, R. and Boyde, A., (1985a). The tandem scanning reflected light microscope. Part 1 - The principle, and its design. Proc. Roy. Microsc. Soc., **20**, 125-129.

Petráň, M., Hadravský, M. and Boyde, A., (1985b). The tandem scanning reflected light microscope. Scanning, **7**, 97-108.

Petráň, M., Hadravský, M., Boyde, A. and Muller, M., (1986). Tandem scanning reflected light microscopy. In: *Science of biological specimen preparation.* SEM Inc., AMF O'Hare Il.

Petráň, M., Hadravský, M., Egger, M. D. and Galambos, R., (1968). Tandem scanning reflected light microscope. J. Opt. Soc. Am., **58**, 661-664.

Sheppard, C. J. R., (1988). Super-resolution in confocal imaging. Optik, **80**, 53-54.

Sheppard, C. J. R., (1989). Axial resolution of confocal fluorescence microscopy. J. Microsc., **154**, 237-241.

Sheppard, C. J. R. and Mao, X. Q., (1988). Confocal microscopes with slit apertures. J. Mod. Opt., **35**, 1169-1185.

Sheppard, C. J. R. and Wilson, T., (1981). The theory of the direct-view confocal microscope. J. Microsc., **124**, 107-117.

Tolansky, S., (1948). Multiple beam interferometry of surfaces and films. pp. 196. Oxford University Press, Oxford.

Wijnaendts-van-Resandt, R. W., Marsman, H. J. B., Kaplan, R., Davoust, J., Stelzer, E. H. K. and Stricker, R., (1985). Optical fluorescence microscopy in three dimensions: microtomoscopy. J. Microsc., **138**, 29-34.

Wilson, T. and Carlini, A. R., (1989). The effect of aberrations on the axial response of confocal imaging systems. J. Microsc., **154**, 243-256.

Xiao, G. Q. and Kino, G. S., (1987). A real-time confocal scanning optical microscope. In: *Scanning imaging technology.* Wilson, T. and Balk, L. (Editors), Proc. S.P.I.E., **809**, 107-113.

Young, J. Z. and Roberts, F., (1951). A flying spot microscope. Nature, **167**, 231.

10. The Confocal Microscope as an Instrument for Measuring Microstructural Geometry

V. HOWARD

I. Introduction

The advent of the confocal scanning light microscope (CSLM) has immediately opened new horizons for quantitative microscopy. The fact that its appearance has coincided with an explosion of activity in the development of a new family of unbiased stereological methods is proving to be a particularly fortuitous juxtaposition. Before CSLM, stereologists were firmly rooted in the two-dimensional world of the histological section. Admittedly, physical sections can be used serially to re-create the third dimension or in pairs to form a three-dimensional probe, as will be discussed, but with any method that necessarily destroys the specimen as part of the preparative technique the sampling opportunities are restricted, and can, for any particular region, be performed only once. In confocal microscopy we now have a non-destructive three-dimensional probe. The same specimen will now be able to be sampled on multiple occasions, either in a time series or with different geometrical probes, or indeed with the same probe from a number of orientations. This is already leading to new ideas in stereology and more will follow.

Microscopists must have always had a wish, subconscious at the very least, to be able to look "inside" objects at high resolution. For biologists in particular, the concept of being able to observe living processes in four dimensions is most exciting. By combining three-dimensional measurement techniques with "vital" CSLM there is the prospect of gaining insight into many physiological processes at the cellular level.

It is never good to overstate a case, and there are still severe technical barriers to the realisation of the aspirations outlined above. Most commercially available CSLMs are darkfield microscopes. The features of interest must therefore either reflect or fluoresce if they are to be imaged confocally. This is not always possible. Transmission CSLMs exist but are not yet readily available. They also have some drawbacks, mainly concerning the upper limit of specimen mass. However, we are clearly at the beginning of a wave of technological development which is moving very rapidly. In the rest of this chapter I will discuss the prospects for unbiased three-dimensional measurements using CSLM, in the context of recent advances in stereological thinking.

II. Dimensions and probes

Measurements can be made of geometrical properties of features in three-dimensional space by "throwing" random geometrical probes, of various dimensions, into the space and recording the way in which they intersect with structures of interest. The choice of probe will depend on the dimensions of the feature being scrutinised. Let us consider this further.

We live in a three-dimensional world and are all quite used to thinking three-dimensionally. It is only when sitting at the microscope that this habit of a lifetime will be totally disregarded by a considerable proportion of users. On a macroscopic scale, consider the room that you are sitting in to be the equivalent of a histological block and the objects therein to be features of interest. If the room was suddenly filled with a regular three-dimensional lattice of points each, say, 1cm apart, then it should be clear that they would fall within objects in strict proportion to their volume (i.e. the grandfather clock would contain more points within it than the ink pot!). We can make this statement more general by adding that neither the shape of the objects nor their orientation will have any effect on this result, only their volume.

In standard histology this imaginary feat is actually achieved in two stages. First, we cut a section; second, we apply a grid of points to the section, the latter act being rather "two-dimensional" in nature. Those who are thinking in three dimensions, however, will immediately see that this is analogous to "throwing" some points into a volume. The histological section therefore becomes the "vehicle" for, in this case, a zero-dimensional probe.

Let us now consider surfaces. If we fill our room with parallel lines travelling between one pair of opposite walls, and again arranged in a regular lattice, we can imagine them cutting the surface of objects. In fact they will be hitting the objects, in this instance, in proportion to their area projected normal to the lines. We can also see how the orientation properties of the surfaces will affect the number of times they are "hit" by the lines. A sheet of paper has a large surface-to-volume ratio but, if we held it so that the sides of the paper were perfectly parallel to the lines in our room, there would be no hits. Conversely, if it was held normal to the lines there would be a large number of hits. Therefore, if we are to use linear probes in the estimation of a surface, their orientation properties must be taken into account.

The most common geometrical probe used in microscopy is the two-dimensional section, though it is rarely thought of as such. In your room, imagine a "false ceiling" that can be moved to any position, with uniform random probability, between the floor and the real ceiling, while remaining parallel to these surfaces. For a number of random positions it is easy to imagine that the plane would hit the grandfather clock rather more frequently than the ink pot. Moreover, if the grandfather clock were to be placed on its side, it would be hit less frequently. In other words, microscopical sections hit things in proportion to their height normal to the plane of the section. Orientation is also important. When dealing with structures that have length but are not particulate (e.g. tubules in biology or pipes in your room) they will be cut in proportion to their length per unit volume. The strictures concerning orientation remain important.

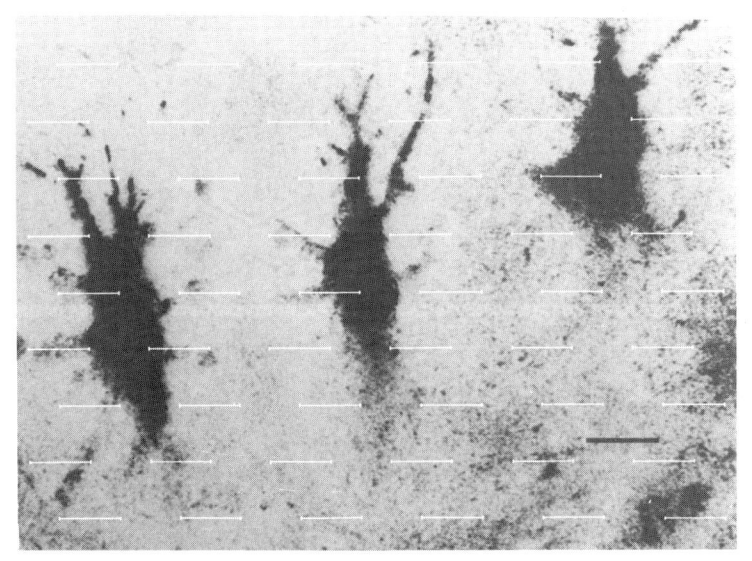

Figure 10.1. A reflectance image of osteocyte lacunae in mineralised human jaw bone imaged confocally on a Biorad MRC500 with an N.A. 1.4 oil immersion objective. The scale bar (black) is 10μm. A grid of lines with associated points has been superimposed on the image. *Volume density V_V estimation.* There are 108 points in the reference space. 11 points fall within profiles of lacunae. Therefore by Delesse's principle an unbiased estimate of $V_V = 11/108 = 10\%$. *Surface density S_V estimation.* Assume that this section is IUR. The total length of test line present in real units is 410μm. The number of intersections of test line with the boundary of lacunar profiles is 21. Therefore the number of intersections per unit length of test line $I_L = 21/410 = 0.05$ per μm. If the test system is isotropic then $S_V = 2I_L = 0.1\mu\text{m}^2/\mu\text{m}^3$.

Now consider that you wish to count the number of chairs in the room. In three dimensions, this is not difficult. Chairs are each assigned a "weight" of "1" if they are present and of "0" if they are not. Thus they are selected with uniform probability. Note that it doesn't matter about the shape of any particular chair (i.e. it isn't necessary to assume that each chair is a sphere!) nor is there any problem associated with orientation. It is easy to imagine how difficult it would be to try to estimate chair number from a single slice through the room. As stated above, chairs would be hit in proportion to size and orientation would be significant.

There are some properties that can only be measured in three dimensions. Lower-dimensional probes will give no useful information about the spatial distribution or shape of objects. Furthermore, they will provide no solutions to determining connectivity patterns. Real feature shape and size distribution measurements are possible only in three dimensions [De Hoff, 1983].

The reader may by now be wondering where all this is leading! The concepts outlined above are of fundamental importance for those considering using confocal microscopes for making geometrical measurements. In the subsequent sections I will use these ideas freely.

In summary, on the subject of dimensions, when a three-dimensional object is cut by a thin section there is a dimensional reduction, as shown below:

3-D Object	2-D Slice
Volume (L^3)	Area (L^2)
Surface (L^2)	Boundary (L^1)
Length (L^1)	Point or profile number (L^0)

Table 10.1.

If we interrogate a space with geometrical probes they will intersect with features according to their dimensionality as follows:

Probe	3-D Feature Selected	Anisotropy Sensitive?
Point (L^0)	Volume (L^3)	No
Line (L^1)	Surface (L^2)	Yes
Area (L^2)	Height or length (L^1)	Yes
Volume (L^3)	Zero-dimensional quantities (L^0)	No

Table 10.2.

Note that the dimensions of the probe and of the feature sampled sum to three, the dimension of the world in which they exist.

III. Volume estimation

This can be dealt with briefly. To estimate the volume proportion of a particular phase in a three-dimensional object with a confocal microscope, the optical section is placed at a uniform random depth within the specimen. If the image has adequate contrast, it can be segmented automatically in two dimensions by simple thresholding. The ratio of the area of the phase of interest to the area of the containing space (reference volume) is an unbiased estimate of the volume density, V_V, by Delesse's principle (1847). If automatic segmentation has been performed, this can be done by pixel counting. If this is not possible, it can be performed by simple point-counting on a grid overlaid onto the micrograph, Figure 10.1. Stable estimates can be obtained by repeated sampling with optical sections at uniform random depths. In a well-designed procedure it should never be necessary to count more than 200 points in any specimen [Gundersen and Jensen, 1987]. For an exhaustive treatment of point-counting methods, consult Weibel (1980).

The estimation of the actual volume of individual particles can be made in confocal microscopy with great ease, using Cavalieri's principle [Gundersen and Jensen, 1987]. The particle is serially sectioned in a uniform random manner from any arbitrary direction so that there are about 10 equally spaced sections through it. The product of the sum of the areas of the profiles of the particle and the distance between sections is an unbiased estimate of the particle's volume. Once again, area estimation can be performed by manual point-counting or automatic segmentation. For a full discussion of Cavalieri's estimator, its efficiency and error analysis, see Gundersen and Jensen (1987).

Clearly the above estimates are instantly available by performing image processing on the three-dimensional image directly. This will be discussed below.

IV. Number estimation

Because the moving optical section sweeps through three-dimensional space, if the reference volume is definable and small enough (e.g. a single cell) and the number of objects to be counted is not too great (e.g. the number of mitochondria within that cell) then it will be possible to count these objects directly using confocal microscopes. However, even at this stage of the discussion, warning bells should be ringing because no consideration has yet been made about how that particular cell (or any other relevant object) might have been chosen from the hundreds of others present. If it were to have been selected from a single "random" optical section, then that object itself would have been selected from a sub-population with probability proportional to its height normal to the optical section. This is not a uniform sampling scheme and would clearly have an effect on the results, for "large" objects would have a higher chance of being sampled.

On most occasions we will wish to count in systems where the reference space will be too large to count everything within it (e.g., the whole brain) or undefinable (e.g., a piece of rock). Then it will be necessary to sample with the use of an unbiased counting rule, of which there are several varieties [Gundersen, 1986].

Let us first consider an unbiased two-dimensional counting rule used commonly in microscopy, which was first proposed by Gundersen (1977). The problem is to unbiasedly "capture" a number of two-dimensional profiles with a counting grid, as illustrated in Figure 10.2. It must be possible to move the grid in a two-dimensional tesselation across the field without counting any profile more than once. By using the "forbidden line" as illustrated, this is easily achieved.

For counting in three dimensions the two-dimensional tiling rule can be extended to a three-dimensional "bricking" rule [Gundersen, 1981; Howard et al., 1985]. You will recall that a random two-dimensional probe, like a histological section, hits things in proportion to their height. If we want to count or make measurements on particles sampled with uniform probability, it is essential to use a three-dimensional probe. In the case of the "room" cited above we only attempted to count things wholly within the three-dimensional probe. In actual microscopy the probe, or "brick", will be positioned within the specimen randomly and some particles will only partially intersect the brick (in the room analogy this would be similar to the position

Measuring Microstructural Geometry

of the steam engine in Salvador Dali's painting 'Time transfixed').

In the first place, we can state that particles can be sampled with uniform probability by simply using a pair of planes, an unknown distance apart [Cruz-Orive, 1987]. The rule that is used was first proposed by Sterio (1984) and consists of sampling particles, profiles of which appear in one of the planes and not the other. By sampling in this way, particles will only be included in the sample if their "top" is in the space between the planes. As each particle has only one "top", in one of the two directions normal to the sections, then the sampling scheme must be uniform. In the confocal microscope, this will be equivalent to placing the optical section at a random height within the specimen, discarding all particles giving rise to profiles appearing at that height, racking the optical section through the specimen a certain distance and only sampling particles which have newly appeared during the racking of the optical section. These particles will have been sampled uniformly, and it is of fundamental importance that

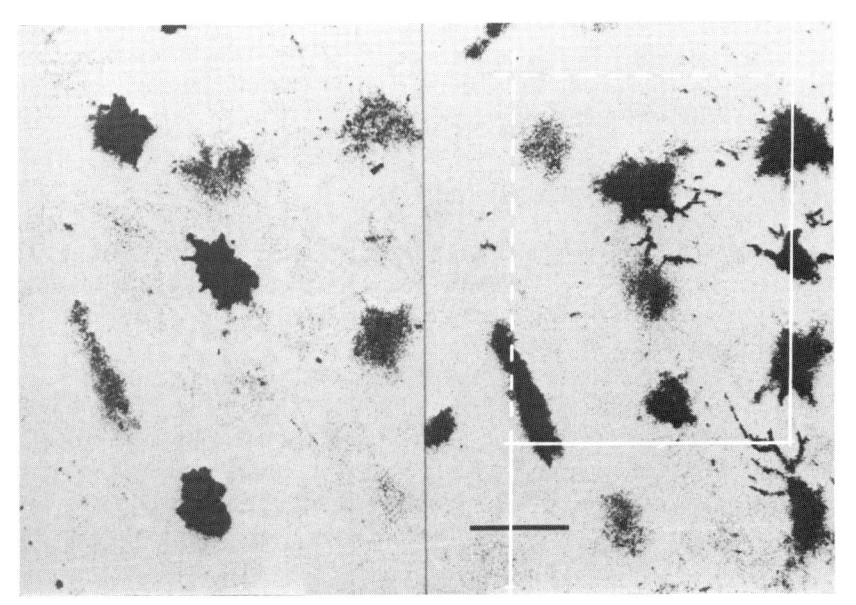

Figure 10.2. Two images similar to that in Figure 10.1. The scale bar (black) is 25μm. On the right hand image is a two-dimensional unbiased counting frame. Profiles are selected provided that they intersect the quadrant but not the fully drawn "forbidden" line. Therefore, in this case 5 profiles would be rejected and 4 would be selected. The left hand image is separated in the z-axis from the right hand one by 6μm. Of the 4 profiles selected on the right, only one (the lowest) does *not* appear on the left, and therefore the "disector" count, Q^-, in this example is 1. The area of the quadrant in real units is 8850μm^2. Therefore the volume of the "disector" $V_{dis} = 8850 \times 6 = 53100\mu$m^3. Therefore an unbiased estimate of the numerical density $N_V = 1/53100 = 1.9 \times 10^{-5}\mu$m^{-3}.

those wishing to use confocal microscopes to make objective and unbiased measurements realise that they cannot simply put the optical section at a random level in the specimen and start measuring. Those objects would of course have been selected from the height-weighted distribution of size.

For actual estimates of numerical density, N_V (number per unit volume) the "disector" [Sterio, 1984] can be used very efficiently. In this case the distance between the pair of planes must be known. Profiles are selected on one of the planes, called the reference plane, using a two-dimensional sampling frame and then counted if they *don't* appear on the other "look up" plane, Figure 10.2.

To progress from a density estimate to one of total particle number, it is necessary to know the total volume in which the population resides. This can be performed using Cavalieri's principle. For a lucid example of the use of the "disector/Cavalieri" combination to estimate total numbers of neurons in brain nuclei, see Pakkenberg and Gundersen (1988).

This is a convenient time to point out that the optical sectioning properties of high N.A. oil immersion objectives can be used for disector counting in conventional microscopes [Braendgaard *et al.*, 1990] as well as confocal machines.

V. Surface and length estimation

Consider briefly the surface of a sphere. It has rather special properties because it is isotropic. What this means is that if we take a random element on the surface, the normal to that element will have a uniform probability of pointing in any direction over a solid angle of 4π. Let us imagine that we have a globe of the world in our "room". The parallel lines running across the room will hit the globe with the same probability, no matter what the orientation of the globe, because it always presents the same aspect from any random direction. Under such circumstances the probability of a line intersecting a surface is related purely to the size of the globe and the density of lines in space. It has been proved that [Weibel, 1980]

$$S_V = 2\, I_L \qquad (10.1)$$

where S_V is the surface per unit volume (surface density) and I_L the number of intersections per unit length of test line with the surface. For perfectly anisotropic surfaces (i.e. where all surfaces are parallel) the numerical constant 2 changes to 1. In reality, we usually have partial anisotropy of an unknown degree. Under such conditions it is safer to invert the approach to this problem and make no assumptions about the orientation properties of the surface under scrutiny if we are to ensure unbiasedness in our estimations. By this I mean that we should make the linear probe isotropic in Euclidean space. Therefore, just as we could use lines of arbitrary orientation to measure the isotropic surface of the sphere, we can employ isotropic lines to unbiasedly estimate surfaces of arbitrary orientation. This is called a design-based approach.

Isotropic lines in space can be generated in two ways in microscopy. Isotropic uniform random (IUR) planes, in other words sections, can be taken and used as a vehicle for lines with a uniform random orientation in the plane between 0 and π. IUR planes in space can be defined by the orientation of their normals. These normals can be created as directions in space by the creation of pairs of random angles, see for example Baddley *et al.* (1986).

It is easy to see how IUR optical sections might be sampled in a confocal microscope with the use of a universal goniometer stage. However, major practical problems abound. The short working distances of high N.A. objective lenses means that for all but the smallest of specimens, tilting would only be possible to a limited extent, and many orientations would therefore not be available to sample. Possibly a more promising approach is to collect the binarised three-dimensional image in a computer and then "interrogate" it with IUR linear probes in software (*vide infra*). There is another problem here, though, which is that the axial resolution of the confocal microscope is at least three times worse than lateral resolution. For unbiased estimation of metric quantities, equiaxial resolution is desirable, though not absolutely essential. It could be obtained from suitably merged tilt series of mutually orthogonal stacks of optical sections. However, this raises the same problems of specimen size mentioned above, and we find ourselves in something of a circular argument. Improvement of the quality of the captured three-dimensional image is something which is justifiably the subject of intensive research at present.

I mentioned above that there was a second method of creating IUR lines in space, which is by the use of Baddeley's "vertical sections". Because of poor z-axis resolution problems, it is clearly advantageous at present to view boundaries (i.e., the one-dimensional trace of two-dimensional surfaces in the section) in x-y rather than any other (synthetic) projection. "Vertical sections" probably provide us with the best approach at present. Baddeley (1984) proposes the production of IUR lines in space as a two-step procedure. First, sections are taken with one uniform random degree of freedom to rotate about a vertical axis which is normal to a defined (arbitrary) horizontal plane. This now qualifies as a "vertical section" and the surface can be unbiasedly estimated by counting intersections of boundaries with lines whose length are proportional to the sine of a uniform random angle defining its orientation to the "vertical" direction (which must be known). This can be very easily achieved by using a cycloidal test system. For a lucid explanation of "vertical sections" the reader is referred to Baddeley et al. (1986). Vertical sections could be sampled, for example, by fixing the specimen on a needle at right angles to the optical axis of the microscope. The needle must be free to rotate around its own longitudinal axis, and then, for uniform random rotations around that axis, optical sections could be taken, each one qualifying as a "vertical section". After intersection-counting with a cycloidal test system, the formula given above can be used for the estimation of surface density, Figure 10.3.

A new and powerful surface estimator has been described by Sandau (1987) using the "spatial grid". An application of this method to the measurement of surfaces of individual osteocyte lacunae has been given by Howard and Sandau (1990).

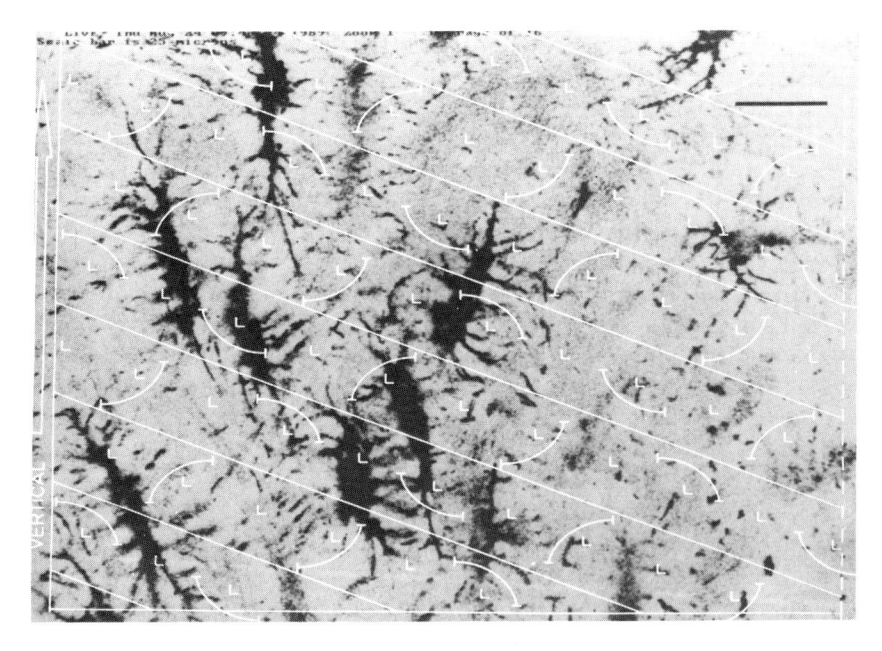

Figure 10.3. An image similar to that in Figure 10.1. The scale bar (black) is 25μm. This field has been sampled as a "vertical" section and has a cycloidal test system superimposed. The vertical direction is indicated. *Surface density estimation.* The total length of cycloidal arc in the reference space in real units is 790μm. The number of intersections of test arc with the boundaries of spaces is 53. Therefore the number of intersections per unit length of test line $I_L = 53/790 = 0.067\mu\text{m}^{-1}$. Under this sampling regime $S_V = 2I_L = 0.13\mu\text{m}^2/\mu\text{m}^3$.

Length cannot be unbiasedly measured on "vertical sections" and IUR sections are, therefore, the only option. If linear probes can be used to estimate surface, then *ipso facto* surfaces (i.e. planes) can be used to estimate the metrical properties of linear structures. If we section a structure with IUR planes, then the length of a structure such as a tubule or filament per unit volume (length density, L_V) will be proportional in some way to the number of times that structure cuts the plane. This is measured as the density of profiles of the structure per unit area of section (Q_A). The formula is directly analogous to that for surface [Weibel, 1980].

$$L_V = 2\,Q_A \qquad (10.2)$$

where the numerical constant 2 assumes isotropy of either the orientation distribution of the structure or of the sections. Unbiased sampling of profiles in two dimensions is performed with a tile [Gundersen, 1977] as for the disector.

If the structure is highly anisotropic — for example, a highly orientated tissue like muscle — then the results will be very sensitive to even slight departures from IUR sampling. A method of minimising the sampling variance (and therefore the amount of work required!) is to sample with ORTRIPS, **O**rthogonal **TRI**iplet **P**robe**S**, as proposed by Mattfeldt *et al.* (1985).

A very new idea for measuring length in projections through "vertical slabs" has just been presented by Gohkale (1990). This is potentially a very powerful technique in confocal microscopy.

VI. Estimation of particle size

It is possible to measure mean particle size indirectly using a combination of the stereological size and number estimators. For example, the mean volume of particles could be calculated from estimates of V_V and N_V

$$\bar{v}_N = V_V \,/\, N_V \qquad (10.3)$$

where $\bar{v}_N$ is the mean particle volume in the number-weighted distribution. Similarly

$$\bar{s}_N = S_V \,/\, N_V \qquad (10.4)$$

where $\bar{s}_N$ is the mean particle surface in the number-weighted distribution. What is meant by *number-weighted distribution*? First we must consider the process by which we *select* particles on which we will subsequently make *measurements*, for it is always a two-stage process. Furthermore, it is important to note that the two processes, selection and measurement, should be *independent* in a statistical sense.

We have mentioned above that by using probes of different dimensions we can hit particles with different probabilities. On a group of particles sampled from any one of these four weighted distributions, we could, in theory at least, consider marking measurements of size. We are all accustomed to measuring, living as we do in three-dimensional space, such quantities as length, surface and volume. Thus it is possible to describe 12 measures of size, as shown in Table 10.3 [Gundersen, 1986].

Although in model-based techniques linear measurements of size have been the norm, e.g., mean particle "diameter" of spheroids, it is difficult to imagine the functional meaning of such measures in a biological setting. Volume and surface on the other hand have some fairly obvious functional correlates. Volume is a measure of "biomass" which is related to such activities as protein production, total oxidative capacity, etc. Surface is correlated with transport phenomena, etc.

Weighted Distribution	Size Measure		
	Height	Surface	Volume
Number	h_N	s_N	v_N
Height	h_H	s_H	v_H
Surface	h_S	s_S	v_S
Volume	h_V	s_V	v_V

Table 10.3.

The three number-weighted measures are all estimable by indirect methods, as described above. We must now consider direct methods of measuring particle size. The first, and most obvious, is to select particles on the "disector" principle and then measure the volume of each particle thus sampled by Cavalieri, which will lead to a direct estimate of v_N and also provide a measure of size distribution. This approach requires a knowledge of shrinkage. It should be noted that such a method implemented on particles sampled with single sections would lead to an estimate of $\bar{v}_H$.

There exists an extremely efficient direct method for measuring $\bar{v}_V$, the particle mean volume-weighted volume, from single sections of IUR or "vertical" orientation. This has been developed by Gundersen and Jensen (1985). The technique usually requires between 50 and 100 measurements and takes less than 10 minutes per sampling unit when the micrographs are to hand.

The volume-weighted distribution of size is otherwise known as the "sieving" distribution. The first step is to sample particles in proportion to their volume. This is done by "throwing" points into three-dimensional space and only measuring a particle if it is hit by a point. The second step is to measure the length of an isotropic line passing through the sampling point between the intersections with the boundary of the particle.

In practice, this may be achieved by taking an IUR section and placing a grid of points associated with lines onto the micrograph. The direction of the lines must be uniformly random in direction about π to complete the process of creating an IUR line in space. Given IUR sections, this is a staightforward procedure. The intercept length, l_0 (for it is measured with the use of a zero-dimensional probe) is raised to the power of 3. Gundersen and Jensen's (1985) beautifully simple finding (the result of very complex integral geometry analyses) is that the mean of these cubed intercept lengths multiplied by $\pi/3$ is an unbiased estimator of $\bar{v}_V$, irrespective of particle shape, orientation or spatial distribution.

$$\bar{v}_V = \pi/3 \; \bar{l}_0^3 \qquad (10.5)$$

For efficient methods of measurement, see the above papers as well as Braendgaard and Gundersen (1986).

Because this measurement can be made on single IUR or "vertical" optical sections, it is becoming widely used. The majority of studies so far have been at the LM level [Nielsen et al., 1986; Howard, 1985] in the field of diagnostic histopathology. The method can be implemented in "vertical" sections, though clearly it will not be possible to use cycloid arcs to measure intercept length. In this case we have to weight the orientations of straight lines by the sine of the angle from the vertical using an orientating frame attributed to Gundersen and illustrated by Howard (1985) and Cruz-Orive et al. (1986). A low-cost apparatus is available for capturing images and applying test grids at sine-weighted angles by computer graphics [Moss et al., 1989], which assures that the design of the sampling is adequately controlled.

Another interesting development is the "nucleator" [Gundersen, 1988]. Instead of taking a random point within the particle, as in the method described above, it is possible to take a uniform random point within a uniquely definable subspace of the particle — e.g., the nucleolus of a cell. This has to be selected with uniform probability using the "disector" principle. The distance to cell boundary is then measured in an independent direction. This distance raised to the third power and multiplied by $4\pi/3$ is an unbiased estimate of v_N. Because in a cell the nucleus or nucleolus is usually near the centre, this means that the cell membranes are cut non-obliquely, which greatly aids identification. Furthermore it means that thin "caps" are not missed. The "nucleator" is in fact a more general statement following on from the "selector" principle described by Cruz-Orive (1987). Space in this chapter does not allow a full description of these techniques, but the two immediately abovementioned papers are strongly recommended. Nucleator sampling will also lead to information on particle number and spatial distribution in certain situations [Evans and Gundersen, 1990].

VII. Other zero-dimensional quantities

Most previous efforts in the field of spatial statistics have been directed toward two-dimensional problems, see for example Diggle (1983). The reason that three-dimensional studies have previously been unknown is largely because three-dimensional data sets have not been available, as the effort required for their collection was prohibitive. In microscopy at least, confocal machines are rapidly changing this state of affairs.

Baddeley *et al.* (1987) have published a report on the spatial distribution of osteocyte lacunae in bone using a three-dimensional k-function and demonstrated a hard core packing. This will clearly be the first of many such papers on the subject.

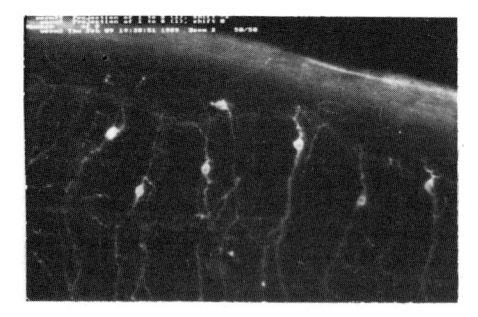

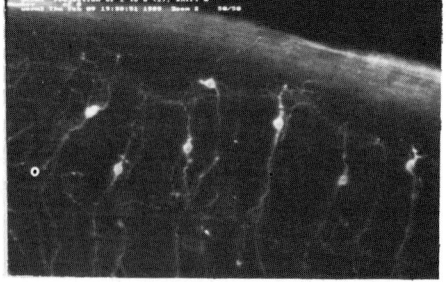

Figure 10.4. Stereopair image of serotonergic neurons immunofluorescently stained with a standard anti 5-HT streptavidin/boitin/FITC regime in an intact specimen of *Grillotia erinaceus*, a shark parasite. The field width is 300μm and the depth of field 80μm. The specimen was viewed epiflourescently with a 20X 0.75 N.A. air objective. The connectivity of the network is clearly visible. (This work was performed in collaboration with Dr. George McKerr, University of Ulster.)

Another zero-dimensional property of immense interest is connectivity. In biology, the connectivity of neurons in the central nervous system is likely to be a key parameter. In the field of petrology, connectivity of pore spaces in sandstone rocks is of major interest. Figure 10.4 shows how useful confocal microscopes will be in providing connectivity data, which can only be seriously approached in three dimensions.

VIII. Future developments

A number of research groups and commercial concerns are now developing image-processing algorithms which operate directly on the three-dimensional data set. As with image processing in two dimensions, scene segmentation is always a major hurdle. However, confocal microscopes give high contrast images which should make their output more amenable to image analysis than those from conventional light microscopes. Many of the techniques described above require manual counting methods using grids of different dimensions because there is no satisfactory alternative at present. However, these stereological techniques are simply waiting to be combined with image-processing algorithms, as has already happened in this laboratory.

For image display purposes the whole data set is required, which puts great pressure on mass storage devices. For the purposes of measurement, data compression of many thousandfold can be achieved, reducing the requirement for mass storage capacity though retaining the need for adequate temporary buffer memory.

Just as digitiser tablets were (and still are, unfortunately!) used very inefficiently in two dimensions to measure profile area and boundary, giving very biased measures of quantities such as surface-to-volume ratio, there is little doubt that similarly unreliable measurements will be presented to the consumer as the three dimensional image-processing movement becomes a headlong rush amongst the commercial image-analysis software companies. For example, surface can only be estimated unbiasedly with the use of lower-dimensional linear probes. Yet "tiling" algorithms are being offered as the solution to this problem, even though tiling will invariably underestimate surface area, sometimes severely.

As an indication of the probable direction of future events, I will conclude with an example of the work of Stephane Gesbert, a student from the Paris School of Mines who did a three-month final-year project in this laboratory in collaboration with Jean Serra and Dominique Jeulin under the COMET scheme. His task was to take a three-dimensional binary data set from the Biorad MRC500 microscope and develop image transformations which operate in three-dimensions. Two-dimensional image processing machines generally work on digitisations achieved with a lattice of quadratic or hexagonal form. In three dimensions a "digital ball" such as the centre-faced-cube or rhombododecahedron can be employed, the latter giving seven independent directions in space from the central voxel. Thus nine basic structuring elements are available as tools for the construction of three-dimensional "mathematical morphology" [Serra, 1982] algorithms. During his project, Gesbert wrote programs for performing dilation, erosion, opening, closing, labelling and the distance and frontier functions all working in three dimensions on binarised images.

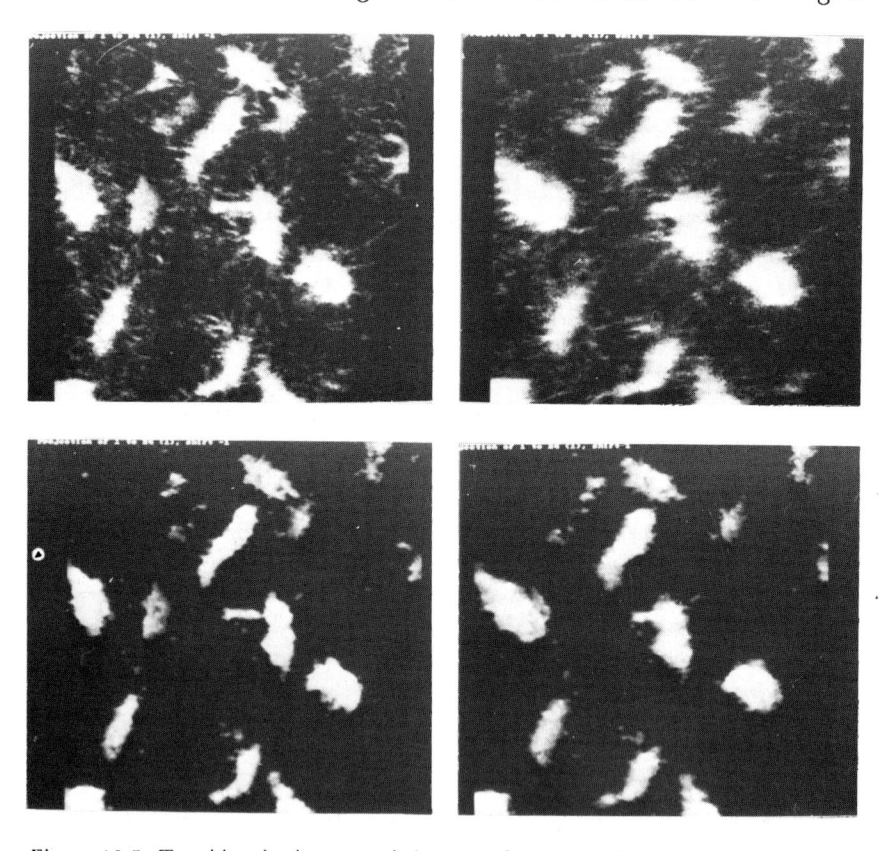

Figure 10.5. Two binarised stereopair images of osteocyte lacunae shown in reverse relief to those in Figures 10.1-10.3. The initial image was obtained in reflectance as a stack of 24 serial optical sections. The field width is 300μm and the field depth 100μm. The image was digitised using a rhombododecahedral structuring element. The upper image is unfiltered after binarisation. Many canaliculi can be seen. The lower image has undergone a three-dimensional opening and this has achieved a segmentation of the lacunar spaces in three dimensions. (This work was performed in collaboration with S. Gesbert, D. Jeulin and J. Serra of the Ecole des Mines de Paris.)

With the distance and frontier functions, measures of particle number and their individual volumes and surfaces become available. Furthermore, measures of spatial distribution can easily be computed. Figure 10.5 shows stereopair binary images of osteocyte lacunae before and after a three-dimensional "opening".

Confocal scanning light microscopes are truly three-dimensional microscopes. We should, therefore, regard purely two-dimensional measurements made in the plane to be a thing of the past. These machines can be used most efficiently to make stereological measurements, where the three-dimensional quantity has statistical properties. There is now major activity underway to develop software that will make direct measurements on digitised three-dimensional images produced by confocal machines. These are exciting times for those who use microscopes to measure!

References

Baddeley, A. J., (1984). Vertical sections. Stochastic geometry, geometric statistics, stereology. Ambartzumian, R.V. and Weil, W. (Editors). pp. 43-52. Proc. Conf. at Oberwolfach, 1983. Teubner Texte zur Mathematik, nr 65, Teubner, Leipzig.

Baddeley, A. J., Gundersen, H. J. G. and Cruz-Orive, L. M., (1986). Estimation of surface area from vertical sections. J. Microsc., **142**, 259-276.

Baddeley, A. J., Howard, C. V., Boyde, A. and Reid, S., (1987). Three dimensional analysis of the spatial distribution of particles using the tandem scanning reflected light microscope. Acta Stereol., **6**, 87-100.

Braengaard, H., Evans, S. M., Howard, C. V. and Gundersen, H. J. G., (1990). The total number of neurons in the human neocortex unbiasedly estimated using optical disectors. J. Microsc., **157**, 285-304.

Braendgaard, H. and Gundersen, H. J. G., (1986). The impact of recent stereological advances on quantitative studies of the nervous system. J. Neurosci. Meth., **18**, 39-78.

Cruz-Orive, L. M., (1987). Particle number can be estimated using a disector of unknown thickness: the selector. J. Microsc., **145**, 121-142.

Cruz-Orive, L. M. and Hunziker, E. B., (1986). Stereology for anisotropic cells: application to growth cartilage. J. Microsc., **143**, 47-80.

De Hoff, R. T., (1983). Quantitative serial sectioning analysis: preview. J. Microsc., **131**, 259-263.

Delesse, M. A., (1847). Procede mechanique pour determine la composition des roches. C. R. Acad. Sci. (Paris), **25**, 544.

Diggle, P. J., (1983). *Statistical analysis of spatial point patterns.* Academic Press, New York.

Evans, S. M. and Gundersen, H. J. G., (1990). Unbiased and general stereological estimators of three-dimensional distributions of distances and densities using the nucleator. To be published.

Ghokale, A., (1990). Unbiased estimation of curve length in 3-D using vertical slices. J. Microsc., **159**. In press.

Gundersen, H. J. G., (1977). Notes on the estimation of the numerical density of arbitrary profiles: the edge effect. J. Microsc., **111**, 219-223.

Gundersen, H. J. G., (1981). Stereologi: eller hvordan tal for rumlig form og indhold opnas ved iagttagelse af structureer pa snitplaner. Laegeforeningens Forlag, Copenhagen.

Gundersen, H. J. G., (1986). Stereology of arbitrary particles. A review of unbiased number and size estimators and the presentation of some new ones, in memory of William R. Thompson. J. Microsc., **143**, 3-45.

Gundersen, H. J. G., (1988). The nucleator. J. Microsc., **151**, 3-21.

Gundersen, H. J. G. and Jensen, E. B., (1985). Stereological estimation of the volume-weighted mean volume of arbitrary particles observed on random sections. J. Microsc., **138**, 127-142.

Gundersen, H. J. G. and Jensen, E. B., (1987). The efficiency of systematic sampling in stereology and its prediction. J. Microsc., **147**, 229-263.

Howard, V., (1985). Rapid nuclear volume estimation in malignant melanoma using point sampled intercepts in vertical sections: a review of the unbiased quantitative analysis of particles of arbitrary shape. In: *Quantitative image analysis in cancer cytology and histology.* Mary, J.Y. and Rigaut, J.P. (Editors). pp. 245-254. Elsevier Science Publishers, New York.

Howard, V., Reid, S., Baddeley, A. J. and Boyde, A., (1985). Unbiased estimation of particle density in the tandem scanning reflected light microscope. J. Microsc., **138**, 203-212.

Howard, V. and Sandau, K., (1990). Measuring the surface area of a cell by the method of the "spatial grid" with a CSLM – an application. J. Microsc.

Mattfeldt, T., Mobius, H-J. and Mall, G., (1985). Orthogonal triplet probes: An efficient method for unbiased estimation of length and surface of objects with unknown orientation in space. J. Microsc., **139**, 279-289.

Moss, M., Browne, M., Howard, V. and Joyner, D., (1989). An interactive image analysis system for mean particle volume estimation using stereological principles. J. Microsc., **156**, 79-90.

Nielsen, K., Clostrup, H., Nilsson, T. and Gundersen, H. J. G., (1986). Stereological estimates of nuclear volume correlated with histopathological grading and prognosis of bladder tumour. Virchows Arch. Cell Pathol., **52**, 41-54.

Pakkenberg, B. and Gundersen, H. J. G., (1988). Total number of neurons and glial cells in human brain nuclei estimated by the disector and the fractionator. J. Microsc., **150**, 1-20.

Sandau, K., (1987). How to estimate the area of a surface using a spatial grid. Acta Stereol., **6**, 31-36.

Serra, J., (1982). *Image analysis and mathematical morphology.* Academic Press, New York.

Sterio, D. C., (1984). The unbiased estimation of number and sizes of arbitrary particles using the disector. J. Microsc., **134**, 127-136.

Weibel, E. R., (1980). *Stereological methods. Practical methods for biological morphometry. Vol. 1.* Academic Press, New York.

11. Confocal Microscopy of Ocular Tissue

B. R. MASTERS

I. Introduction

Confocal microscopy has enhanced our ability to image the unstained, living, *in vivo* eye. The application of this technology permits the acquisition of images of high spatial resolution and contrast as compared to conventional microscopy. This chapter addresses the following topics:

(1) The development of optically sectioning microscopes in ophthalmology.

(2) The current applications of confocal microscopy in ophthalmology.

(3) The development of a real-time confocal system for clinical ophthalmology.

(4) The future applications of confocal microscopy in ophthalmology.

While there are numerous applications of confocal microscopy in cell biology, anatomy, and solid state physics, it is in clinical ophthalmology that major diagnostic applications will be developed in the next few years.

II. The structure and optical properties of the eye

In order to understand the problems associated with confocal imaging of ocular tissue, it is imperative to elucidate its unique structure and optical properties. The human eye has evolved into a wonderful light detection system: it has the capability of detecting single photons of visible light, and can produce images from a wide dynamic range of intensities. However, the unique structural and functional properties of the eye as an optical system lead to great difficulty when one attempts to image the component parts of the eye. The important properties of transparency [McCally and Farrell, 1989] and limited scatter of visible light, the properties that permit us to see the external world, present a great hindrance when one attempts to use visible light to image the eye. The optical problem is how to image a transparent or semi-transparent object with very little intrinsic contrast.

In order to image the eye, it is necessary to consider the optical and structural properties of the entire tissue from the tear film to the retina. Each of the layers of the eye presents different structures with unique optical properties. The light that enters the eye is transmitted across the tear film, the cornea, the aqueous humor, the ocular lens and the vitreous humor, and is finally blocked at the rear of the eye after traversing the thickness of the retina. With the exception of the rear of the retina, all of these tissues and fluids have two major optical properties: transparency due to little scatter of visible light, and low contrast. In order to visualize cellular structure within the eye there are two requirements: transparency, to permit the light to enter the eye and to permit the reflected light to leave the eye, and enhanced contrast to permit image formation.

The structure of the eye is uniquely adapted for its function: light collection, image formation and detection. The cornea and the ocular lens together with the ocular humors present the minimum contrast in addition to high transparency. The retina exhibits little transparency and enhanced contrast owing to its complex network of neurons and blood vessels.

III. Development of optically sectioning microscopes in ophthalmology

The unique optical properties of the cornea, the ocular lens, and the retina resulted in the development of distinct ophthalmic imaging devices. The slit lamp (biomicroscope) was developed to produce an optical section across the 0.5mm thickness of the human cornea. Endothelial cells comprising a 6-micron layer at the inner layer of the cornea could be visualized by specular reflection. The ocular lens can be imaged with the development of the Scheimpflug camera. The direct and the indirect ophthalmoscopes are used to examine the retina. The fundus camera is a specialized optical instrument which is used in clinical examination of the fundus. The recent development of the laser scanning ophthalmoscope resulted in the development of a new real-time clinical tool to visualize the surface of the retina. The application of video imaging together with digital image processing of ophthalmic images permits enhancement, quantitation, and mass storage of ocular images which was previously impossible.

The structure of the cornea [Klyce and Beuerman, 1988] presents a challenge to the development of optically sectioning microscopes. The transparent, low-contrast structure possesses remarkable anatomical symmetry. A tear film (7 microns) coats the surface of the cornea and provides an optically smooth surface. Directly under the tear film is the epithelium which is composed of 5-6 layers of cells. The epithelial layer is usually divided into three regions: the superficial cell layer, the wing cell layer, and the basal cell layer. The innermost layer is the basal epithelium where cellular mitosis occurs. These cells eventually migrate to the corneal surface and slough off into the tear film. Under the basal epithelial cells is a basement membrane and Bowman's membrane which separates the corneal epithelium from the stromal region. Bowman's membrane is present in the human cornea but not in the cornea of the rabbit. The stroma contains many keratocytes and nerve fibers in the anterior region. Between the stroma and the endothelial cells is Descemet's membrane. The innermost region of the cornea is composed of the 6-micron thick layer of endothelial cells which are critical for the maintenance of corneal transparency.

In order to image this remarkable transparent structure, several instruments were developed to produce the specular image from the regions of the cornea with large differences of refractive index, i.e., the interface between the endothelial cells and the aqueous humor. Maurice has the credit for the development of the specular microscope which is based on specular reflection of light from the cornea [Maurice, 1968]. While the initial development was for a laboratory specular microscope, several other investigators have modified the specular microscope for clinical diagnostics [Laing, 1980]. Today the specular microscope is an important clinical tool to evaluate endothelial morphology. The optical principle is as follows. A slit of light is projected through one half of the microscope objective and the specularly reflected light is collected through the other half of the objective. This development permitted the imaging of the corneal endothelial interface with the aqueous humor as well as surface epithelial cells and the interior of the stroma.

A problem arose from the fact that the depth of the optical section was a function of the slit height. If the slit height was reduced, an optical section with a smaller depth of field resulted but the number of cells in the field of view was dramatically reduced. Several approaches have been developed to overcome this problem.

Maurice developed a scanning slit microscope (SSM) which employed a second slit which was parafocal with the first slit (confocal design) [Maurice, 1974]. The film in a film carrier was transported in synchrony with the lateral motion of the eye under the microscope objective [Gallagher and Maurice, 1977]. With the use of a 3-micron slit under the film plane, high-resolution and high-contrast images were obtained on enucleated eyes. The objective was a Nikon 40X water imersion lens with an N.A. of 0.75. The images showed the surface epithelial cells, nuclei of stromal keratocytes deep within the cornea and the interface of the corneal endothelial cells with Descemet's membrane. A typical image is shown in Figure 11.1. The quality of the images obtained with this microscope rivals those obtained today with the latest confocal microscopes. The major limitations were the time required to obtain the image and the fact that it could only be used on enucleated eyes.

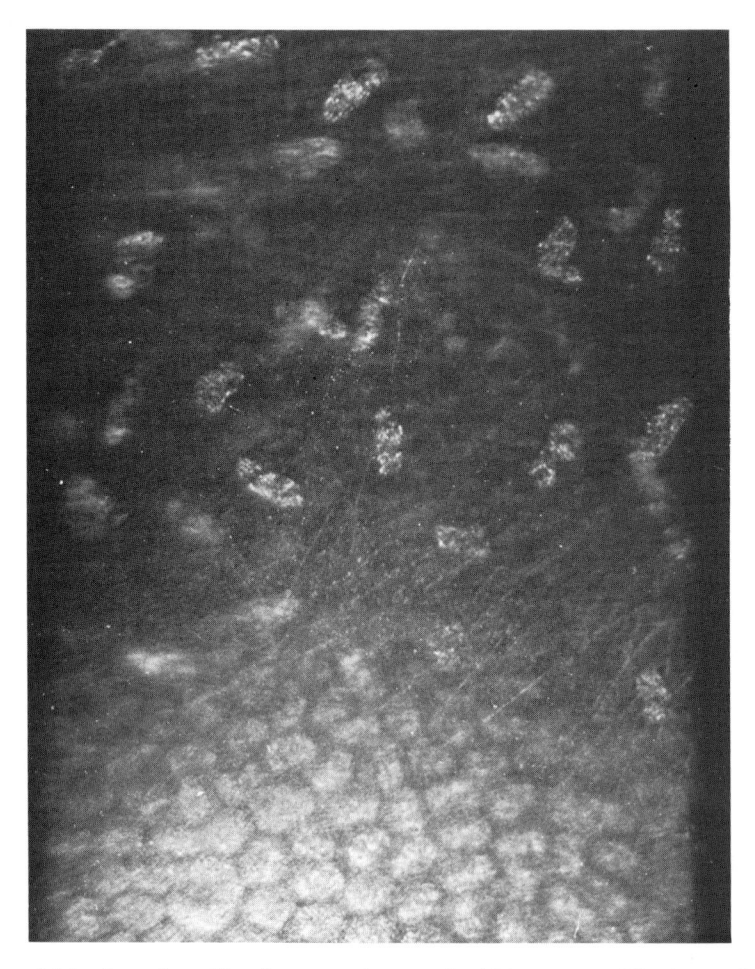

Figure 11.1. Scanning slit microscope image of rabbit cornea. The image is an oblique view through the posterior corneal surface. The lower region shows the distinct cell boundaries of the interface between the endothelial cells and the aqueous humor. The middle region shows the diffuse cell borders of the endothelial cells adjacent to Descemet's membrane, the filaments being assocaiated with Descemet's membrane. The upper region shows the nuclei of the stromal keratocytes. This image represents a montage of thin optical sections in the film plane. Figure provided by Dr. D. Maurice, Department of Ophthalmology, Stanford University.

Several recent developments in clinical wide-field specular microscopy have provided real-time instruments that can image endothelial cells in a field of one square millimeter. A wide-field scanning specular microscope was developed by Koester. The narrow optical section was obtained with the use of two narrow slits and a vibrating prism which rapidly scans the image of the slit over the object [Koester, 1980; Koester et al., 1980]. This instrument has been used on enucleated eyes to image the corneal epithelium, the stroma, and the endothelium, and has been developed into a clinical optically-sectioning confocal microscope. The present limitation is the low numerial aperture of the dipping cone (0.3 N.A.). Another wide-field specular microscope has been developed and marketed as the Keeler-Konan wide-field specular microscope. It is a non-scanning system and uses a set of graded refractive index materials to match the refractive index of the cornea and the instrument optics, thereby minimizing the strong specular reflection from the tear/air interface. Both of these instruments produce images of the endothelial cell patterns from human eyes [Mayer, 1984]. They have not been fully exploited as optically sectioning systems to image thin optical sections in the eye [Mizushima, 1988].

IV. Applications of confocal microscopy to ophthalmology

The developments of optically-sectioning confocal microscopes provided investigators with a tool to image living tissue in real time. This technique could provide high-resolution, high-contrast images from ocular tissue without the use of stains or fixatives, and would thereby reduce the artefacts which make interpreting the images difficult. The tandem scanning reflected light microscope was used by Petráň et al. to image the cornea (epithelium, stroma, and endothelium), the ocular lens and the layers of the retina. A similar microscope modified to operate in a horizontal mode has been used to image the full-thickness cornea [Lemp et al., 1986; Jester et al., 1988, 1989]. These authors imaged exfoliating epithelial cells, the surface of Bowman's membrane, interdigitating collagen plate in the stroma, stromal keratocytes and the endothelial cells of rabbit and human enucleated eyes.

In addition to these confocal microscopes, which are based on the Nipkow disk (tandem scanning), there are new developments which are being used to image ocular tissue. A no-moving-parts video-rate laser beam-scanning confocal reflected/transmission microscope has been developed by Goldstein. The operation principle is based on acousto-optic beam deflectors to scan the object (the eye). The image is produced on an image dissector tube which is raster-scanned coincident with the congruent image of the laser beam incident on the object [Goldstein, 1989].

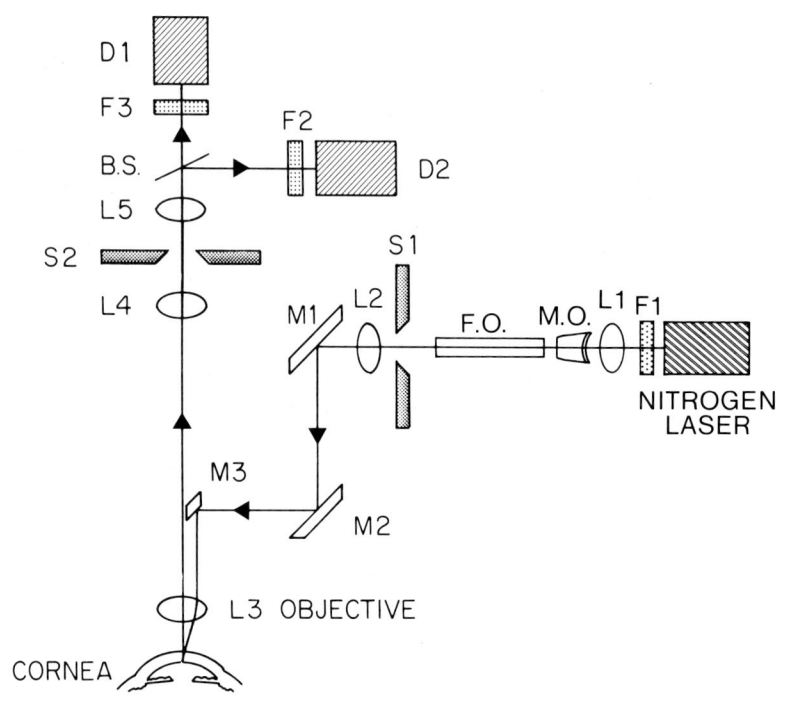

Figure 11.2. Schematic diagram of the scanning optical sectioning microscope showing a light ray path. The source is either a laser or a mercury arc lamp connected to the slit by a quartz fiber optic. F1 and F2 are narrow-band interference filters to isolate the excitation wavelengths. F3 is a narrow-band interference filter to isolate the emission light. The conjugate slits are labelled S1 and S2. M1, M2 and M3 are front surface mirrors, and B.S. is a quartz beam splitter. L3 is the scanning objective 50X, N.A. 1.00. The piezo-electric device scans the objective along the optic axis of the eye. A drop of bicarbonate Ringer's is the optical coupling fluid between the surface of the cornea and the objective.

A confocal scanning optical microscope (CSOM) was developed by Xiao and Kino using a Nipkow disk system in which the light rays illuminating the specimen and the light rays from the specimen pass through the same holes [Xiao and Kino, 1987; Xiao, Corle and Kino, 1988]. This greatly simplifies the optical system of the microscope as compared to the tandem scanning microscope developed by Petráň and Hadravský. This is a real-time confocal system and has been applied to image ocular tissue.

The examples discussed in this section use enucleated eyes. The next section will discuss applications of confocal microscopy to human ocular imaging. An optically sectioning fluorescence microscope was developed by Masters for use in biophysical studies of corneal metabolism [Masters, 1984, 1988, 1989]. The microscope measures the fluorescence and the reflectance of optical sections along the optical axis of the cornea. The fluorescence is from the naturally occurring reduced pyridine nucleotides in the mitochondria of the cells in the corneal epithelium and the corneal endothelium. The technique of redox fluorometry is a noninvasive optical method to assess the oxidative metabolism of the cells in the cornea [Masters, 1984]. The instrumentation is shown in Figure 11.2. The basic instrument is a modification of the specular microscope developed by Maurice. There are two modifications: the use of a second conjugate slit in the eyepiece to produce the optical sectioning or depth discrimination, and a floating objective lens which is scanned on the optic axis by a computer-controlled piezoelectric driver [Masters, 1988]. Figure 11.3 shows the principle of the depth discrimination of the microscope. The depth resolution of

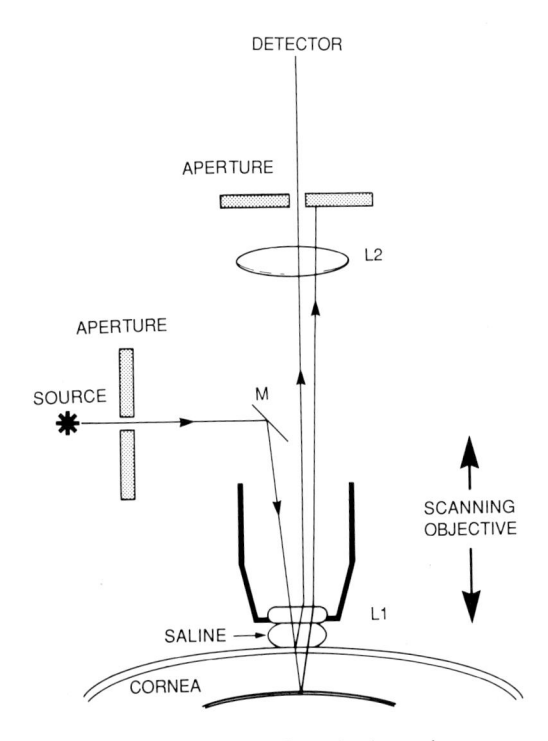

Figure 11.3. The principle of the optical sectioning microscope. A simplified ray path is shown. Physiological saline or bicarbonate Ringer's solution is the optical coupling fluid between the surface of the cornea and the scanning objective of the microscope. The position of the objective determines the volume of the object that is optically sectioned. There are two conjugate slits, one adjacent to the source and one at the detector. The light ray from the upper surface of the cornea (epithelium) is detected after passing through the detector slit; the ray from the lower surface (endothelium) impinges upon the conjugate detector slit and is not detected.

this instrument is a function of the objective and the slit height. For a slit width of 100 microns and an objective with a N.A. of 1.00 (100X water imersion), the depth resolution is about 18 microns. Figure 11.4 shows the fluorescence and reflectance scans made from a cornea in a living rabbit. The peaks in the reflectance signal clearly correspond to large changes in the refractive index. The fluorescence signals from the corneal epithelium (40 microns) and the corneal endothelium (6 microns) are clearly separated. This type of one-dimensional confocal microscope does not produce a two-dimensional image, but it optically sections the cornea in a living animal and measures the fluorescence and the reflectance signals along the optic axis of the cornea. The intensity scans illustrated in Figure 11.4 can be deconvolved if the instrument response function is independently determined. The scans represent the convolution of the instrument response function and the biological structure.

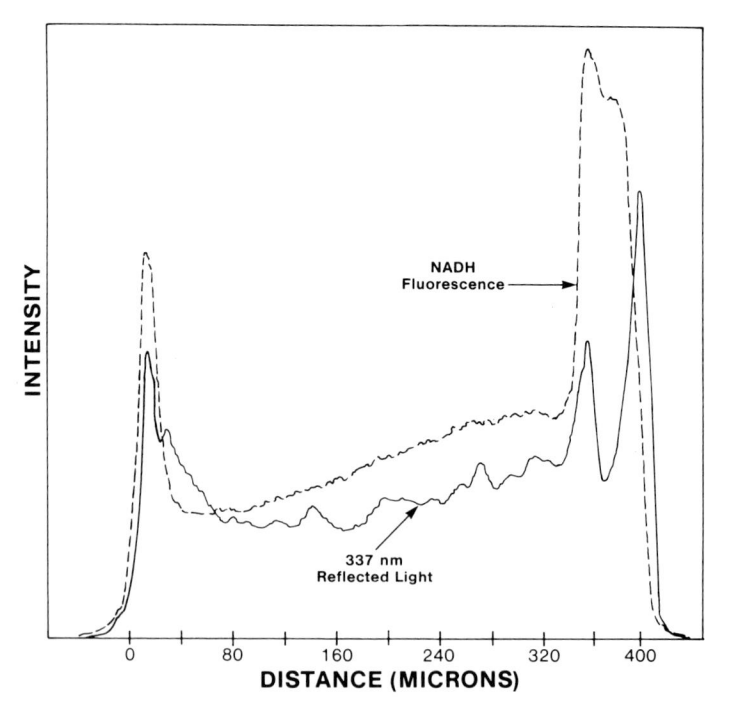

Figure 11.4. An optical section through a live rabbit cornea illustrating the range resolution for the reflected light (solid line) and the 460nm fluorescence emission. The intensity of the reflected light is 10 times that of the fluorescence. The ordinate is relative intensity and the abscissa is distance into the cornea. The tear film is on the right side and the aqueous humor is on the left side of the figure. The fluorescence from the region between the endothelial and the epithelial intensity peaks (stromal region) is non-specific fluorescence.

The following section will illustrate the applications of imaging confocal microscopes to morphological studies of the eye. Before describing the various types of confocal microscopes and illustrating the images made with them, it is valuable to discuss the significance of these morphological studies.

What is the significance of confocal microscopes to the study of the eye? The development of tissue-specific devices which optimize certain parameters will be available: laser scanning ophthalmolscopes [Webb et al., 1987] and confocal microscopes, tandem scanning [Petráň et al., 1968, 1974, 1984, 1985] and non-tandem scanning microscopes for the study of the cornea and the ocular lens and the retina. From a device-independent point of view, what are the significant properties of these optical devices? The two main advantages of a confocal microscope as compared to a conventional microscope are the following:

(1) Enhanced depth or range resolution.
(2) Enhanced lateral resolution.

A natural consequence of the first property is that the images are not degraded in contrast owing to out-of-plane contributions from scattered light. A striking result is that the confocal microscope produces good images in high contrast from the transparent tissues of the cornea and the ocular lens. Some of the images of the cornea and the lens are reminiscent of those produced with an electron microscope. This results in the ability to image unstained transparent, low-contrast ocular tissue and to observe subcellular components in high resolution and high contrast. This results in the ability to image the eye from the tear film through the cornea to the endothelial cell layer. In addition, the ocular lens and the retina may be optically sectioned and subjected to quantitative morphological analysis. This means that the ophthalmologist is no longer limited to specularly reflected images from the corneal endothelium and low-contrast images from the corneal epithelium. In particular, semi-transparent regions of the cornea and the ocular lens may be optically sectioned and viewed, while prior to the advent of confocal microscopy this was not readily done. In summary, confocal microscopy gives the ophthalmologist a tool to image transparent, low-contrast, and semi-opaque ocular tissue and to produce high-resolution, high-contrast images of ocular tissue.

We will now briefly describe several types of confocal microscopes and give some illustrative examples of images of ocular tissue that were made with these systems. The images shown from the cornea were made on enucleated rabbit eyes. The ocular lens was removed from the rabbit eye prior to imaging. The image of the retina was made on a region of rabbit retina that was removed from the globe. All of the images were made in reflected light using a Leitz 50X water immersion objective with a numerical aperture of 1.0 and a working distance of about 0.7mm. The sequences of images follows the anatomical progression from the tear side of the cornea, through the cornea to the endothelial cells, to the ocular lens, and finally to the retina. Each of these images shows structure and contrast that has been previously observed only with electron microscopy or with the scanning slit specular microscope developed by Maurice.

Figure 11.5 is an image made with a CSOM of exfoliating epithelial cells with the focal plane inside the cell nuclei.

The next six confocal images, Figures 11.6-11.11, were made on the Bio-Rad laser scanning confocal microscope [White et al., 1987]. The light source was the combined output from an argon ion laser (488nm and 514nm light). The image required several seconds to be displayed on the monitor. The high amplification required to detect the images resulted in a strong reflection artefact in the center of the image. Several images have a black marker placed over the reflection artefact to indicate its presence, while in other images the reflection artefact was cropped out in the photographic processing. This is not fundamental to the confocal principle, but rather to a problem with the particular Bio-Rad instrument used.

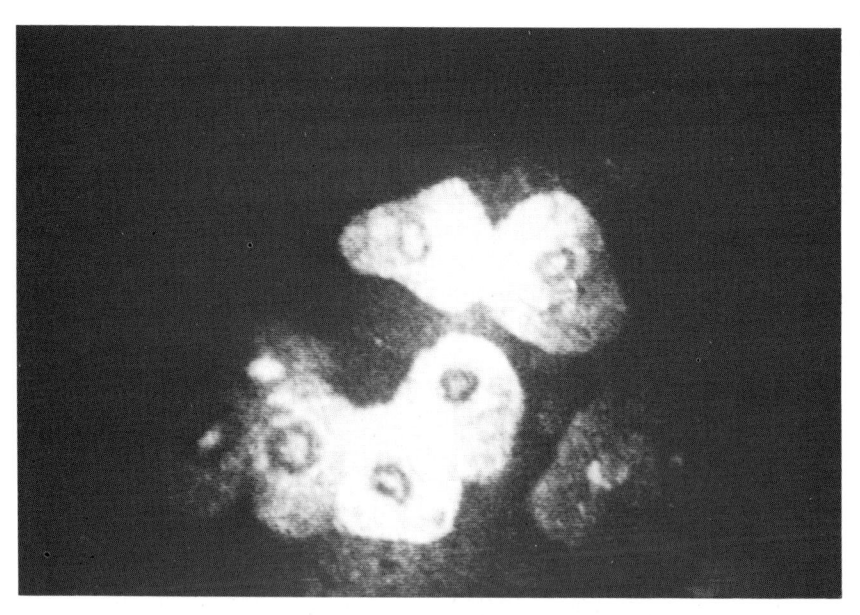

Figure 11.5. Confocal (CSOM) reflected light microscope image of the superficial epithelial cells of a rabbit cornea. The polygonal epithelial cells with their cell borders and nuclei are shown. The focal plane of the image is centered at the center of the nuclei of the cells. The bright cells are on the surface and the darker cells are slightly below the focal plane. Structure within the cell nuclei are shown. The reflected light image was made with a confocal scanning optical microscope (CSOM) described in the text. This image was made together with Drs. G.Q. Xiao and G.S. Kino at Stanford University.

Confocal Microscopy of Ocular Tissue 315

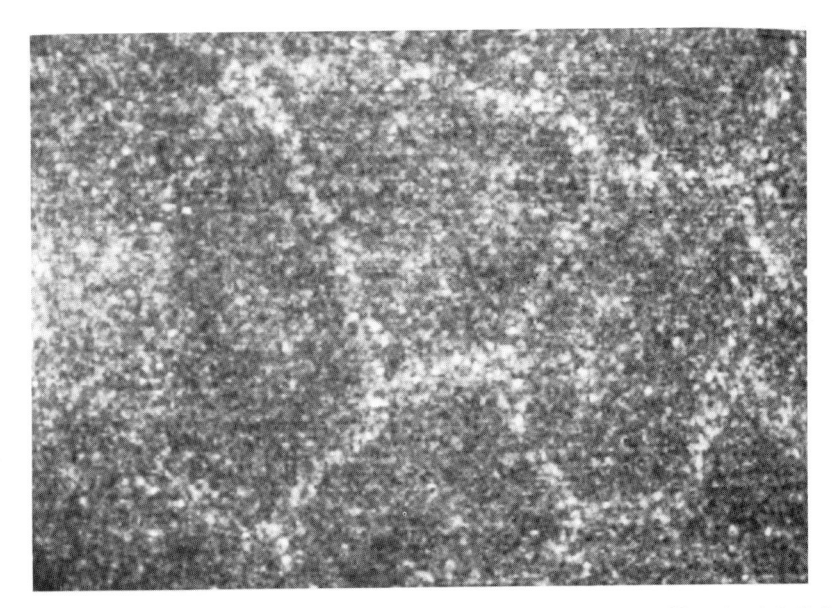

Figure 11.6. Confocal (CLSM) reflected light microscope image of basal epithelial cells of the corneal epithelium of a rabbit. These cells are about 40 microns below the corneal surface. The focal plane is centered at the center of the nuclei of the epithelial cells. These cells are about 20 microns thick and about 10 microns in diameter. They are the only cells that undergo mitosis. This image was made in collaboration with Dr. S. Paddack, IMR, University of Wisconsin.

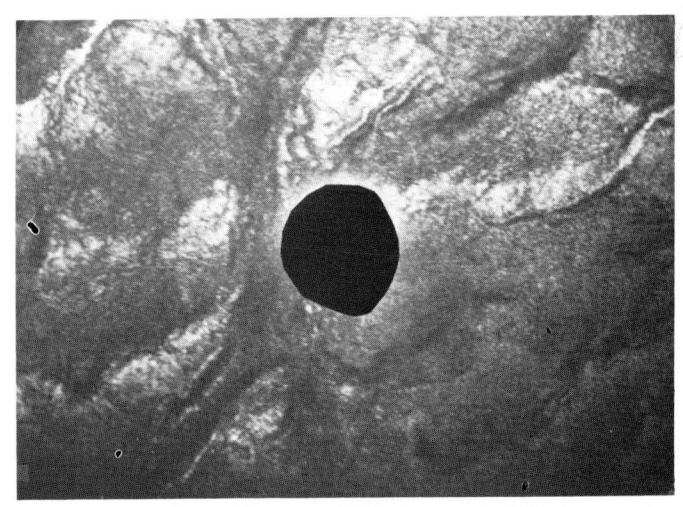

Figure 11.7. Confocal (CLSM) reflected light image of rabbit basement membrane showing folds. This membrane is at the base of the basal epithelium and separates it from the stromal region. The image shows a section covering 75 microns. This image was made in colloboration with Dr. S. Paddack, IMR, University of Wisconsin.

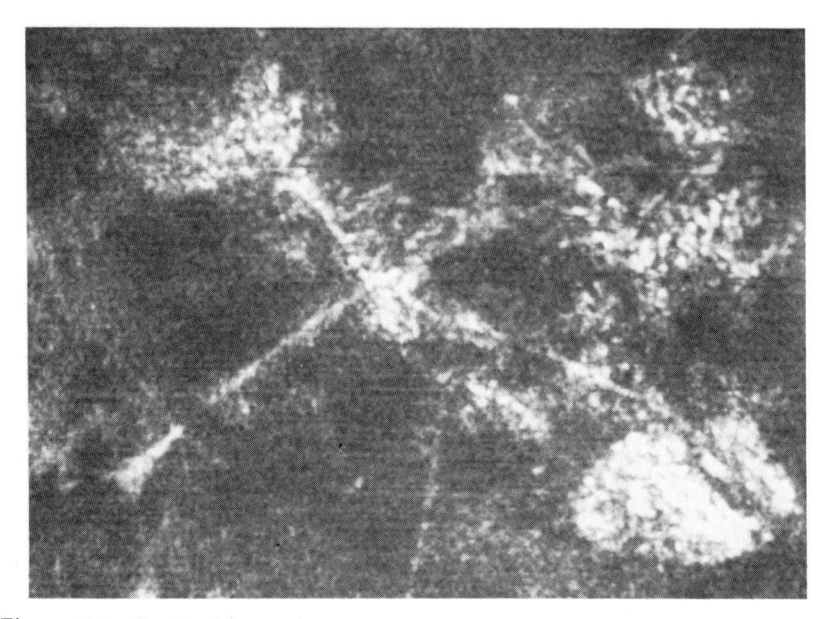

Figure 11.8. Confocal (CLSM) reflected light image of a subepithelial nerve plexus in the anterior region of a rabbit stroma. The linear fibers are nerves and the nuclei of stromal keratocytes are shown. This image was made in collaboration with Dr. S. Paddack, IMR, University of Wisconsin.

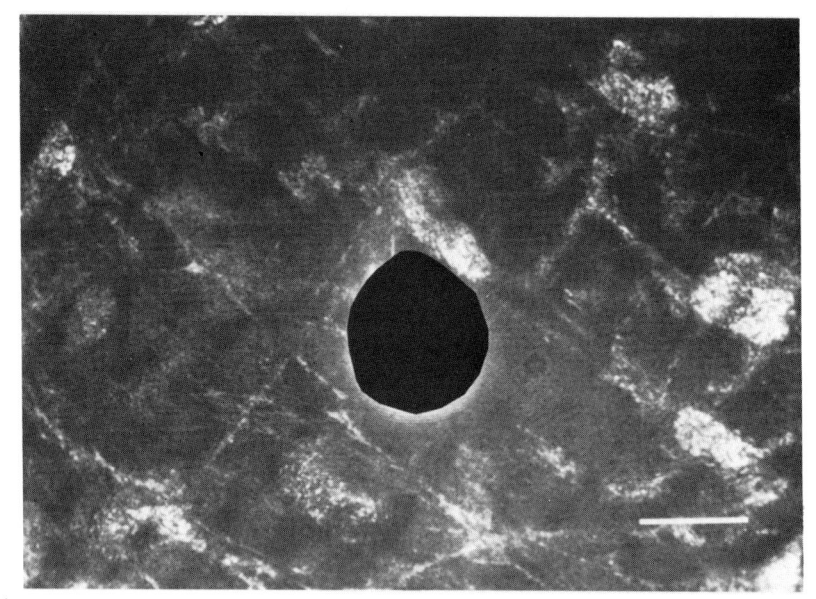

Figure 11.9. Confocal (CLSM) reflected light image of stromal keratocytes and an orthogonal array of filaments in the stroma of a rabbit cornea. This image was made in collaboration with Dr. S. Paddack, IMR, University of Wisconsin.

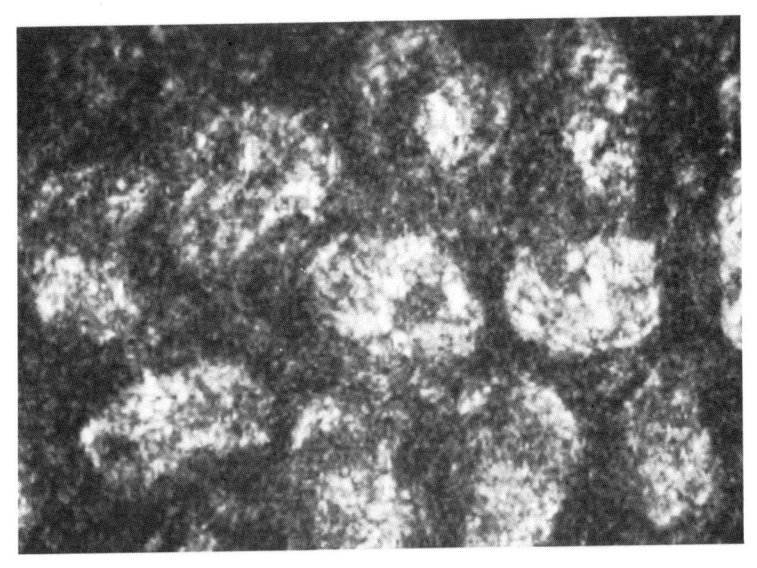

Figure 11.10. Confocal (CLSM) reflected light image of endothelial cells of a rabbit cornea. The focal plane is at the center of the cell nuclei which appear as dark disks within the cells. This image was made in collaboration with Dr. S. Paddack, IMR, University of Wisconsin.

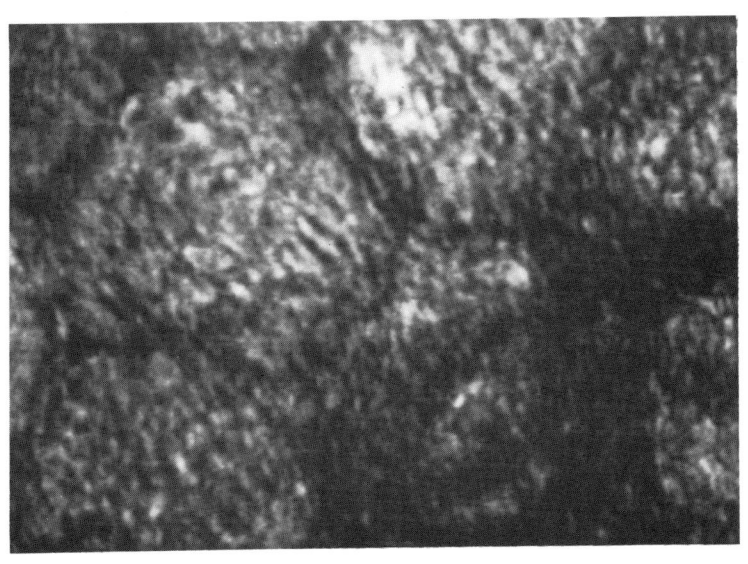

Figure 11.11. Confocal (CLSM) reflected light image of endothelial cells of a rabbit cornea. The focal plane is at the surface of the endothelial cells and shows the presence of microvilli at the interface between the cells and the aqueous humor. This image was made in collaboration with Dr. S. Paddack, IMR, University of Wisconsin.

In addition to the cornea, it is possible to image the ocular lens as well as the component layers of the retina with confocal microscopes. The ability to focus on a particular ocular tissue is limited by the free working distance of the objective used in the microscope. Of course the laser scanning instruments (i.e., the laser scanning ophthalmoscope) can focus through the full thickness of the human eye in real time. Figure 11.12 shows the image of the rabbit ocular lens made on a freshly excised lens. The CSOM real-time confocal microscope was used to produce this image. The Bio-Rad laser confocal microscope was used to produce the images shown in Figures 11.13 and 11.14 from a freshly excised piece of rabbit retina.

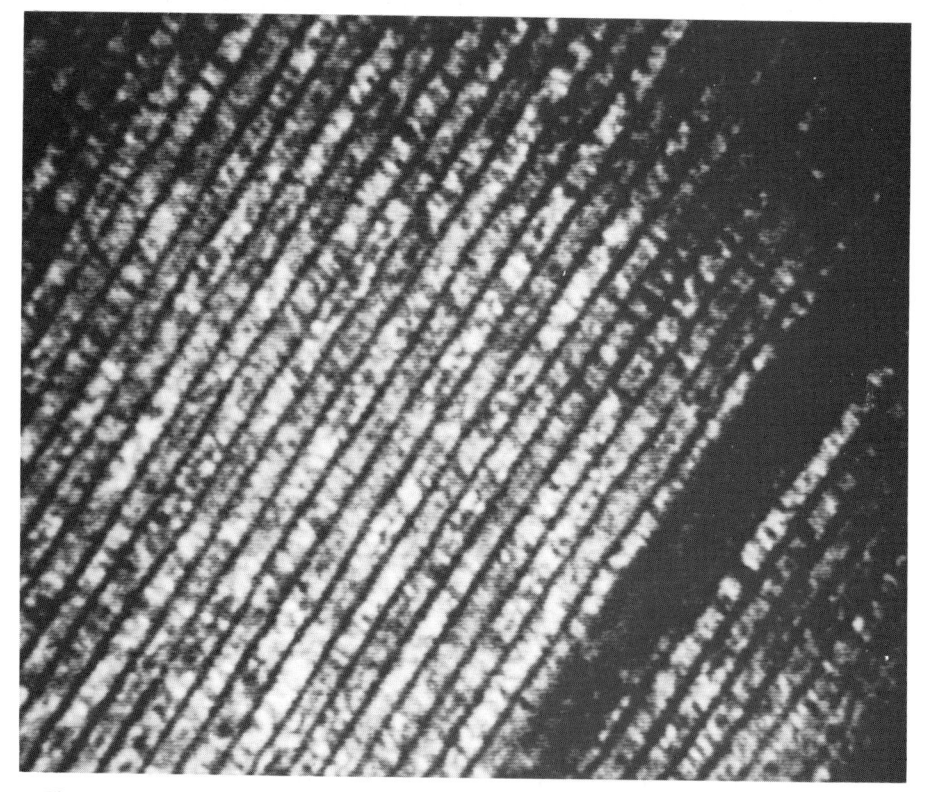

Figure 11.12. Confocal (CSOM) reflected light image of the ocular lens from a freshly excised rabbit lens which shows the lens fibers. The separation between the adjacent fibers is about two microns, and the submicron transverse bands are shown. The dark object is the nucleus. The image is made from the bow zone of the lens, about 200 microns below its surface. This image was made together with Drs. G.Q. Xiao and G.S. Kino at Stanford University.

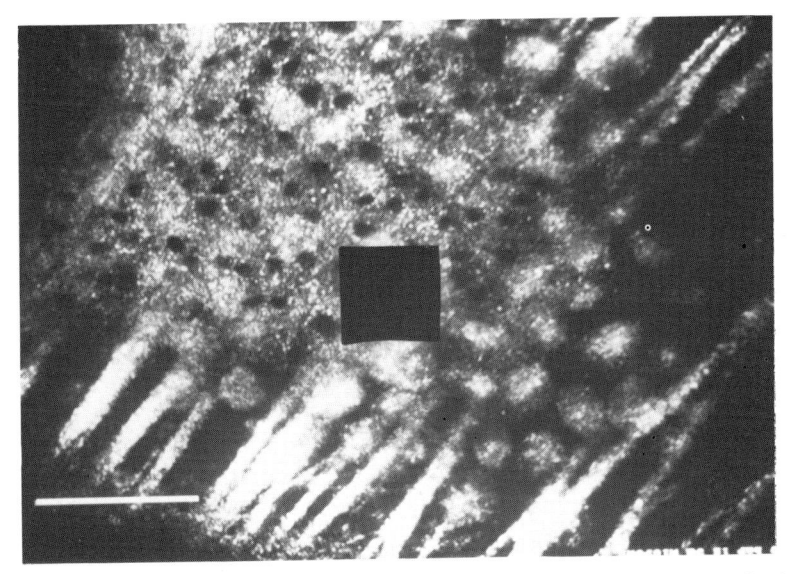

Figure 11.13. Confocal (CLSM) reflected light image of a freshly excised piece of rabbit retina cut perpendicular to the plane of the retina. The bar represents 50 microns. The image shows the rod inner segments in the lower region of the image, and the outer nuclear layer (nuclei of photoreceptors) in the middle and upper regions.

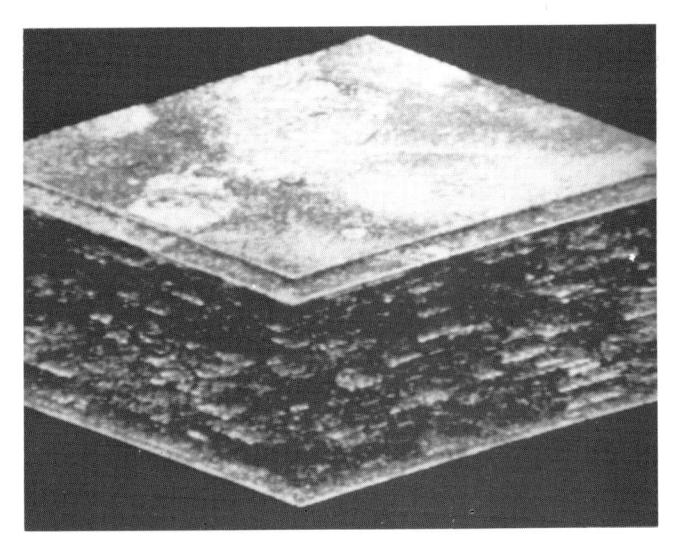

Figure 11.14. Three-dimensional reconstruction of the rabbit cornea. The data set consisted of three micron sections obtained with a BioRad laser confocal microscope. This image was made in colloboration with Dr. S. Paddack, IMR, University of Wisconsin.

V. Development of a real-time confocal system for clinical ophthalmology

The observer of the images of ocular tissue produced with various types of confocal microscopes would naturally inquire whether the technology could be developed into a clinical microscope for diagnostic microscopy. In order to address this question, the desired characteristics of such a clinical microscope are discussed. In addition, there is a discussion of the potential limitations.

The choice of light source will depend on the intended use of the microscope. If the goal is a morphological examination of the cornea and the ocular lens, then a suitable source could be a xenon arc lamp. A set of interference filters could be selected to remove the ultraviolet and infrared components and to produce quasi-monochromatic light. This would reduce the effect of chromatic aberrations of the objective, which result in different wavelengths focusing on different planes. An alternative source could be a HeNe laser or even a solid-state diode laser which could be tuned to various wavelengths. The advantage of the red light is that the light scatter is minimized as compared to blue light, and there is better penetration of the light into the cornea and the ocular lens. For fluorescence studies of the ocular tissue there is the problem that the excitation and the emission wavelengths differ. Therefore, the image detected will be degraded in range resolution.

A set of applanating objectives with different powers and therefore different working distances would be desirable. All of these objectives should have a large numerical aperture. It would be an important design consideration to have a system of internal focusing, so that the tip of the objective could rest on the surface of the cornea and remain stationary. An internal lens system would permit focusing at different depths within the cornea or ocular lens.

The development of a clinical confocal microscope for diagnostic ophthalmology requires careful attention to two other potential limitations. The first is light damage to the eye and the second is eye movement. The use of a confocal system with a Nipkow disk [Nipkow, 1884] requires high-intensity light sources since only a small fraction of the light incident on the disk reaches the object. There are numerous studies on the effect of light on the cornea, lens and retina, and these guidelines must be followed [James et al., 1988; Sliney and Wolbarsht, 1980]. Ultraviolet light incident on the cornea is mostly absorbed by the adult ocular lens and there is thus little danger to the retina. However, blue light is readily transmitted through the cornea and the ocular lens and can present a danger to the retina [Sliney and Wolbarsht, 1980]. The use of high-speed photographic film or microchannel plate solid-state intensifiers in front of the video camera would permit minimal light levels to be incident on the ocular tissue.

A more severe potential problem is the natural motion of the eye within the head. The cardiac cycle results in a pressure pulse at a frequency of about 1 Hz. In addition there are saccadaic movements of the eye. These movements occur in bursts with quiescent periods between them. Several methods are available to minimize the effects of eye motion. There are real-time eye trackers which could monitor the motion of the eye and gate the detectors for the quiet periods. The video image of the ocular structures could be recorded onto high-resolution (500 horizontal lines) video tape. Upon playback of the video tape, the frames of the quiet eye could be selected as the in-focus images for further processing and storage. A major factor in the development of a clinical confocal microscope for ocular examination will be the development of a new set of applanating objectives. The surface of the cornea is round, and therefore, to obtain wide fields of in-focus cells with a confocal microscope, it is important to flatten the cornea. This is performed by first placing a drop of anesthetic onto the surface of the cornea to inhibit pain, then gently touching the surface of the applanating lens to the surface of the cornea. Sometimes an index-matching gel or solution is placed on the tip of the objective. A similar procedure is routinely employed in the use of the wide-field clinical specular microscope. The use of an applanating lens has several advantages:

(1) The lateral movement of the eye is diminished owing to the friction of the eye and the tip of the applanating objective.

(2) The surface of the eye and the tip of the objective are optically coupled by the index-matching fluid or gel.

(3) The distances between the tip of the eye and focal planes on the optic axis are stabilized.

(4) A wide field of cells are in the same focal plane.

What can we propose about future developments of confocal microscopy in ophthalmology? The advances in our understanding of aging, wound healing, and the normal and pathological development of the eye will be made through the use of confocal microscopy. The first major area is that of morphological studies. The use of a clinical confocal microscope will permit full-section serial sections to be made of ocular tissue. The use of digital image processors will be used both for noise removal of the individual sections, but more importantly for three-dimensional reconstructions [Agard, 1984] of the cornea and the ocular lens. The study of such reconstructions would permit investigations of innervation, topology, and morphology on living eyes without the use of mechanical sectioning, staining, fixing and electron microscopy, which destroys the tissue. Morphological changes could be followed over long periods of time on the same eyes. This is important for studies of growth factors, and other pharmacological agents. The result would be a reduction in the number of animals needed for a given pharmacological investigation.

Another important application is the use of fluorescent monoclonal antibodies to label specific molecules in ocular tissue [Amos et al., 1987]. The optical sectioning characteristics of confocal microscopes would result in sharp images of these fluorescent labels.

Finally, the clinical use of confocal microscopy would provide a useful tool to examine ocular tissue that is slightly opaque, i.e., the endothelial cells under a swollen cornea, or the regions in the ocular lens behind an opacity. The future is bright for the research and clinical applications of confocal microscopy in ophthalmology. The early work of researchers on confocal imaging systems for biological structures [Nipkow, 1884; Baer, 1970; Davidovits and Egger, 1971; Petráň et al., 1968] has led to the recent developments described in this paper.

Acknowledgements

This work was supported by a grant from N.I.H. EY-06958, and a departmental grant from Research to Prevent Blindness, Inc. Dr. Steve Paddack of the Integrated Microscopy Resource for Biomedical Research, University of Wisconsin, Madison, Wisconsin, collaborated in obtaining the laser confocal images. Drs. G.Q. Xiao and G.S. Kino of the Edward L. Ginzton Laboratory, W.W. Hansen Laboratories of Physics, Stanford University, Stanford, California collaborated in obtaining the CSOM images of the corneal epithelium and the ocular lens.

References

Agard, D.A., (1984). Optical sectioning microscopy: cellular architecture in three dimensions. Ann. New York Acad. Sci., 191-219.

Amos, W.B., White, J.G. and Fordham, M., (1987). Use of confocal imaging in the study of biological structures. Appl. Opt., 26, 3239-3243.

Baer, S.C., (1970). Optical apparatus providing focal-plane-specific illumination. U.S. Patent No. 3,547,512.

Davidovits, P. and Egger, M.D., (1971). Scanning laser microscope for biological investigations. Appl. Opt., 10, 1615-1619.

Gallagher, B. and Maurice, D., (1977). Striations of light scattering in the corneal stroma. J. Ultrastructure Res., 61, 100-114.

Goldstein, S., (1989). A no-moving-parts video-rate laser beam scanning type 2 confocal reflected/transmission microscope. J. Microsc., **153**, Pt. 2, RP-1.

James, R.J., Bostrom, R.G., Remark, D. and Sliney, D.H., (1988). Hand-held ophthalmoscopes for hazards analysis: an evaluation. Appl. Opt., **27**, 5072-5076.

Jester, J. V., Cavanagh, H. D., Dilly, P. N. and Lemp, M. A. (1989). Confocal microscopic imaging of the eye and the tandem scanning confocal microscope (TSCM). In: *Non-invasive Diagnostic Techniques in Opthalmology.* Ed. Masters, B. R. Springer-Verlag, New York.

Jester, J.V., Cavanagh, H.D. and Lemp, M.A., (1988). In vivo confocal imaging of the eye using tandem scanning confocal microscopy. Proc. 46th annual meeting of the Electron Microscopy Society of America, San Francisco. Bailey, G.W. (Editor), 56-57.

Klyce, S. D. and Beuerman, R. W. (1988). Anatomy and physiology of the cornea. In: *The Cornea.* Ed. Kaufman, H. E., Barron, B. A., McDonald, M. B. and Waltman, S. R. Churchill Livingstone, Inc., New York, 3-54.

Koester, C.J., (1980). Scanning mirror microscope with optical sectioning characteristics: applications in ophthalmology. Appl. Opt., **19**, 1749-1757.

Koester, C.J., Roberts, C.W., Donn, A. and Hoefle, F.B., (1980). Wide-field specular microscopy: clinical and research applications. Ophthalmol., **87**, 849-860.

Laing, R. A. (1980). Specular microscopy. In: *Current Topics in Eye Research*, **3**, Ed. J. Zadunaisky and H. Davson. Academic Press, (New York), 157-218.

Lemp, M.A., Dilly, P.N. and Boyde, A., (1986). Tandem-scanning (confocal) microscopy of the full-thickness cornea. Cornea, 4, 205-209.

Masters, B.R., (1984). Noninvasive corneal redox fluorometry. In: *Current topics in eye research, Vol. 4.* Zadunaisky, J. and Davson, H. (Editors), 139-200. Academic Press, New York.

Masters, B.R., (1984). Noninvasive redox fluorometry: how light can be used to monitor alterations of corneal mitochondrial function. Curr. Eye Res., **3**, 23-26.

Masters, B.R., (1988). Effects of contact lenses on the oxygen concentration and epithelial mitochondrial redox state of rabbit cornea measured noninvasively with an optically sectioning redox fluorometer microscope. In: *The cornea: transactions of the world congress on the cornea III.* Cavanagh, H.D. (Editor), 281-286. Raven Press, New York.

Masters, B.R., (1988). Optically sectioning ocular fluorometer microscope: applications to the cornea. Proc. S.P.I.E., **909**, Time-resolved laser spectroscopy in biochemistry. 342-348.

Masters, B.R., (1989). Confocal microscopy of the eye. Proc. S.P.I.E., **1161**, 350-365.

Maurice, D., (1968). Cellular membrane activity in the corneal endothelium of the intact eye. Experientia, **24**, 1094-1095.

Maurice, D., (1974). A scanning slit microscope. Invest. Ophthalmol. Vis. Sci., **13**, 1033-1037.

McCally, R.L. and Farrell, R.A., (1989). Light scattering from cornea and cornea transparency. In: *Noninvasive diagnostic techniques in ophthalmology.* Masters, B.R. (Editor). Springer, New York.

Mayer, D. J. (1984). *Clinical Wide-field Specular Microscopy.* Bailliere Tindall, London.

Mizushima, Y. (1988). Detectivity limit of very small objects by video-enhanced microscopy. Appl. Opt., **27**, 2587-2594.

Nipkow, P., (1884). German Patent No. 30,105. Published Jan. 15.

Petráň, M., Hadravský, M., Egger, M.D. and Galambos, R., (1968). Tandem-scanning reflected-light microscopy. J Opt. Soc. Am. **58**, 661-664.

Petráň, M. and Sallam, S.M., (1974). Microscopical observations of the living (unprepared and unstained) retina. Physiologia Bohemoslov., **23**, 369.

Petráň, M., Hadravský, M., Benes, J. and Boyde, A., (1984). In vivo microscopy using the tandem scanning microscope. Ann. New York Acad. Sci., 440-447.

Petráň, M., Hadravský, M., Benes, J., Kucera, R. and Boyde, A., (1985). The tandem-scanning reflected light microscope. 1. The principle and its design. Proc. Roy. Microscop. Soc., **20**, 125-129.

Petráň, M., Hadravský, M. and Boyde, A., (1985). The tandem scanning reflected light microscope. Scanning, **7**, 97-108.

Petráň, M., Hadravský, M., Boyde, A. and Mueller, M., (1985). Tandem scanning reflected light microscopy. In: *Science of biological specimen preparation for microscopy and microanalysis.* Bhatt, S.A., Muller, M., Becker, R.P., Boyde, A. and Wolosewick, J.J. (Editors), 85-94. SEM Inc., AMF O'Hare, Chicago.

Sliney, D., Wolbarsht, M. (1980). *Safety with Lasers and other Optical Sources,* Plenum Press, New York.

Webb, H., Hughes, G.W. and Delori, F.C., (1987). Confocal scanning laser ophthalmoscope. Appl. Opt., **26**, 1492-1499.

White, J.G., Amos, W.B. and Fordham, M., (1987). An evaluation of confocal versus conventional imaging of biological structures by fluorescence light microscopy. J. Cell Biol., **105**, 41-48.

Xiao, G.Q. and Kino, G.S., (1987). A real-time confocal scanning optical microscope. Proc. S.P.I.E., **809**, Scanning Imaging Technology. Wilson, T. and Balk, L. (Editors), 107-113.

Xiao, G.Q., Corle, T.R., and Kino, G.S., (1988). Real-time confocal scanning optical microscope. Appl. Phys. Lett. **53**, 716-718.

Xiao, G.Q., Kino, G.S. and Masters, B.R., (1989). Observation of the rabbit cornea and lens with a new real-time confocal scanning optical microscope. Scanning, June, 1990.

12. Biological Perspectives of Confocal Microscopy

V. SARAFIS

I. Introduction

The field of confocal microscopy is a rapidly developing one. The advantages of this new type of microscopy are the possibility of increased lateral resolution, an increase in depth discrimination — also much improved in fluorescence — the ease of using spatial filters, a high signal-to-noise level compared to conventional microscopy and an image acquisition mode which is digitised and can be easily manipulated by software applications to construct two- and three-dimensional images of desired attributes [Boyde, 1985, 1987, 1988; Brakenhoff, 1984; Cox, 1988; Petráň et al., 1985; Sheppard, 1987; Wilson and Carlini, 1988; Wilson and Sheppard, 1984].

The techniques of confocal microscopy were originally developed at the University of Oxford and the University of Amsterdam in the mid- to late 70s [Brakenhoff, 1979; Brakenhoff et al., 1979]. User environments propelled the developers along different lines of interest: reflection microscopy for surface examination was mainly developed at Oxford, while transmission and, later on, fluorescence confocal microscopy, especially for biological specimens, was developed in Amsterdam [van der Voort et al., 1985, 1987a, 1987b, 1987c].

As with all new techniques, the value and application of confocal microscopy depends on an understanding of the limitations as well as the advantages of the instrumentation. Confocal microscopy is chosen as a technique because it produces exceptionally clear views of biological systems, especially in three dimensions, where it can be utilised as an *essential* tool in solving particular biological problems.

The theoretical background had originally emphasised edge reflecting specimens such as microchips, where the focussed spot meets the specimen at only one level of focus without focusing through intervening layers [Wilson and Sheppard, 1984]. More recently, a full theory for resolving planes, lines and points has been developed, applicable to transmission as well as reflection systems [van der Voort, 1989].

II. Applications of reflection microscopy

II.1. Reflection from phase specimens

If we consider a transparent specimen with planar phase steps within its volume, a reflection image will be stronger than a transmission one [Wilson and Sheppard, 1984; Sheppard, 1987].

Let us consider what kind of biological specimens of the above type could thus be expected to be imaged well. Organisms such as scaly flagellates and other naturally coated microorganisms could be examined in the live state by reflection, as could plant surfaces, spores, fungal colonies and animal tissue cells. Differential confocal phase contrast can be applied to these specimens. The technique is described by Benschop (1987).

Imaging specimens where a large phase step occurs just beneath a smooth surface, e.g., euglena, or striated muscle fibres, lend themselves to confocal imaging. One would expect that most of the signal would be provided by the reflected wave from the phase steps.

However, when a specimen containing several layers with strong phase steps is imaged in reflection, interference fringes appear between the reflected wavefront(s) and the incident one(s). For example, when imaging the staminal hairs of *Tradescantia*, the cuticular ornamentation of the staminal hairs and the phase step provided by the cell wall cytoplasm boundary produce interference patterns which detract from image quality. Upon observing water-imbibed spores of the moss *Dawsonia superba* by reflection in an aqueous mount, strong interference fringes appear for spores located close to the cover slip. The ease with which interference fringes in the reflection mode can be made to appear is exploited in highly sensitive interference reflection contrast microscopy, used, for example, in studying attachment sites of fibroblasts to substrates.

Use has been made of the strong reflection signal when looking at spores of the moss *Dawsonia superba* immersed in immersion oil. These spores are 5-6μm in diameter with very fine wall ornamentation approximately 0.2μm apart. In transmission microscopy, using the classical mode, the wall is almost invisible, but in reflection confocal microscopy a strong signal due to the ornamentation on the spore wall, which is close to the optical resolving limit, gives rise to an image in which the ornamentation appears bright on a dark background. When imaging oil-immersed spores with larger-sized wall ornamentation, such as those of the moss *Polytrichum commune*, the strong reflection signal obtained consists of a mixture of minute interference fringes, suggesting interference from different planes of the spore wall. The reflection signal obtained from spores of *Dawsonia superba* is sufficiently strong for one to be able to perform a through-focus series in reflection, and get a high-resolution image of both the wall closest to the objective and the one farthest away, with almost all the detail of the farthest wall being as good as that of the front wall.

II.2. Reflection from gold immunolabelled material

Gold immunolabelling of cell components, e.g., microtubules with or without silver enhancement, offers an excellent method of utilising the reflection mode of the confocal microscope. Gold gives an extremely strong red reflection signal owing to its selective reflectivity in the spectral range of the red [Bajer et al., 1986; Mole-Bajer and Bajer, 1988] for particles whose size are greater than the wavelength. For sub-wavelength particles, however, Rayleigh scattering is enhanced in the green. An even stronger signal-to-noise ratio might be available through adapting the Nikon Epipolarizing illumination pod for confocal use, with the attendant gains which the confocal microscope gives in resolution and depth discrimination [Hoefsmit et al., 1986].

II.3. Autoradiography

Autoradiography using silver halide emulsions [Baker, 1988] also offers an area of use for the reflection mode in confocal microscopy, with the possibility of clearly discriminating the emulsion layer from the object layer.

III. Transmission microscopy

III.1. Transmission in brightfield

Because reflection mode imaging can result in problems due to interference, transmission microscopy may offer a better means of examining thick specimens, assuming that the specimen is not a strong multi-layer absorber. Specimens such as bacteria and flattened animal cells or the epidermis of plant leaves fall into this category. However, in preparing to examine organs which heavily scatter light because of strong phase boundaries — for example, the mesophyll tissues of a leaf or endolithic algae — the resolution degrades rapidly beyond a depth of about 20 microns, owing to the air spaces which act as very strong scatterers and result in the scanning spot losing its Airy disc-like character, with a consequent loss in resolution and signal-to-noise ratio. In specimens which do not have such strong scatterers but which may instead have small, highly refractive lens-like elements, such as small oil droplets, a degradation of the scanning spot with increasing depth is also evident. Thus, in assessing a structural investigation problem, one should try to minimise, whether in reflection or in transmission, any feature under the experimenter's control which may increase the scanning spot size with increasing depth.

III.2. Correcting for spherical aberration

The spherical aberration which may be present in aqueous mountants has been shown by Brakenhoff to be compensated by the shape of the specimen (bacterial cells of *E.coli*) [Brakenhoff *et al.*, 1980]. However, this may not be the case for other kinds of specimens. A correction collar, such as is available with the water immersion 1.25 numerical aperture Neofluar objective of C. Zeiss, could be used to correct for spherical aberration caused by differing refractive indices within the specimen, or by the effects of a mountant of a refractive index different from the design specifications of the objective.

The high contrast possible in transmission with the very much improved signal-to-noise ratio has enabled a study of the chromosomal material during division in *E.coli* to be monitored *in vivo* at a resolution hitherto unattained. *E.coli*, when fixed by a chemical fixative, has always shown some change from the *in vivo* state [Woldringh and Nanninga, 1985; Valkenburg, 1985; Valkenburg *et al.*, 1985]. Thus, the confocal microscope offers the opportunity of studying the replication of the complex DNA conformation in living bacterial cells. Similarly, we have investigated the structure of chloroplasts in transmission using the absorbance of chlorophyll in the blue or red within living cells of mesophyll tissues with a low level of artefact [Sarafis *et al.*, 1987]. Hitherto, electron microscopical techniques have been responsible for giving a well-preserved but possibly artefactual appearance [Goodchild, 1971; Spencer, 1971].

IV. Fluorescence confocal microscopy

IV.1. Fluorescence microscopy

In the fluorescence mode, the confocal microscope really comes into its own. Here the resolution of extended objects is improved considerably in comparison to the classical incoherent microscope [Cox *et al.*, 1982]. Little out-of-focus information or veiling glare reduce the contrast of the picture. Because in this mode the various scan planes in the vertical direction are well discriminated, very *clean* pictures may be constructed, including three-dimensional views. Fluorescence imaging generally suffers from fading and quenching, but in the confocal mode, owing to the very short period of time in which the scan spot dwells upon an object element, bleaching is much reduced. Image integration is, however, necessary owing to the relatively low amounts of signal emitted, but it is possible, by appropriate choice of excitant dye wavelength and irradiance strength, to mitigate such problems, as well as by the appropriate choice of fluorophore and its delivery, e.g., by liposomes [Truneh and Machy, 1987].

IV.2. Fluorescence microscopy with thick scattering and absorbing specimens

It should be borne in mind that resolving power, whether in two- or three dimensions, is worse when viewed in fluorescence in thick specimens than in thin ones. First, the resolution is reduced by scattering and spherical aberration within the specimen in the exciting beam. Second, it is reduced by scattering and spherical aberration within the specimen in the emitted beam. In addition, quenching of fluorescence occurs as well as losses owing to the inner filter effect and reabsorption [Tanke et al., 1981] and may need to be compensated by increasing the exciting illumination strength as one traverses a specimen with a high extinction coefficient. Recently, software solutions to the problem of decreasing brightness with increasing depth have been used to enhance imaging of cyanophytes embedded in calcareous rock. In a recent study, we have shown the power of confocal fluorescence microscopy in analysing the living structure of chloroplasts as an example of the visualisation of a plant cell structure at the limits of optical resolution [van Spronsen et al., 1989].

Three-dimensional reconstruction of structures which may suffer distortion upon fixing and embedding, or of structures which are inaccessible, are particularly well studied by confocal fluorescence microscopy. Complex three-dimensional structures are also usefully studied by this means [Brakenhoff et al., 1986, 1987; Stelzer et al., 1986].

IV.3. Application of confocal scanning light microscopy in fluorescence

Some examples of the use of fluorescence confocal microscopy are in the study of food emulsions [Heertje et al., 1987] and the study of nerve cells [Carlsson et al., 1985; Carlsson and Åslund, 1987; Frankson et al., 1987].

A range of fluorescent probes renders the study of three dimensions in cells and tissues by confocal microscopy particularly useful. Some examples of such probes and their potential uses are: Mithramycin in the study of nuclear structure [Brakenhoff et al., 1985] and immunolabelled material within cells [Busby and Gunning, 1988a, 1988b; Hoefsmit et al., 1986; Koch et al., 1987; Mineyuki et al., 1988]. The study of the Golgi apparatus by means of C_6-MBD-Ceramide [Lipsky and Pagano, 1985] has opened the possibility of looking at Golgi activity in living cells.

Some examples of other probes that may be used in living cells are: DAPI in the study of plastid DNA configuration [Coleman, 1985], new DNA fluorescent dyes excitable in the red [Shapiro and Stephens, 1985], a PH probe used in fluorescence ratio imaging [Bright et al., 1987], calcium probes for the study of calcium distribution in cells [Grynkiewicz et al., 1985], and a thiol reagent for the study of SH groups [Olive et al., 1982]. A range of other probes and their applications are discussed by Hawkes (1988).

Relatively inaccessible material which has been investigated includes bone and teeth [Boyde, 1985, 1987, 1988; Boyde et al., 1983; Jones and Boyde, 1987; Petráň et al., 1985; Watson and Boyde, 1987].

A very recent application of confocal fluorescent microscopy has been the study of digestion in young fish with relatively transparent bodies by using fluorescent labelled food particles [Lim, 1988]. Applications to sphingolipid sorting have been described by van Meer et al. (1987).

Interesting suggested areas of application, amongst many that could be made, are the study of the three-dimensional structure of bacterial colonies [Shapiro, 1988], striated and cardiac muscle fibre reconstruction [Sarafis and Brakenhoff, 1988], plasmodesmatal connections between plant cells [Carr, 1976], lichen thallus construction [Ahmadjian and Hale, 1973], chloroplasts of shade plants [Lee, 1983], algae in mussel shells such as *Ostreobium* and *Conchocelis* [Oltmanns, 1922], the substomatal alga *Stomatochroon* in its habitat of the substomatal space in leaves [Oltmanns, 1922], the algae *Trichophilus* and *Cyanoderma* found in the hairs of sloths [Oltmanns, 1922], mycorrhizae in roots [Harley, 1969], cellular slime mould slugs [Vardy et al., 1986], *Sacculina*, a parasitic crustacean in crabs [Barnes, 1987], and giant pheromone-secreting cells in noctuids [Clearwater and Sarafis, 1971].

IV.4. Advantages of confocal fluorescence microscopy for increasing resolving power

The above studies and possible applications demonstrate some of the useful features of confocal fluorescence microscopy. Yet it should be realised that the resolution claimed for an infinitely thin specimen does not hold for a strong absorbing specimen, which may require specialised spatial filters and three-dimensional tomography by specimen rotation to overcome this problem. Spatial filters may be used to enhance lateral resolution or, in apodisation, to extend the depth of field or to reduce it [Magiera and Magiera, 1987; Hegedus and Sarafis, 1986; Hegedus, 1989]. They may be combined to achieve a particularly desirable optical probe.

Fluorescence microscopy should be a prime target for the use of spatial filters, because the optical transfer function for confocal fluorescence microscopy, whilst cutting the abscissa at an optical frequency twice as high as conventional fluorescence microscopy, closely hugs the abscissa from the classical limit up to the spatial frequency cut-off [Cox et al., 1982]. Spatial filters raise the optical transfer function height from the abscissa quite considerably [Cox et al., 1982] and would enable easier visualisation of details at the new lateral resolution limit.

Histochemistry offers new avenues for confocal microscopy. Several errors are known from classical microscopy. They relate to the unequal distribution of dye in particulate high extinction units which results in the underestimation of the chemical being measured. Another problem is that the depth discrimination of classical microscopy is so poor that it is difficult to obtain an accurate assessment of chemical concentrations. Confocal microscopy does not suffer from these limitations, and, in addition, identifies material from a much smaller volume than in classical microscopy. It is therefore the method of choice, particularly when fluorescence techniques are to be used. This method lends itself to mathematical software manipulation, with a view to displaying rates of change of tissue and cell components, indicating an average of levels above a particular set level.

V. Application to stereology

Another application of confocal microscopy is in the area of stereology. It is a known limitation of this technique that section thickness contributes to error [Weibel, 1969]. The apparent thickness from which the signal is derived in the confocal microscope comes only from the plane of focus, with virtually no contributions from other planes. The accuracy of stereology in confocal microscopy is therefore expected to be higher.

VI. Confocal microscopy applications for real-time imaging

In the future, confocal microscopy may increasingly become a method of choice, not only because of its unique ability to solve problems but because of the likely development of user-friendly instruments. A serious difficulty to overcome is imaging in real time. This may be achieved by speeding up the scanning stages or by using honeycomb scanning lens systems in which the scanning field is dissected, so that only a small proportion of the honeycomb is dedicated to scanning a particular part of the object. This would imply that only a small scan amplitude would be needed to image the entire field, and hence the speed of image acquisition would be considerably increased. This would necessitate new optical designs which could come into being only if demand for such instrumentation increases dramatically. Beam-scanning devices, whilst recently developed into real-time scanning systems, often suffer from uncorrectable off-axis image defects. However, the ease of real-time imaging should be balanced against the better optical characteristics of object scan and, at high magnifications, the decay in the point spread function of highly corrected objectives near the axis is probably tolerable.

Immobilizing the objects in calcium alginate [Draget et al., 1988] offers a means of examining otherwise mobile organisms. A recently developed, more powerful method can be used to immobilize cells and intracellular organelles as well as to manipulate them by using focussed infrared laser beams [Ashkin et al., 1987].

VII. White light confocal microscopy

Some scientists would very much like to see things in white light and thus benefit from colour viewing, having every possible chromatic cue in the image. With one laser this is certainly not easily possible, but with three laser lines, red, green and blue, traversing the appropriate dichroic mirrors, the specimen will be scanned simultaneously by three different colours at the same juncture, and the three different colours emerging from the objective can be divided to arrive at three different detectors to synthesize a three-colour image. A commercial instrument based on this principle is available from Lasertec Corporation.

Scanning tandem light microscopy can utilise white light directly, but has lower lateral and depth resolution [Petráň et al., 1985; Wilson and Sheppard, 1984].

VIII. Extensions of confocal microscopy methodology

Another useful area for confocal microscopy is polarisation imaging. For example, the technique of combining condenser and objective, as in classical microscopy would open up new possibilities for Nomarski [Horikawa et al., 1987], phase contrast and darkfield microscopy.

The technique of fringe-scan flow cytometry [Mullikin et al., 1988] may also be improved by implementing confocal techniques. This should result in higher sensitivity and permit smaller specimens to be analysed.

Extensions of confocal scanning microscopy could be made into the short wavelength domain by, for example, the use of quartz optics for the ultraviolet or zone plates for the soft x-ray zone. Scanning electron microscopy could also be converted to the confocal mode. Application of such methods would allow very much higher resolutions than are currently available from the nonconfocal versions of these microscopes.

References

Ahmadjian, V. and Hale, M.E., (1973). *The lichens.* Academic Press.

Ashkin, A., Dziedzic, J. M. and Yamane, T., (1987). Optical trapping and manipulation of single cells using infrared laser beams. Nature, **330**, 769-771.

Bajer, A. S., Sato, H. and Mole-Bajer, J., (1986). Video microscopy of colloidal gold particles and immuno-gold labelled microtubules in improved rectified DIC and epi-illumination. Cell Struct. Funct., **11**, 317-330.

Baker, J. R. J., (1988). Autoradiography - past, present and future. Microscopy and Analysis, **7**, 43-45.

Barnes, R. D., (1987). *Invertebrate Zoology.* 5th Edition. CBS Saunders.

Benschop, J. P. H., (1987). Confocal differential phase contrast in scanning optical microscopy. Proc. S.P.I.E., **809**, 90-96.

Boyde, A., (1985). Stereoscopic images in confocal (tandem scanning) microscopy. Science, **230**, 1270-1272.

Boyde, A., (1987). Applications of tandem scanning reflected light microscopy and 3-dimensional imaging. Ann. New York Acad. Sci., **483**, 428-439.

Boyde, A., (1988). Confocal optical microscopy. Microscopy and Analysis, Jan. 1988, 7-13.

Boyde, A., Petráň, M. and Hadravský, M., (1983). Tandem scanning reflected light microscope (TSRLM). J. Microsc., **146**, 137-142.

Brakenhoff, G. J., (1979). Confocal scanning light microscopy (CSLM). J. Microsc., **117**, 233-242.

Brakenhoff, G. J., (1984). Confocal scanning light microscopy. In: *Analysis of organic and biological surfaces.* Echlin, P. (Editor), 285-300. J. Wiley.

Brakenhoff, G. J., Binnerts, J. S. and Woldringh, C. L., (1980). Developments in high resolution confocal scanning light microscopy (CSLM). In: *Scanned image microscopy.* Ash, E. A. (Editor), 183-200. Academic Press.

Brakenhoff, G. J., Blom, P. and Barends, P., (1979). Confocal scanning light microscopy with high aperture immersion lenses. J. Microsc., **117**, 219-232.

Brakenhoff, G. J., van der Voort, H. T. M., van Spronsen, E. A., Linnemans, W. A. M. and Nanninga, N., (1985). Three-dimensional chromatin distribution in neuroblastoma nuclei shown by confocal scanning laser microscopy. Nature, **317**, 748-749.

Brakenhoff, G. J., van der Voort, H. T. M., van Spronsen, E. A. and Nanninga, N., (1986). Three-dimensional imaging by confocal scanning fluorescence microscopy. Ann. New York Acad. Sci., **483**, 405-415.

Brakenhoff, G. J., van der Voort, H. T. M., van Spronsen, E. A. and Nanninga, N., (1987). Three-dimensional imaging of biological structures by high resolution confocal scanning laser microscopy. Scanning Microscopy, **2**, 33-40.

Bright, G. R., Fisher, G. W., Rogowska, J. and Taylor, D. L., (1987). Fluorescence ratio imaging microscopy: temporal and spatial measurements of cytoplasmic pH. J. Cell Biol., **104**, 1019-1033.

Busby, C. H. and Gunning, B. E. S., (1988a). Establishment of plastid-based quadripolarity in spore mother cells of the moss *Funaria hygrometrica*. J. Cell Sci., **91**, 117-126.

Busby, C. H. and Gunning, B. E. S. (1988b). Development of the quadripolar meiotic cytoskeleton in spore mother cells of the moss *Funaria hygrometrica*. J. Cell Sci., **91**, 127-137.

Carlsson, K. and Åslund, N., (1987). Confocal imaging for three-dimensional digital microscopy. Appl. Opt., **26**, 3232-3238.

Carlsson, K., Danielsson, P. E., Lenz, R., Liljeborg, A., Majlöf, L. and Åslund, N., (1985). Three-dimensional microscopy using a confocal laser scanning microscope. Opt. Lett., **10**, 53-55.

Carr, D. J., (1976). Plasmodesmata in growth and development. In: *Intercellular communication in plants: studies on plasmodesmata*. Gunning, B. E. S. and Robards, A. W. (Editors), 243-289. Springer-Verlag.

Clearwater, J. R. and Sarafis, V., (1973). The secretory cycle of a gland involved in pheromone production in the noctuid moth, *Pseudaletia separata*. J. Insect Physiol., **19**, 19-28.

Coleman, A. E., (1985). Diversity of plastid DNA configuration among classes of eukaryote algae. J. Phycol., **21**, 1-16.

Cox, G., (1988). Confocal scanning optical microscopy. Australian EM Newsletter, **20**, 3-7.

Cox, I. J., Sheppard, C. J. R. and Wilson, T., (1982). Super-resolution by confocal fluorescence microscopy. Optik, **60**, 391-396.

Draget, K. I., Myhre, S., Evjen, K. and Ostgaard, K., (1988). Plant protoplasts immobilized in calcium alginate: a simple method of preparing fragile cells for transmission electron microscopy. Stain Tech., **63**, 159-164.

Franksson, D., Liljeborg, A., Carlsson, K. and Forsgren, P-O., (1987). Confocal laser microscope scanning applied to three-dimensional studies of biological specimens. Proc. S.P.I.E., **809**, 124-129.

Goodchild, D., (1971). C.S.I.R.O. Division of Plant Industry, Canberra. Personal communication.

Grynkiewicz, G., Poenie, M. and Tsien, R. Y., (1985). A new generation of Ca indicators with greatly improved fluorescence properties. J. Biol. Chem., **260**, 3440-3450.

Harley, J. L., (1969). *The biology of mycorrhiza.* 2nd Edition. Leonard Hill.
Hawkes, C., (1988). Subcellular localization of macromolecules by microscopy. In: *Plant molecular biology - a practical approach.* Shaw, C. H. (Editor), 103-130. IRL Press.
Heertje, I., van der Vlist, P., Blonk, J. C. G., Hendricks, H. A. C. M. and Brakenhoff, G. J., (1987). Confocal scanning laser microscopy in food research: some observations. Food Microstructure, **6**, 115-120.
Hegedus, Z. S., (1989). Pupil filters in confocal imaging. This volume.
Hegedus, Z. S. and Sarafis, V., (1986). Superresolving filters in confocally scanned imaging systems. J. Opt. Soc. Am., **3**, 1892-1896.
Hoefsmit, E. C. M., Korn, C., Blijleven, N. and Ploem, J. S., (1986). Light microscopical detection of single 5 and 20nm gold particles used for immunolabelling of plasma membrane antigens with silver enhancement and reflection contrast. J. Microsc., **143**, 161-169.
Horikawa, Y., Yamamoto, M. and Dosaka, S., (1987). Laser scanning microscope: differential phase images. J. Microsc., **148**, 1-10.
Inoué, S., (1986). *Video microscopy.* pp. 584. Plenum Press, New York.
Jones, S. J. and Boyde, A., (1987). Scanning microscopic observations on dental caries. Scanning Microsc., **1**, 1991-2002.
Koch, G. L. E., Macer, D. R. J. and Smith, M. J., (1987). Visualization of the intact endoplasmic reticulum by immunofluorescence with antibodies to the major ER glycoprotein, endoplasmin. J. Cell Sci., **87**, 535-542.
Lee, D. W. (1983). Unusual strategies of light absorption in rain-forest herbs. In: *On the economy of plant form and function.* Givnish, T. J. (Editor), 105-129. Cambridge University Press.
Lim, M., (1988). Zoology Department, National University of Singapore, Singapore. Personal communication.
Lipsky, N. G. and Pagano, R. E., (1985). A vital stain for the Golgi apparatus. Science, **228**, 745-747.
Magiera, A. and Magiera, L., (1987). Two-point resolution in conventional and confocal microscopes with apodizers. Atti. Fond. Giorgio Ronchi, **42**, 517-523.
Mineyuki, Y., Wick, S. M. and Gunning, B. E. S., (1988). Preprophase bands of microtubules and the cell cycle: kinetics and experimental uncoupling of their formation from the nuclear cycle in onion root-tip cells. Planta, **174**, 518-526.
Mole-Bajer, J. and Bajer, A. S., (1988). Relation of F-actin organization to microtubules in drug treated haemanthus mitosis. Protoplasma, **1**, 99-112.
Mullikin, J., Norgren, R., Lucas, J. and Gray, J., (1988). Fringe-scan flow cytometry. Cytometry, **9**, 111-120.
Olive, P. L., Biaglow, J. E., Varnes, M. E. and Durand, R. E., (1982). Characterization of the uptake and toxicity of a fluorescent thiol reagent. Cytometry, **3**, 349-353.

Oltmanns, F., (1922). Morphologie und Biologie der Algen. 2 Aufl., 3 Bde. G. Fischer.
Petráň, M., Hadravský, M., Benes, J., Kucera, R. and Boyde, A., (1985). The tandem scanning reflected light microscope. Part 1: The principle and its design. Proc. Roy. Microscop. Soc., **20**, 125-129.
Sarafis, V., van Spronsen, E. A. and Brakenhoff, G. J., (1987). Unpublished.
Sarafis, V. and Brakenhoff, G. J., (1988). Unpublished.
Shapiro, H. M. and Stephens, S., (1985). Flow cytometry of DNA content using Oxazine 750 or related laser dyes with 633nm excitation. Cytometry, **7**, 107-110, (1986).
Shapiro, J. A., (1988). Bacteria as multicellular organisms. Scientific Amer., 62-69.
Sheppard, C. J. R., (1987). Scanning optical microscopy. Adv. in Opt. and Electron Microscopy, **10**, 1-98.
Spencer, D., (1971). C.S.I.R.O. Division of Plant Industry, Canberra. Personal communication.
Stelzer, E. H. K. and Wijnaendts-van-Resandt, R. W., (1986). Applications of fluorescence microscopy in three dimensions: microtomoscopy. Proc. S.P.I.E., **602**, 63-70.
Tanke, H. J., van Oostveldt, P. and van Duijn, P., (1981). A parameter for the distribution of fluorophores in cells derived from measurements of the inner filter effect and reabsorption phenomenon. Cytometry, **2**, 359-369.
Truneh, A. and Machy, P., (1987). Detection of very low receptor numbers on cells by flow cytometry using a sensitive staining method. Cytometry, **8**, 562-567.
Valkenburg, J. A. C., (1985). The nucleoid of *Escherichia coli* as visualized by confocal scanning light microscopy, phase contrast microscopy and electron microscopy. Ph.D. Thesis, University of Amsterdam.
Valkenburg, J. A. C., Woldringh, C. L., Brakenhoff, G. J., van der Voort, H. T. M. and Nanninga, N., (1985). Confocal scanning light microscopy of the *Escherichia coli* nucleoid: comparison with phase-contrast and electron microscope images. J. Bacteriol, **161**, 478-483.
van der Voort, H. T. M., Brakenhoff, G. J., Valkenburg, J. A. C. and Nanninga, N., (1985). Design and use of a computer controlled confocal microscope for biological applications. Scanning, **7**, 66-78.
van der Voort, H. T. M., Brakenhoff, G. J. and Janssen, G. C. A. M., (1987a). Determination of the three-dimensional optical properties of a confocal scanning laser microscope. Optik, **78**, 48-53.
van der Voort, H. T. M., Brakenhoff, G. J., Janssen, G. C. A. M., Valkenburg, J. A. C. and Nanninga, N., (1987b). Confocal scanning laser fluorescence and reflection microscopy: measurements of the three-dimensional image formation and application in biology. Proc. S.P.I.E., **809**, 138-143.
van der Voort, H. T. M., Brakenhoff, G. J. and Baarslag, M. W., (1987c).

Three-dimensional visualization methods for confocal microscopy. J. Microsc., **151**.

van der Voort, H. T. M., (1989). University of Amsterdam, Department of Molecular Cytology, Netherlands. Personal communication.

van Meer, G., Stelzer, E. H. K., Wijnaendts-van-Resandt, R. W. and Simons, K., (1987). Sorting of sphingolipids in epithelial (Madin-Darby canine kidney) cells. J. Cell Biol., **105**, 1623-1635.

van Spronsen, E. A., Sarafis, V., Brakenhoff, G. J., van der Voort, H. T. M. and Nanninga, N., (1989). Three-dimensional structure of living chloroplasts as visualized by confocal scanning laser microscopy. Protoplasma 148/1. To be published.

Vardy, P. H., Fisher, L. R., Smith, E. and Williams, K. L., (1986). Traction proteins in the extracellular matrix of dictyostelium discoideum slugs. Nature, **320**, 526-529.

Watson, T. F. and Boyde, A., (1987). Tandem scanning reflected light microscopy: application in clinical dental research. Scanning Microsc., **1**, 1971-1981.

Weibel, E. R., (1969). Stereological principles for morphometry in electron microscopic cytology. Int. Rev. Cyt., **26**, 235-299.

Wilson, T. and Carlini, A. R., (1988). Three-dimensional imaging in confocal imaging systems with finite sized detectors. J. Microsc., **149**, 51-66.

Wilson, T. and Sheppard, C. J. R., (1984). *Theory and practice of scanning optical microscopy*. Academic Press, London.

Woldringh, C. L. and Nanninga, N., (1985). Structure of nucleoid and cytoplasm in the intact cell. In: *Molecular cytology of Escherichia coli*. Nanninga, N. (Editor), 161-197. Academic Press.

13. Semiconductor Metrology

R. W. WIJNAENDTS-VAN-RESANDT

I. Introduction

In modern semiconductor manufacturing, structure geometries are shrinking while chip complexity, number of processing steps, wafer size and number of mask levels are increasing. This change has been taking place in the last decades and is expected to continue. New tools continue to be developed to meet the requirements of modern, highly sophisticated device processing. Measurement and inspection methods become more and more important for tight process control and quality assurance. The size of today's smallest structures has now reached the wavelength of visible light, and classical microscopy methods no longer satisfy the resulting sensitivity requirements. Furthermore, in many cases the aspect ratio of the structures to be measured is rapidly increasing, and a single value no longer accurately describes such a structure's size. Measurement and inspection equipment is currently in transit from operator-intensive to fully automated systems. This means fully robotic systems, including handling, measurement and decision. Only these robots will be able to overcome the limitations of the human operator. The end of this evolution will be the fully automatic wafer-processing facility.

Analytical microscopy of two-dimensional structures of known geometry is in principle not limited by optical resolution, and size measurements of very high accuracy and precision are possible with a proper analysis of the reflected intensity signals. However, in the typical case of resist structures on some substrates, a number of additional parameters influence the final resulting wave form, and a straight forward inversion is no longer possible. Confocal microscopy possesses a real three-dimensional response which can be used to obtain some of the necessary additional information. Because of this property, confocal microscopy has been able to extend the applicability of optical methods to sub-micron metrology and inspection, and in the past few years a number of systems have been developed especially for in-process semiconductor measurements [Kleinknecht and Meier, 1985; Hamilton and Wilson, 1982; Zapf and Wijnaendts-van-Resandt,

1986; Lindow et al., 1985]. An estimated total of approximately 30 to 35 systems are presently in use. The performance of most of these systems is still being tested in pre-production lines, and more information is needed in order to determine the exact application region of the confocal microscope in semiconductor metrology. The most promising areas seem to be in-process linewidth and overlay measurements of VLSI circuits down into the sub-micron region, linewidth and size measurements for reticle and mask processing, and general microscopy for circuit inspection and defect detection. In addition, three-dimensional microscopy of larger structures can be used for a tighter inspection of various other process steps in semiconductor manufacturing. In the area of sub-micron metrology, a strong competition currently exists between sophisticated optical, confocal techniques and low resolution electron-microscopical methods. The purpose of this chapter is to review the current status of confocal optical metrology. The basic requirements of VLSI metrology will be discussed, followed by a short paragraph on the most important features of confocal optical size measurements. The current state of the art will be shown with the help of some recent results, and a short comparison will be made between the low voltage SEM (Scanning Electron Microscope) and the confocal microscope.

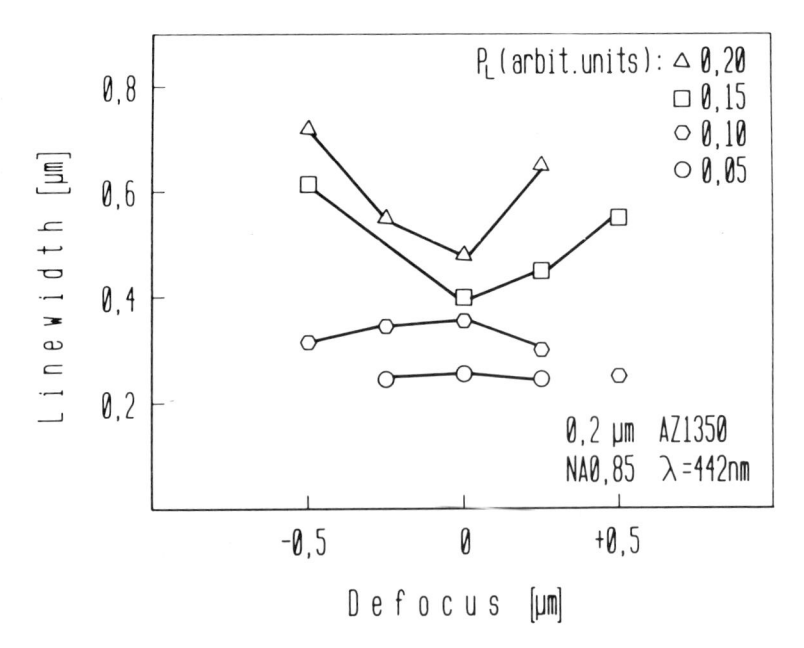

Figure 13.1. Linewidth results for various focus levels at different exposure energies, P_L, obtained with a laser-scanning lithographic tool [Ulrich et al., 1987]. The measurements were made with the Heidelberg Instruments LPM.

Semiconductor Metrology 341

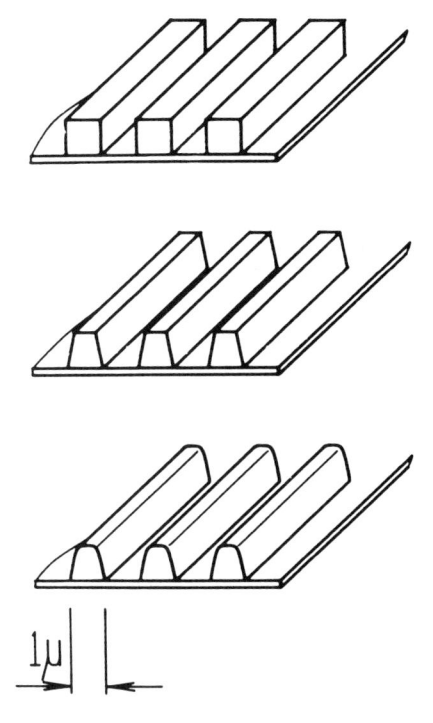

Figure 13.2. Schematic diagram of a 1-micron line of 1-micron-thick resist at various conditions. All these structures should yield the same linewidth measurement result, because the bottomwidth is the same.

II. VLSI metrology

The most important purpose of metrology in semiconductor processing is the control of the optical lithographic process [Harris *et al.*, 1983]. Two major critical dimensions are of interest: linewidth measurements and overlay measurements.

II.1. Linewidth measurements

A strong dependence exists between the dimensions of the pattern produced and focus level and illumination energy. A practical example of this dependence is shown in Figure 13.1. Figure 13.2 shows schematically a typical structure of a one-micron resist line. The resist thickness is usually also of the order of one micron, and it is obvious that a single measurement is not able to specify structure size. However, for the final etching process, which succeeds the resist development, the bottom linewidth is the crucial parameter.

II.2. Registration measurements

Overlay measurements are of crucial importance in checking the registration performance of the lithographic tool. Continuous shrinking of the overlay budget requires measurement procedures with very high accuracy and precision. Measurements are usually performed on special structures. Figure 13.3 shows such a structure as well as the overlay measurement results from the measured offset of the inner box with respect to the outer box. The boxes are printed in different lithography cycles, and hence lie at different focal planes of the microscope system.

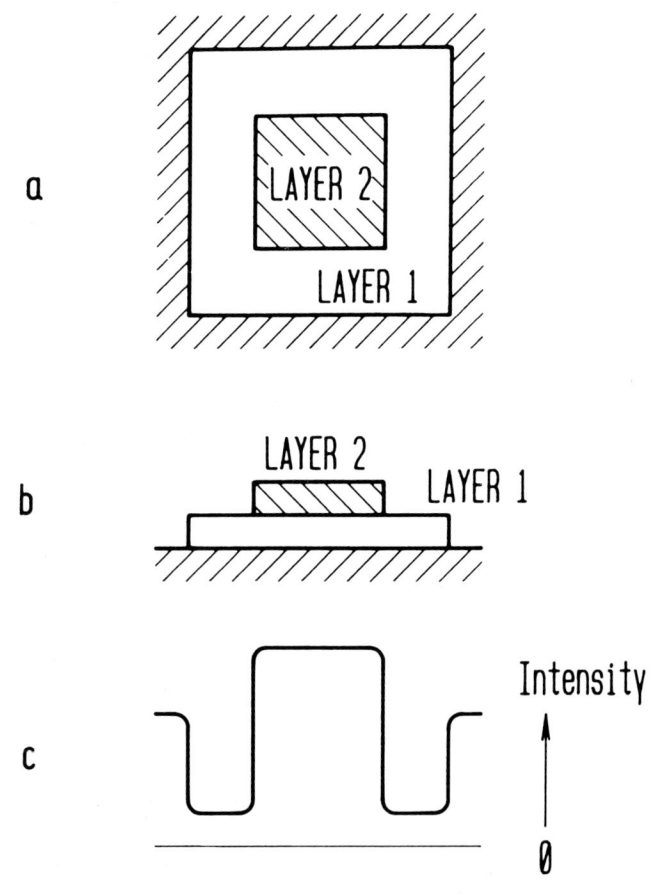

Figure 13.3. Schematic diagram of a typical overlay measurement structure. The two structures are at different focal levels, and the relative position of the inner box must be determined with respect to the outer box. The extended focus intensity signal is shown below. The contrast is usually obscured by additional layers.

II.3. Defect inspection

The reticle is one of the most sensitive elements in the lithography cycle, and small defects should be detected as soon as possible. Since one reticle usually contains two or more identical image fields (dies), repeating defects can be found by directly comparing the images resulting from these two fields. In the future, also, direct comparison of a field to a database will be needed. This is shown schematically in Figure 13.4. The detection of small structural differences in the printed pattern requires a very different measurement strategy. Pattern inspection on wafers is still mainly done manually, which results in long turn-around times and limited defect detection sensitivity. In addition, as structures become smaller and die sizes increase, the number of allowable defects per cm^2 requires that large areas be inspected. This is impossible manually, which means that fast, sensitive, automatic systems are needed.

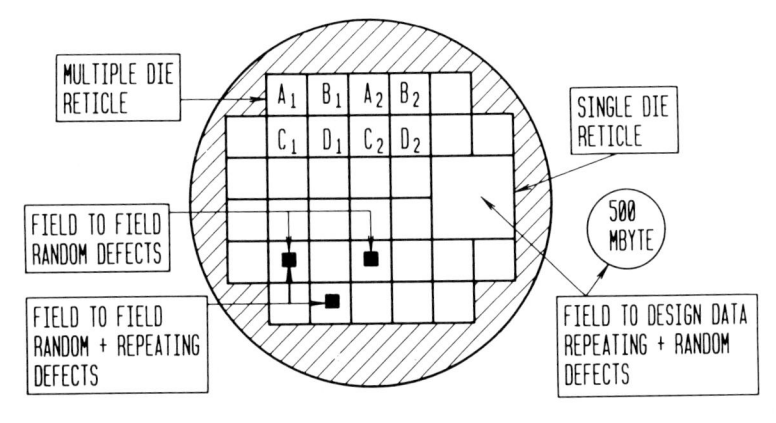

Figure 13.4. An example of a defect detection scheme. A number of possible ways of detecting defects in an image-to-image comparison are shown. As long as one reticle contains more than one identical field, image-to-image comparison can yield random process-dependent defects (A1 to A2, B1 to B2, etc.) and repeating mask-dependent defects by comparing A1 to B1, C1 to D1 etc. In the near future, database or "reference" image comparison will be needed.

II.4. Critical dimensions of masks and reticles

In addition to critical dimension measurements on wafers, similar requirements exist for masks and reticles. In this case, the object is optically better defined as the three-dimensional character is less pronounced. The measurement of resist structures before etching is also important. The final measurements of etched structures are twofold:

the measurement of linewidth variations in the mask or reticle field and the measurement of absolute dimensions over larger distances. For defect inspection, high resolution image comparison is necessary, including the direct comparison to the design data. Both tasks should be suitable for specially adapted confocal systems. Figure 13.5 reviews the current and expected near-future requirements for wafer and mask metrology.

VLSI METROLOGY REQUIREMENTS	CURRENT < 1990	FUTURE > 1990
Min. linewidth on wafer:	1 - 0.7 μ	< 0.5 μ
ACCURACY:		
Registration	0.05 - 0.02 μ	< 0.02 μ
Linewidth:	0.10 - 0.05 μ	< 0.05 μ
REPEATABILITY		
Registration	0.01 - 0.005 μ	< 0.005 μ
Linewidth:	0.02 - 0.01 μ	< 0.005 μ
APPLICATIONS:		
WAFERS:	High aspect ratio resist structures on various substrates and substrate combinations	
MASKS:	Before and after resist stripping and etching 1X, 5X and 10X reticles	

Figure 13.5. Overview of expected near-future requirements for VLSI metrology.

II.5. Other optical inspection techniques

There is a large number of other optical inspection methods which are being used in semiconductor manufacturing, such as macro-inspection of whole wafers in colour and flatness, layer thickness measurements, particle detection on structured and non-structured wafers, endpoint detection for etchers, general microscopy, etc. Most of these techniques have not been the subject of current confocal methods and will not be treated in detail.

III. Quantitative confocal microscopy

The typical confocal microscope consists of a laser source which illuminates a pinhole. The laser can be in the ultraviolet, near infrared or in the visible, depending on the particular application. The source pinhole is imaged onto the structure to be inspected by a high numerical aperture microscope objective. The reflected light is collected by the same objective and imaged onto the detection pinhole. This principle is shown in Figure 13.6. A sensitive optical detector is placed behind the collection pinhole and records only the reflected light intensity at the pinhole. This system measures the reflectivity of a single point of the object, and only by some means of scanning can an image be generated. Thus, the optical system does not directly produce an image, and the complete optical system is being used to obtain the information from a single pixel. This property makes possible the very special contrast modes which are needed to satisfy the requirements of measurements on complex objects which produce phase and amplitude information. One of these properties is real resolution along the optical axis, which has been discussed on many occasions and is the main subject of this book. In the case of VLSI metrology, this means that a full three-dimensional data set of the structure to be inspected can be used for size analysis, and the results are independent of the system's focus conditions. Furthermore, owing to the fact that the microscope is in principle a double system (one microscope for illumination and one for detection), the impulse response is the product of the two single responses. This results in a slightly better resolution and, which is more important, the suppression of the usual edge

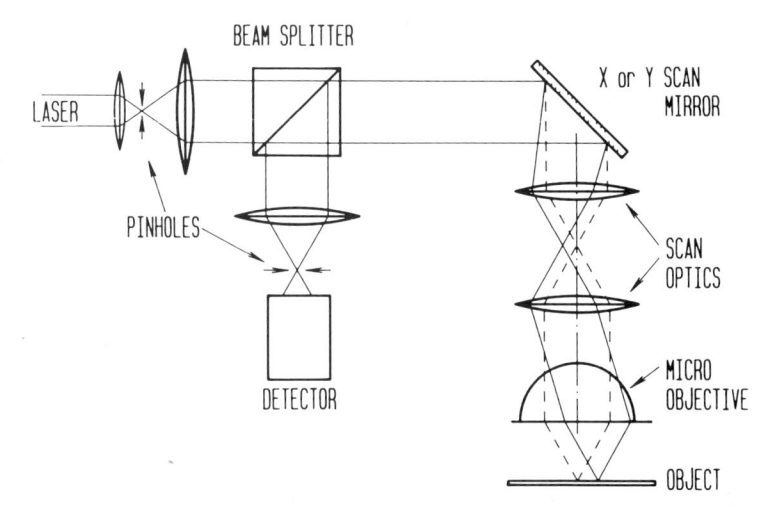

Figure 13.6. Schematic diagram of a confocal beam scanning reflection microscope.

ringing of a coherent system. The signals of a simple two-dimensional structure can be computed by integration of the two-dimensional response function, and are shown in Figure 13.7 for a coherent system and a confocal system. Finally, there is a property which is also very important for sub-micron defect detection. A very high signal-to-noise ratio can be achieved through the use of a laser for illumination and a very efficient detector for signal measurement.

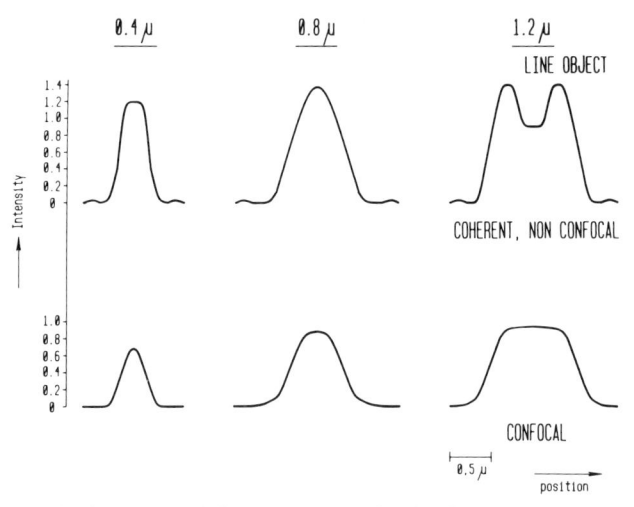

Figure 13.7. Calculated signal from a non-confocal coherent scanning microscope and a confocal scanning microscope. The signals have been computed for ideal lines (the closest to this is, for instance, chrome on glass) in reflection mode. For this calculation: wavelength = 488nm and N.A. = 0.85. Note the strong edge ringing in the non-confocal system.

III.1. The z-response

The most important property of confocal microscopy is the z-response, which should be treated in somewhat more detail. The theoretical intensity for image points along the optical axis can be approximated by:

$$I(u) = I_o \left[\frac{\sin(u/4)}{u/4} \right]^2 \quad (13.1)$$

with $u = kz \sin^2 \alpha$ [Born and Wolf, 1986], for the case of a plane wave incident on the microscope objective used. In Figure 13.8 this is compared to a practical measurement of the response function, and a rather severe discrepancy can be observed. Detailed analysis of this effect at the University of Heidelberg shows that the difference is not due to the approximations which lead to Equation (13.1) but are the result of technical limitations of the high numerical objectives [Hell et al., in preparation].

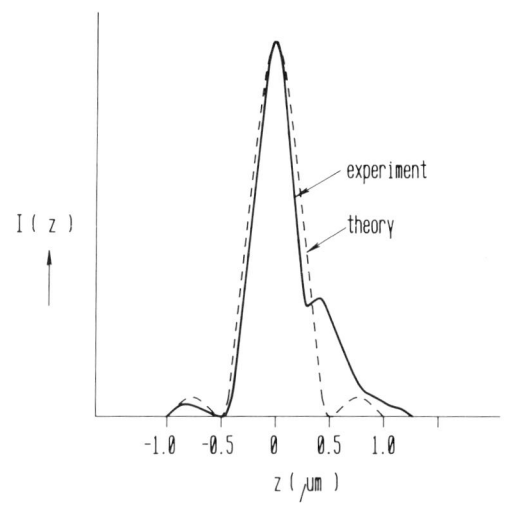

Figure 13.8. Comparison of the theoretical z-response function and the experimentally measured response. The conditions were: wavelength =498nm and N.A. = 0.95.

III.2. Edge-response, object scan versus beam scan

The best measurement of resolution for a metrology tool is the response of the system to an edge. The measurement of the signal resulting from an "infinitely" sharp edge is more possible than that from an infinitely small line. In addition, the result of this measurement can be used as a quality parameter for microscope performance. Since a beam scanning confocal microscope makes different use of the various optical components, in particular of the micro-objective, it is of interest to compare the results of a beam-scanning and an object-scanning system. The edge response was measured by scanning across the very sharp edge of a razor blade. Figure 13.9 shows the results for both the beam-scanning LPM in the left hand panel and the same result for the object-scanning confocal microscope at the European Molecular Biology Laboratory (EMBL) in Heidelberg [Marsman et al., 1983] in the right hand panel. Both edge responses are compared to the theoretically expected curve [Sheppard and Choudhury, 1977]. Both edge scans have been made under rather different circumstances. At the EMBL, the curve was measured using 280nm light from a frequency-doubled, argon ion pumped dye laser and a Zeiss axiomat ultrafluar 125X 1.25 glycerin immersion objective. The measurements were done in the object-scanning mode, using confocal transmission contrast. At Heidelberg with the LPM, an argon ion laser operating at 488nm, a Leitz 50X 0.85 objective, and beam-scanning reflection contrast were used. In order to ascertain that the results did not vary across the field used, several measurements on different locations in the field of view were averaged. For comparison between the results of the beam-scanning and the object-scanning system, the curves were normalized to each other, and the result is that both curves are identical within the experimental uncertainty. This means that the deviation from theory for both results is also identical, which probably indicates that the assumptions which underlay the theoretical calculations are not quite correct. This observation was also made for the edge response along the optical axis for the case of fluorescence contrast, and probably results from the fact that the small angle approximations used for the calculations do not apply for high numerical aperture microscope objectives. For the purpose of this discussion, the main result is the impressive agreement between the two experimental curves and the confirmation that the properties of a beam-scanning confocal microscope are not inferior to those of a stage-scanning system.

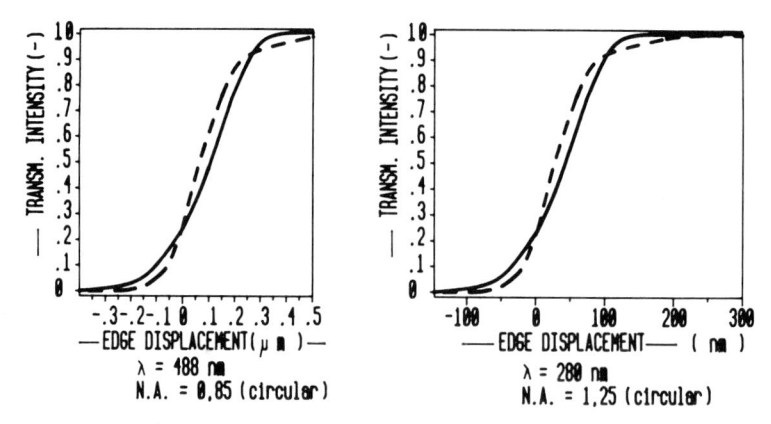

Figure 13.9. Measurement of the edge response for a beam-scanning confocal microscope (left hand panel) and a stage-scanning microscope (right hand panel). In both cases the solid curve is the theoretically expected curve.

III.3. Confocal contrast modes

The measurement of the reflected signal is the most common contrast mode. However, a number of other properties can be used to obtain additional information about the object. For semiconductor metrology, the most important are phase contrast, interference contrast, polarization contrast and darkfield contrast. Fluorescence could also play an important role in the future [Williams et al., 1988]. The major difficulty in the theoretical analysis of the collected data is that most of the usual images are some combination of all of the contrast mechanisms described. It depends very much on the object to be measured which contrast mode is most important and yields the best results for size determination.

III.4. Data quality and analysis

The measurement of a real size from an intensity signal does not only require a very good intensity measurement with a large signal-to-noise ratio, but also a high precision measurement of the position of the laser with respect to the object. Both parameters directly influence the resulting precision (or repeatability) and accuracy. The accuracy of the result presents the possible deviation from the absolute value. As an absolute value, some reference to a standard should be made.

IV. Technical realizations and results

The practical realizations of confocal metrology equipment are quite different depending on the confocal implementation in use. Currently there are two methods in a rather advanced state: the object-scanning method (Siscan Systems) and the beam-scanning method (Heidelberg Instruments). Although confocal microscopy theoretically does not need a laser as a light source, the laser is the only practical, very bright point source. Recently a number of alternative systems have been tested, which have similar properties: the Linnik interference microscope [Davidson et al., 1987] and the quasi-confocal white light system using multiple point sources [Kino et al., 1988]. These, perhaps alternative, methods are beyond the scope of this chapter. We will concentrate on the technical principles of the laser confocal systems.

IV.1. Stage-scanning systems

The early confocal systems developed were of the object-scanning type, since it is attractive to have a single optical path at rest at the center of the optical components used. This seems to relax the need for optical corrections, such as flat field, astigmatism, coma, etc. Indeed, it was shown on several occasions [Brakenhoff et al., 1979; Marsman et al., 1983] that carefully constructed confocal object-scanning microscopes perform at the theoretical limits. However, object-scanning systems have one major drawback. In order to accumulate an image within an acceptable time, the object stage must be mechanically scanned with a rather high frequency. E.g., for the accumulation of a 500 x 500 pixel image in 10 seconds, the stage, including the object, must be scanned with a frequency of 50 Hz. This presents severe limitations on object size, mounting and handling. The result is that stage-scanning confocal microscopes are generally difficult to handle and operate. A major effort is involved in developing a stage which can be moved at high speeds and whose position can be measured with nanometer accuracy, in particular for heavy objects such as large masks and reticles.

IV.2. Beam-scanning systems

As an example of a beam-scanning confocal microscope especially designed and built for semiconductor metrology, the Heidelberg LPM will be discussed in more detail. The principle of the optics of the LPM has been shown in Figure 13.6. In more detail: the laser beam

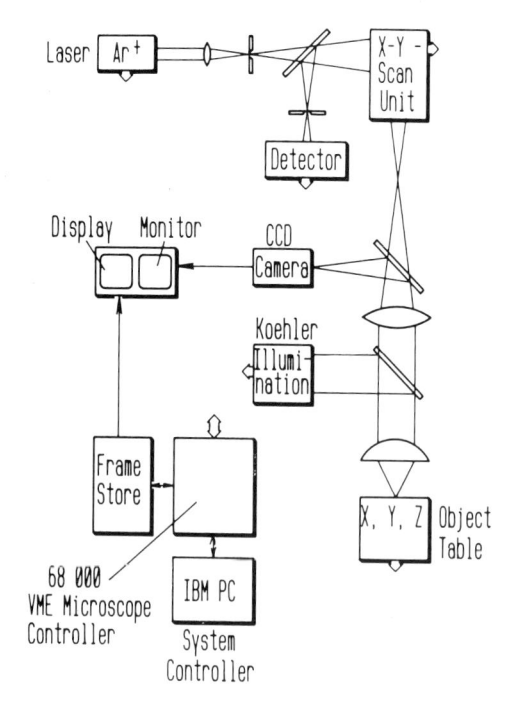

Figure 13.10. Schematic diagram of the main functions of the Heidelberg Instruments LPM. Image collection and processing is performed by the VME/68000 microprocessor and system control is performed by an IBM-PC.

is cleaned with a source pinhole and a parallel beam of a specified diameter is generated. This beam is deflected by two galvanometer mirrors, one for x-scanning and one for y-scanning. The first mirror is imaged into the second by means of standard afocal 4-f optics (see chapter 3). The second scan mirror is imaged in a similar way into the telecentric aperture for the microscope objective. The detector pinhole is situated very close to the excitation pinhole, which ensures a very stable alignment. The size of the source pinhole and the other optics are chosen in such a way that the micro-objective is overfilled and excited by a nearly-plane wave. The detection pinhole is usually somewhat larger, but still satisfies the confocal criterion as shown by Wilson *et al.*, in order to generate fluorescence, polarization and darkfield contrast. In addition to the confocal system, a classical microscope is included. This microscope is used for operator assistance and for automatic alignment and site-finding procedures. The complete block diagram of the system is shown in Figure 13.10. The system includes a wafer-handling system for up to 8 inch wafers, a precision x-y coordinate table for measurement site-positioning and a scanning z-stage with vacuum chuck. An objective revolver allows site-finding with increasing resolution. The system is controlled by a personal computer under a mouse-operated menu. Data collection, image processing and system operation are done by an industrial, high performance, 32-bit microprocessing system. For other applications, the system can be equipped with a general purpose three-dimensional image collection and processing system.

IV.3. Three-dimensional image generation

In principle, the system can be used to collect the entire three-dimensional data set from the xyz-scan of an object. This data can then be used for some form of analysis. An example of the result of such a process is shown in Figure 13.11. However, this process is time- and memory-consuming; and for dimension measurements, a different strategy is chosen. Here an x-z image is collected and analyzed in some form. Next this x-z section is moved along the y-axis as long as specified. In this way, the primary data set is reduced on-line while utilizing the full information in the three-dimensional data. An example of such an x-z image, along with two different processing steps, is shown in Figure 13.12. In this example, the center panel shows the extended focus intensity profile, obtained by adding all information along the z-axis, and the bottom panel shows the height profile, obtained by finding the most likely position of the surface from the raw data.

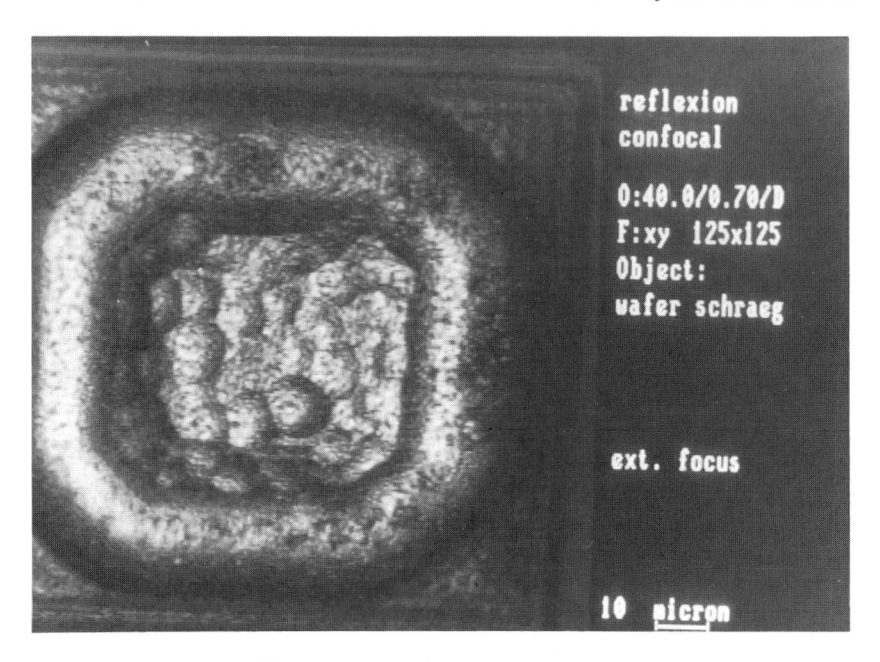

Figure 13.11. Example of the result of a full three-dimensional data collection of a bond pad. The structure is about 25 microns deep, and only through the extended-focus capabilities can all surfaces be shown in focus.

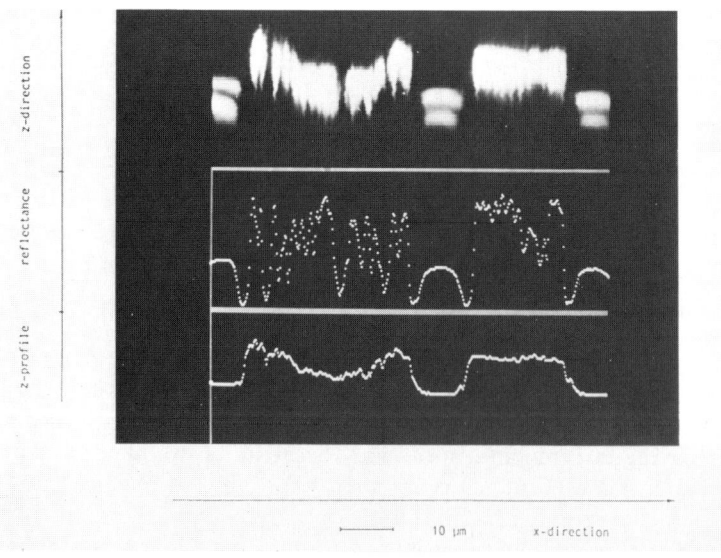

Figure 13.12. An x-z scan through two aluminium lines on silicon. The top image is the rough data set. The center curve is the integrated intensity along z and corresponds to the extended-focus signal. The bottom curve is the calculated surface profile (80X, 0.95 N.A. objective).

IV.4. Data analysis

Semiconductor metrology requires a number of basic critical dimension measurements. Each measurement requires its specific data treatment.

IV.4.1. Pitch measurements

The pitch or distance between two identical structures is measured by selecting the structures in question by means of a double box which is placed over the structure in a scanning x-y image. Next, the three-dimensional dataset inside the box is collected and integrated over the z-axis. The integrated x-y image, which now has an extended depth of focus and is not dependent on the absolute focus conditions of the instrument, is analyzed for pitch. The principle of the data processing is to move the data of the first structure over the data of the second strcuture and to look for optimal cross-correlation between the structures. With this method, the results are not dependent on the detection of certain signal features, and pitch measurements with a very high repeatability can be performed. In principle, this analysis is limited only by the signal-to-noise ratio of the processed data sets. A typical result of such a pitch measurement is shown in Figure 13.13(a).

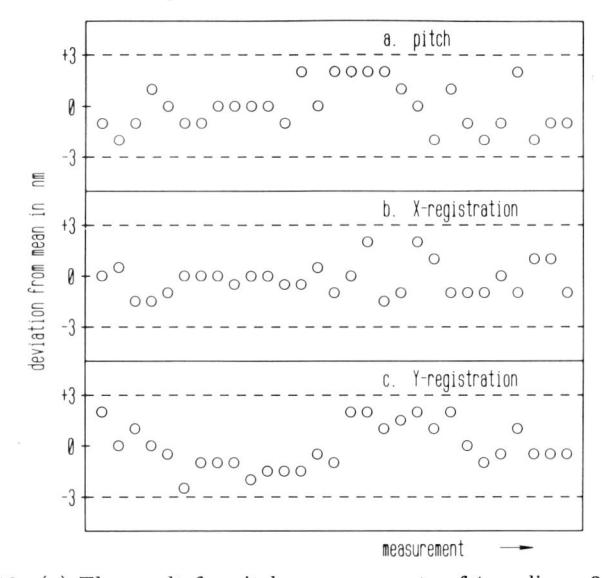

Figure 13.13. (a) The result for pitch measurements of 1μm lines, 2μm pitch of independent measurements at the same site. The 1 sigma standard deviation is 2.1nm. (b) The result for a registration measurement (10μm box in 20μm box) of independent measurements at the same site in the x-direction. The 1 sigma standard deviation is 2.2nm. (c) Same as for (b) for the y-direction. The 1 sigma standard deviation is 2.2nm.

IV.4.2. Registration measurements

The same extended-focus principle is used for registration measurements. For these measurements in particular, this extended depth of focus results in optimal contrast for the edges of the structures in both layers. In normal microscopy, it is impossible to have a sharp image of both layers, which can differ in height level by more than a few microns. In order to be able to measure registration both in x and in y, it is possible to exchange the x-y scanning functions so that the same algorithm can be used for both directions. In one direction, the extended-focus data set is analyzed in terms of symmetry properties between the edges of the overlay structures. This also ensures a specific feature-independent result and an optimal utilization of the information contained in the signals. A typical result of a series of registration measurements is shown in Figures 13.13(b) and 13.13(c).

IV.4.3. Linewidth measurements

It is more diffcult to perform accurate linewidth measurements, for which there are an enormous number of methods and definitions. The main problem results from the fact that for sub-micron linewidth measurements of structures with high aspect ratios, a very complex light scattering problem must be solved [Nyyssonen, 1982]. Confocal microscopy, because of its higher resolution and depth discrimination, can be used to measure sub-micron structures with high repeatability.

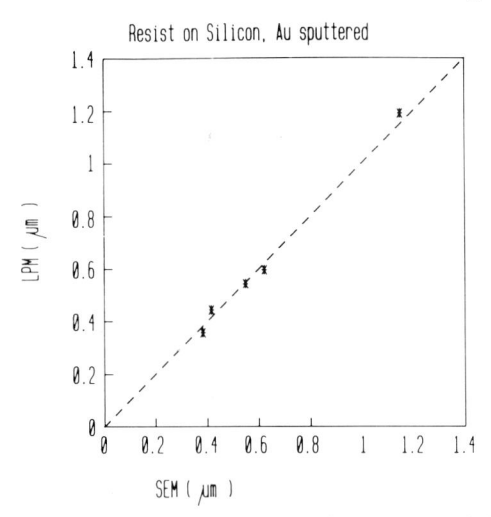

Figure 13.14. Comparison between the results of an SEM and the confocal laser scanning microscope (LPM). The structures are 1.2μm resist on silicon covered with gold.

Semiconductor Metrology 355

The determination of absolute values is, however, very much dependent on the optical properties of the structure to be measured. In most cases, a reference to an SEM image or an up-scaled line will be needed. Further development, both experimental and theoretical, is now taking place. Figure 13.14 shows the comparison between results from the LPM and an SEM for gold-covered resist structures on silicon.

IV.4.4. Height measurements

By measuring across a step and utilizing a x-y or a y-z image, the height of such a step can be measured. The step height is measured in the LPM by comparing the z-level of the maximum in reflectivity left and right of the step. The results are shown in Figure 13.15 and compared to those of a mechanical thickness-measurement tool (Talysurf).

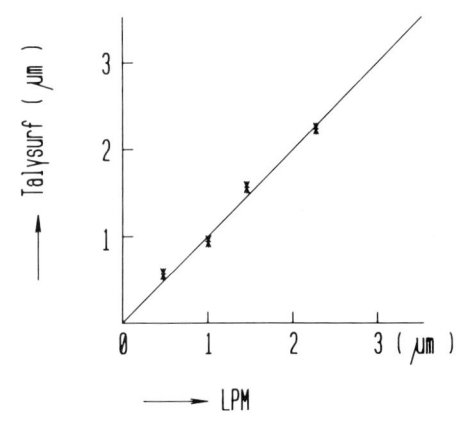

Figure 13.15. Comparison of a height measurement with the LPM and a mechanical stylus measurement system (Talysurf). The objects are etched polysilicon structures.

V. The size measurement of transparent microstructures

The most common application of a critical dimension metrology system is the measurement of resist structures. An exact control of the size of such structures is of crucial importance in the VLSI production process. Depending on the wavelength used, such structures may be highly tranparent, and the actual size measured is very much dependent on the optical properties of these structures. Optical wave guide and interference effects can completely destroy the information being sought. A way out of this problem is to use a wavelength at which the resist is not transparent, such as the near-ultraviolet, or to use dyed resists. A third possibility is to use the effect of the change in refractive index on the polarization components of the light used. This method will be described in somewhat more detail.

V.1. Polarization contrast (QIC)

At the interface between air and resist, a difference exists in the reflectivity between s- and p-polarized light as a function of the incident angle. In the case of high numerical objectives, the incident angles can be quite large. At the Brewster angle the p-polarized component is transmitted 100% into the resist, whereas the s-polarized component is still reflected. The result is that the total reflected polarization measured at the detector is rotated with respect to the incoming beam. Additional depolarization effects, such as birefringence and depolarized light scattering of small objects, also contribute to the final state of polarization of the reflected light. Thus, the contrast

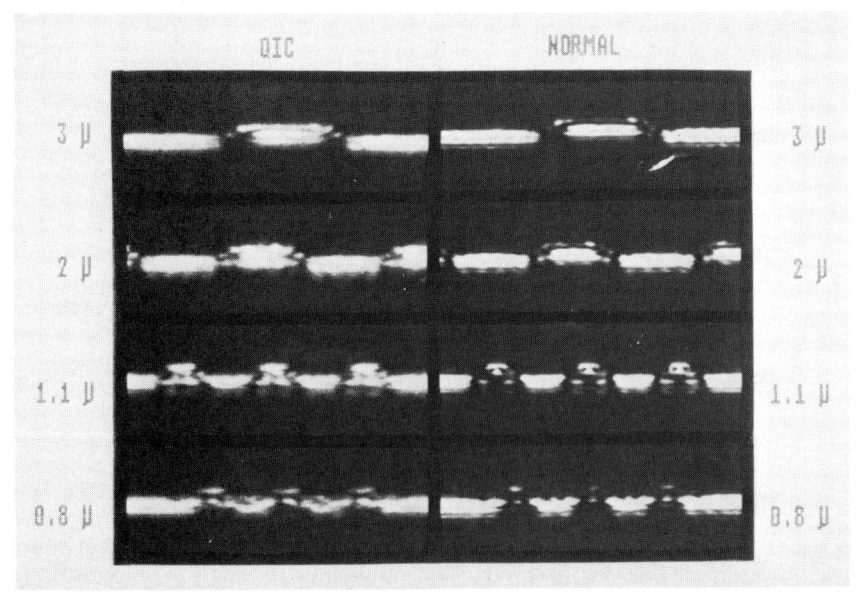

Figure 13.16. Comparison between QIC and normal contrast of 1.2 microns resist on polysilicon for various linewidths. The images are vertical optical sections in the x-z plane.

Semiconductor Metrology 357

based on the polarization of the refractive index change cannot be completely isolated. Due to the quasi-incoherent nature of this effect, the contrast was named QIC (Quasi-Incoherent Contrast). A series of QIC measurements compared to normal contrast is shown in Figure 13.16 between 3 and 0.8 micron lines. In these images, x-z plots are used to show the three-dimensional nature of the structures. Below the x-z sections is the confocal intensity signal summed over z (extended focus). Such sections are normally used for data analysis as they are focus-independent and allow for a full utilization of the three-dimensional features of the signal. From the figure, the difference between the reflectivity of the top resist layer for QIC and normal contrast is obvious. Also, note the very complex pattern for normal contrast due to light guide and interference effects. Such effects destroy the correlation between measured results and actual structure size, as shown in Figure 13.17.

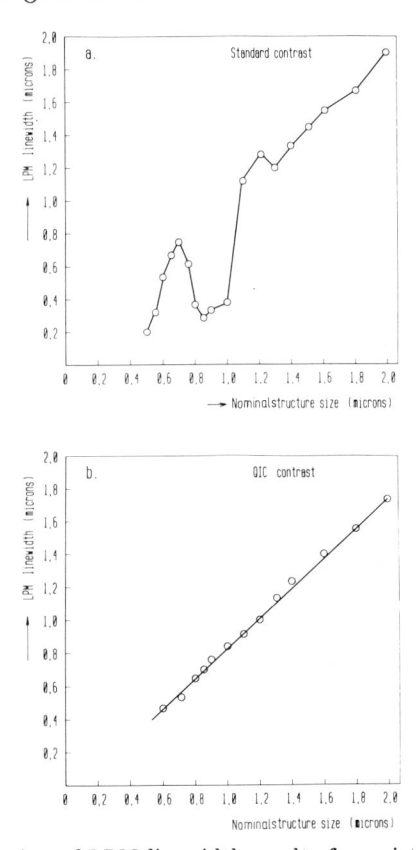

Figure 13.17. Correlation of LPM linewidth results for resist on polysilicon with respect to structure size for normal, Figure 13.17(a), and for QIC contrast, Figure 13.17(b).

V.2. Ultraviolet microscopy

Effects due to the interaction of the light reflected from the top and, through the transparent structure, from the bottom can also be minimized by using ultraviolet light. At a wavelength of below 365nm, the photo resist absorbs strongly and the top signal remains. This also enables a better measurement of the top structure. In this range a slightly better optical resolution is obtained, and, using a smaller numerical aperture, it is easier to measure the "bottom" linewidth. Currently a system exists which can measure at 325nm (HeCd laser). A result of such a measurement at 325nm made with a modified LPM is shown in Figure 13.18.

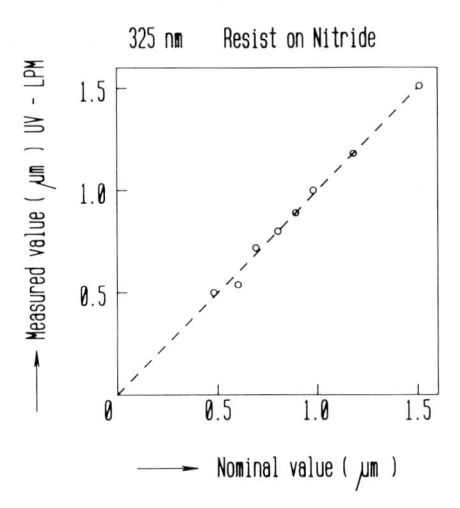

Figure 13.18. Results for linewdith measurements of submicron, 1.2 microns thick resist structures on silicon-nitride.

VI. Confocal microscopy versus scanning electron microscopy

Confocal microscopy is still an optical method, and its resolution relies on high numercial aperture lenses. Moreover, the signals depend on the specific interaction of the optical probe with the sample. Electron microscopy on the other hand utilizes a pencil ray to investigate the object, with which its interaction is quite different. Therefore, in general, it is not possible to make a fair comparison between the two methods. In principle both systems can be used to measure structure sizes, but the results are based on entirely different structural properties. Depending on the application, one method may be more suitable for the purpose than the other. Currently it can be concluded that confocal microscopy yields less information, but of a higher quality. For instance, at structures of high aspect ratio, a single linewidth cannot be defined, and at least three values are needed to describe the structure (topwidth, slope, bottomwidth). Confocal microscopy can yield some indirect information on these parameters, but cannot independently quantify those for any layer-layer combinations. Thus, confocal microscopy can be used as a precise process monitor for linewidth, but an SEM will be needed to inspect the line shapes in detail. A very different situation exists for registration measurements. Here only a relative shift must be determined between two layers and, if signals permit, confocal microscopy can measure such effects with a very high resolution and repeatability. It might be concluded that both optical (confocal) microscopy and electron microscopy are needed in modern sub-micron wafer processing, and should not be considered as alternatives.

VII. Conclusions and outlook

Optical methods in semiconductor metrology continue to remain important in a number of areas. The confocal laser scanning microscope can be used to obtain better and more reliable information. Its slightly higher resolution will make possible measurements in the range down to less than 0.5 micron. A further extension of the resolution can be obtained by using suitable ultraviolet light sources. The additional application of an SEM will be needed in some cases to relate the measured results with the actual geometrical shape. Thus the combination of an SEM and an in-line automatic confocal laser scanning system will be a good solution for in-process metrology. Futher developments of special contrast methods and data-processing algorithms will be needed. Theoretical and experimental studies of each application continue to be the basis for a reliable utilization of the confocal metrology technique.

References

Born, M. and Wolf, E., (1986). *Principles of optics.* Pergamon Press.
Brakenhoff, G. J., Blom, P. and Barends, P., (1979). Confocal scanning light microscopy with high aperture immersion lenses. J. Microsc., **117**, 219-232.
Davidson, M., Kaufman, K. and Mazor, I., (1987). An application of interference microscopy to integrated circuit inspection and metrology. Proc. S.P.I.E., **775**.

Hamilton, D. K. and Wilson, T., (1982). Surface profile measurement using the confocal microscope. J. Appl. Phys., **53**, 5320-5322.

Harris, K., Sandland, P. and Singleton, R., (1983). Wafer inspection automation: current and future needs. Solid State Technology.

Kino, G. S., Corle, T. R. and Xiao, G. Q., (1988). New types of scanning optical microscopes. Proc. S.P.I.E., **921**.

Kleinknecht, H. P. and Meier, H., (1985). Optical profilometer for measuring surface contours of 1 to 150 microns depth. R.C.A. Review, **46**, 34.

Lindow, J. T., Bennet, S. D. and Smith, I. R., (1985). Scanned laser imaging for integrated circuit metrology. Proc. S.P.I.E., **565**, 81.

Marsman, H. J. B., Stricker, R. and Wijnaendts-van-Resandt, R. W., (1983). Mechanical scan system for microscopic applications. Rev. Sci. Instr., **54**, 1047-1052.

Nyyssonen, D., (1982). Theory of optical edge detection and imaging of thick layers. J. Opt. Soc. Am., **72**, 1425-1436.

Sheppard, C. J. R. and Choudhury, A., (1977). Image formation in the scanning microscope. Opt. Acta, **24**, 1051-1073.

Ulrich, H., Wijnaendts-van-Resandt, R. W., Rensch, C. and Ehrensperger, W., (1987). Direct writing laser lithography for production of microstructures. Microelectronic Engineering, **6**, 77-84.

Williams, S., Coates, V. and Ingalls, R., (1988). An improved fluorescence technique for optical measurements below one micron. Proc. S.P.I.E., **921**.

Zapf, Th. and Wijnaendts-van-Resandt, R. W., (1986). Confocal laser microscope for sub-micron structure measurement. Microelectronic Engineering, **5**, 573.

14. Real-time Scanning Optical Microscopes

G. S. KINO AND G. Q. XIAO

I. The tandem scanning optical microscope

The confocal scanning optical microscope has the following major advantages: it yields a very short depth-of-focus, the transverse definition and the contrast of the image is better than with a standard microscope, the device is very well suited to quantitative measurements, and its intensity of illumination with the use of a laser beam can be very high. At the same time, it is possible to choose the wavelength of the illumination very easily [Wilson and Sheppard, 1984; Ash, 1980; Corle et al., 1986; Lemons and Quate, 1973].

The major disadvantages of the scanning optical microscope are associated with the mechanical scan required. This limits the frame time of the image to a few seconds. In all cases, the image is formed too slowly to be directly observable by eye. Thus, image processing in a computer is needed, and the simplicity of the standard microscope is lost. Over twenty years ago, Petráň and Hadravský demonstrated the *Tandem Scanning Optical Microscope* (TSOM), an alternative device used to obtain a real-time image with some of the advantages of the confocal scanning microscope [Petráň et al., 1968; Petráň et al., 1985]. The basic idea was to use, instead of a single pinhole, a large number of pinholes. The pinholes were separated by a distance large enough so that there would be no interaction between the images on the object formed by the individual pinholes. The complete image was formed by moving the pinholes so as to fill in the space between them.

In the original system, the pinholes were drilled in a thin sheet of copper and laid down along a path consisting of a multiple set of interleaved spirals. Several hundred pinholes were illuminated at one time, as illustrated in Figures 14.1 and 14.2 [Petráň et al., 1968]. Typically, since only about one percent of the area of the disk is transparent, a relatively intense light source is required. In the first experiment, the sun was used as a light source; in later experiments, either a mercury arc lamp or a filament lamp were used for the source. As shown in Figure 14.2, in their system, the light passed through the disk to the objective lens of the microscope located a tube-length away from the disk. Thus, an image of each pinhole was formed on the object. A major problem was to eliminate the reflected light from the disk which would cause enough glare so as to obscure the image of the object. This was particularly important to Petráň and Hadravský, and in the later work of Boyde, since they were interested in observing biological objects which have relatively low reflectivity [Petráň et al., 1985]. The reflected light was therefore passed back through a beam inverter and a set of beam splitters and mirrors to a conjugate set of pinholes on the opposite side of the disk and then through a transfer lens to the eyepiece.

Typically, they illuminated a few hundred pinholes at a time, with each pinhole being of the order of 40-50μm in diameter, spaced approximately ten pinhole diameters apart. The disc was spun at a few hundred rpm and a scanned real-time image was obtained. As with any confocal microscope, only the light from the region around the focal plane passed back through the pinholes, while defocused light did not pass back through the pinholes. Consequently, they were able to show good-quality images of biological materials and to carry out *optical cross-sectioning* of these materials. A series of images 7.5μm apart, obtained with a Petráň type of microscope, made by Tracor Northern, are shown in Figure 14.3(a). An extended-focus image, obtained by adding the images together, is shown in Figure 14.3(b).

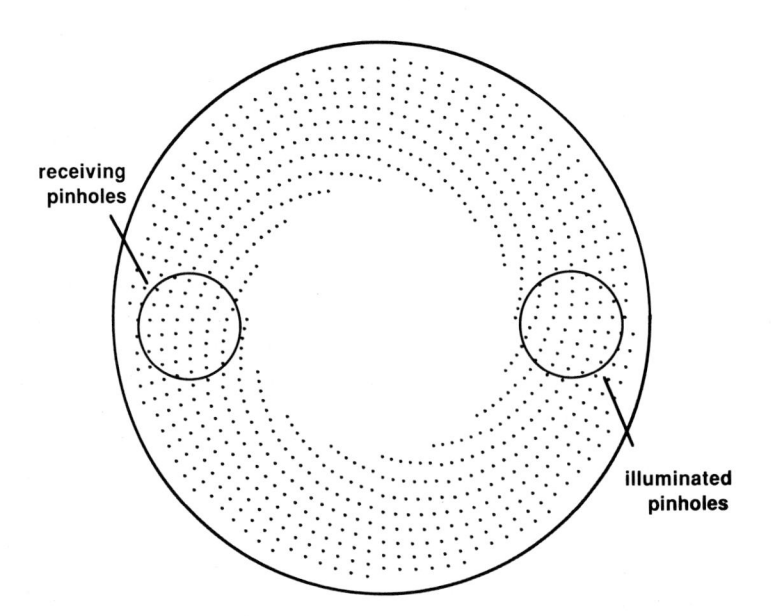

Figure 14.1. A schematic diagram of the Nipkow disk.

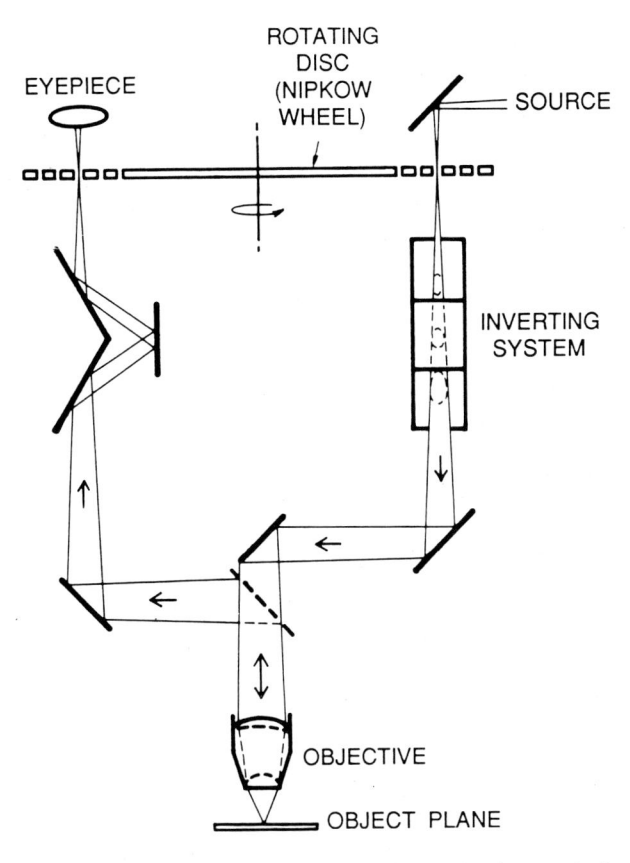

Figure 14.2. A schematic diagram of the tandem scanning optical microscope (TSOM).

The advantages of the microscope are apparent to anyone seeing for the first time. One is its real-time image and good cross-sectioning ability. The disadvantages are a poor light budget, the considerable difficulty in alignment and mechanical complexity. A large number of optical components are required in order to image the reflected light on the correct pinholes, so the mechanical system must be very carefully constructed and aligned in order for the microscope to work properly. For this reason, in the twenty years between its invention and the present time, very few of these microscopes were constructed. Recently, a commercial version of this microscope has been developed. The compromise, made to keep the microscope well aligned, is to use relatively large pinholes so that the alignment is not too critical. For the same reason, the depth-of-focus is somewhat worse than it would be in an equivalent confocal scanning optical microscope.

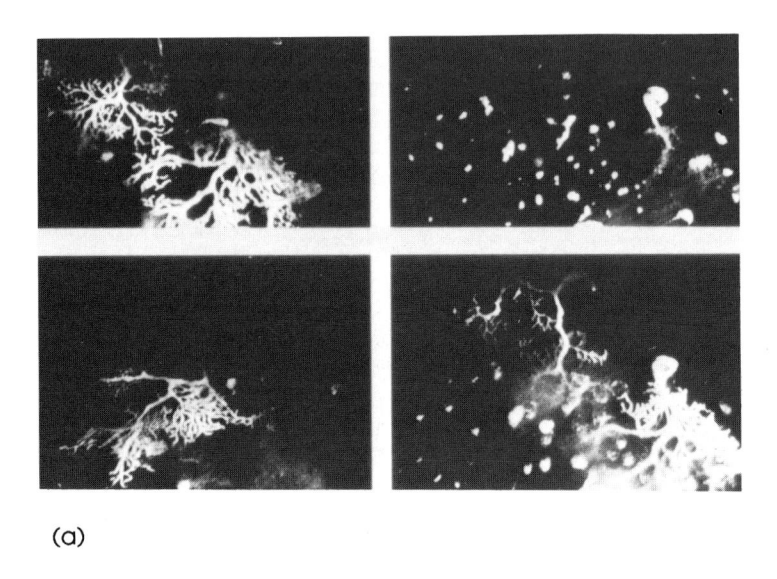

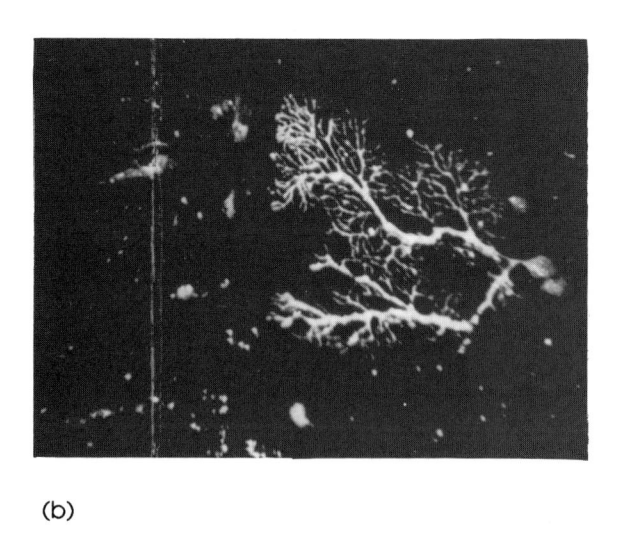

Figure 14.3. (a) Four images, 7.5µm apart, obtained with a Petráň type of microscope. (b) An extended focus image of the same sample.

II. The real-time scanning optical microscope (RSOM)

It was apparent that the improvement needed in the TSOM was to be able to transmit the light through the same pinhole by which it was received. In this case, alignment would not be a problem and a large number of optical components could be eliminated. The problem then was to eliminate the light reflected from the disk. The authors of this work have constructed such a microscope, illustrated in Figures 14.4 and 14.5, which performs well by adopting the following principles [Xiao and Kino, 1987; Xiao et al., 1988]:

(1) The disk is made by photolithographic techniques and consists of black chrome laid down on a glass disk.

(2) As shown in Figure 14.4, the input light is polarized by a polarizer and the light that is received at the eyepiece is observed through an analyzer, with its plane of polarization rotated at right angles to that of the input light. A quarter-wave plate is placed in front of the objective lens so that light reflected from the plane of polarization is rotated by 90 degrees and thus can be observed with the eyepiece.

(3) A stop is placed at the position where the light reflected from the disk is focused, thus further differentiating against the reflected light.

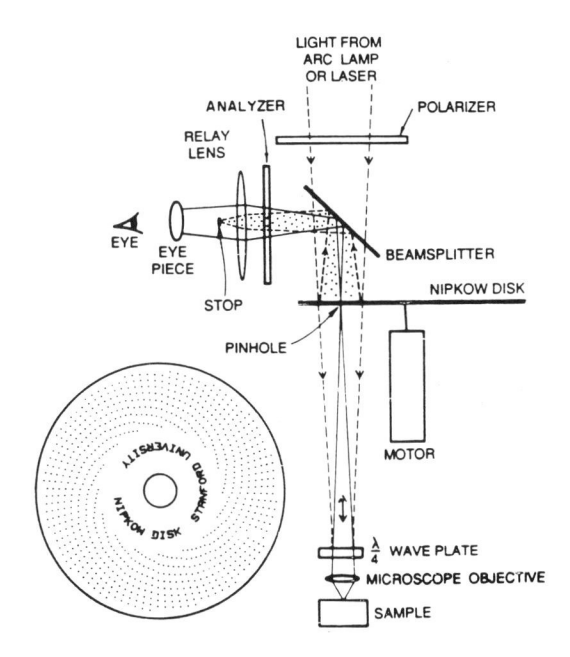

Figure 14.4. A schematic diagram of the real-time scanning optical microscope (RSOM).

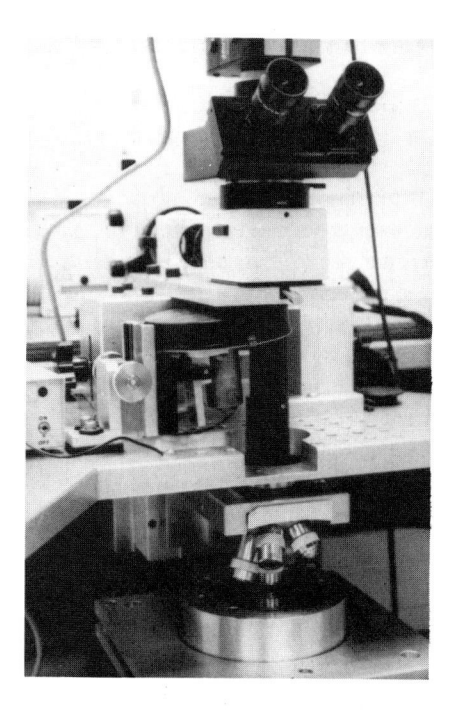

Figure 14.5. A photo of the real-time scanning optical microscope (RSOM).

By using this relatively simple system, the authors were able to construct a disk with 200,000 pinholes, 20 micrometers in diameter, which was rotated at approximately 2000 rpm. This produced a 700 frames/sec, 5000-line image of high quality. The depth-of-focus of this image is as good as the best mechanically-scanned confocal scanning optical microscope (CSOM), as is its transverse resolution. There are, however, some more subtle differences in the performance of the two types of microscopes, because the real-time system uses relatively broadband light.

Later versions of the system employed a tilted disk to still further differentiate against the reflected light, and extra field lenses to optimize the light intensity at the disk and at the pupil of the objective lens. One purpose of these field lenses is like that of Köhler illumination: when the disk is not present, the beam source is focused to a point at the back focus of the objective lens [Born and Wolf, 1975]. With the disk present, this implies that the central axes of the diffracted beams passing through individual pinholes all pass through the center of the back focal plane of the objective lens. Thus the illumination and the definition of the system is made as uniform as possible over the field of view.

III. Theory of the real-time scanning optical microscope

Two important criteria in the design of the real-time scanning optical microscope are the choice of the optimum size and spacing of the pinholes in the disk. If the pinholes are too small, there is a serious loss in light intensity, while if they are too large the range and transverse definitions suffer. With too large a pinhole, the light entering the pinhole is diffracted into a relatively narrow beam which does not fill the pupil of the objective lens, and hence causes it to behave like a lens with an aperture smaller than its full value. Similarly, the beam returning from the objective does not fill the pinhole. So the transverse and range definitions are not as good as for a pinhole of infinitesimal size.

We use scalar theory in the treatment to be given here. However, we have shown that, for instance, the formula for the variation of output intensity with distance z of a plane mirror from the focus can be derived by use of the full vector theory of optics and is identical to the result of the scalar theory. A vector formalism, however, is more complex, and is not worth using most of the time since polarization is not necessarily maintained through the lenses used in the system.

We suppose that the field of the wave incident on the pinhole is uniform and of value Ψ_0 and that paraxial conditions are satisfied on the pinhole side but not on the sample side of the objective lens. We consider the system illustrated in Figure 14.6. It follows from the Rayleigh-Sommerfeld diffraction theory, using the Fraunhofer approximation, that at a distance h_1 from the the pinhole, the field at the pupil plane D_1 an angle θ_1 and radius r_1 from the axis is, for coherent

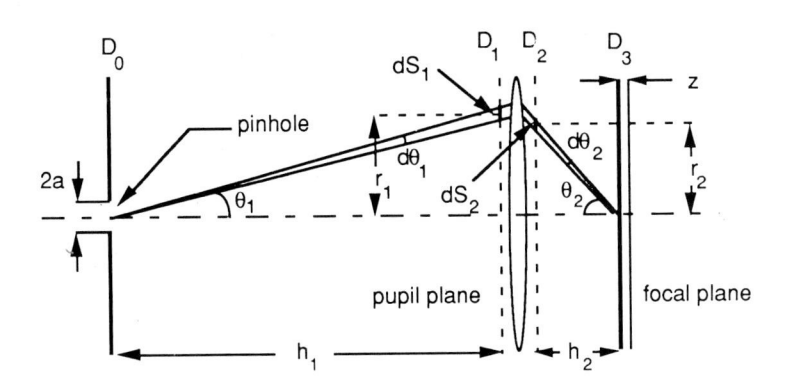

Figure 14.6. A schematic of the system analyzed in the theory.

illumination of the pinhole [Born and Wolf, 1975; Goodman, 1968; Kino, 1987]:

$$\Psi_1(\theta_1) = 2\pi \frac{\Psi_0}{j\lambda h_1} \frac{a^2 J_1(ka\sin\theta_1)}{ka\sin\theta_1} \exp\left[-jk\left(h_1 + \frac{r_1^2}{2h_1}\right)\right] \quad (14.1)$$

where $k = 2\pi/\lambda$, and λ is the optical wavelength in free space.

Using this formula, we may make a rough estimate of the optimum pinhole radius $a(opt)$ by choosing the diameter $d_p(3dB)$ at the half power points of the beam diffracted by the pinhole to be equal to the pupil diameter $2b$ of the objective. This leads to the result [Kino, 1987]:

$$a(opt) = \frac{0.25\lambda h_1}{b} \quad (14.2)$$

where h_1 is the spacing between the pupil and the objective (the tube length). For $h_1 = 180$mm, $\lambda = 540$nm and $b = 2$mm this leads to a pinhole radius $a(opt) \sim 11.6\mu$m which may be compared to the pinhole radius of 10μm used in the present RSOM of Xiao et al. (1987). More exact results will be derived below.

If the pinholes are too closely spaced, there is interference between their images on the object, and speckle effects will be apparent when a narrow-band light source is employed. Furthermore, as the spacing between the pinholes is decreased, the area of the pinholes becomes closer to the total area illuminated. In this case, the intensity of the image no longer falls off rapidly with the defocus distance z, and part of the light reflected from the object can return through alternative pinholes to the eyepiece. For this latter reason, it is advisable to keep the area of the pinholes within the illuminating beam to be less than 1% of the total illuminated area.

Since it is important to obtain as good a definition as possible, which implies the use of large aperture lenses, it is desirable to develop a non-paraxial theory for the design criteria required. This can be done by use of the theory of stigmatic imaging, which leads to the sine condition [Born and Wolf, 1975]:

$$\frac{\sin\theta_2}{\sin\theta_1} = M\frac{n_1}{n_2} \quad (14.3)$$

where M is the magnification of the image, θ_1 and θ_2 are the entrance and exit angles, respectively, of a ray passing through the objective lens, and n_1 and n_2 are the refractive indices of the media on each side of the lens, respectively. We will assume from now on that $n_1 = 1$.

We first determine the range resolution and power efficiency of the microscope when the beam passing through the objective lens is reflected by an ideal plane reflector a distance z from the focus. We consider the power $P(z)$ passing back through the pinhole when the pinhole radius a is finite and determine the ratio $I(z) = P(z)/P_0$. P_0 is the power entering the pinhole from the light source and is given by $P_0 = \pi a^2 I_0$ where I_0 is the intensity of the beam incident on the pinhole. It is assumed that the radius of the pinhole is large enough so that $ka \gg 1$. In this case, it follows from Keller (1957) that the transmission cross-section of the hole is $\sigma = \pi a^2$, i.e., the field across the hole follows closely the form of the incident field, except very close to its edge.

Our procedure is to use scalar theory and find the field radiated from the hole at the input plane to the lens D_1 at a distance h_1 from the pinhole, in the configuration illustrated in Figure 14.6. We determine the reflected field from a perfect reflector. At the pupil plane D_1, the reflected field at h_1, θ_1 is conjugate to the incident field when the reflector is at the focus. When the reflector is a distance z from the focus, the reflected fields at the pupil plane D_1, at an angle θ_1 to the axis, suffer a phase change:

$$\varphi_1'(\theta_2) = 2k_2 z \cos\theta_2 \qquad (14.4)$$

where θ_2 is the angle of a ray on the sample side passing through the point h_1, θ_1 on the pinhole side of the objective lens, $k_2 = 2\pi n_2/\lambda$, λ is the wavelength in free space, and z is the distance of the plane reflector from the focus.

We have already determined the fields at the pupil plane D_1. The next step is to determine the values of the reflected wave fields, which we shall denote by the symbol Ψ'. The reflected wave field at the plane D_1, at the point h_1, θ_1, is given by the relation [Wilson and Sheppard, 1984; Corle et al., 1986; Kino, 1987]:

$$\Psi_1'(h_1, \theta_1) = \Psi_1(h_1, \theta_1) \exp(-2jk_2 z \cos\theta_2) \qquad (14.5)$$

where θ_2 and θ_1 are related by equation (14.3).

We may use the Rayleigh-Sommerfeld theory to determine the reflected wave field at the pinhole, $\Psi_0'(r_0, \phi_0)$, from a knowledge of $\Psi_1'(r_1, \phi_1)$, where r, ϕ are cylindrical coordinates, by writing:

$$\Psi_0' = \frac{1}{j\lambda} \int_0^{2\pi} \int_0^b \frac{\Psi_1' \exp(-jkR_1)}{R_1} \cos\theta_1 \, r_1 \, dr_1 d\phi_1 \qquad (14.6)$$

where $k = 2\pi/\lambda$ and

$$R_1 = \sqrt{r_1^2 + h_1^2 + r_0^2 - 2r_0 r_1 \cos(\phi_0 - \phi_1)} \qquad (14.7)$$

Using the standard paraxial assumptions, we arrive at the result:

$$\Psi'_0(r_0, \phi_0) = -\frac{4\pi^2 \Psi_0 a^2}{\lambda^2 h_1^2} \int_0^b \frac{J_1(ka\sin\theta_1) J_0(kr_0 \sin\theta_1)}{ka\sin\theta_1}$$

$$\times \exp(-2jk_2 z \cos\theta_2) r_1 dr_1 \qquad (14.8)$$

It is convenient to write:

$$r_1 \simeq h_1 \sin\theta_1 \qquad (14.9)$$

and differentiate this expression to show that:

$$dr_1 \simeq h_1 \cos\theta_1 \, d\theta_1 \simeq h_1 \, d\theta_1 \qquad (14.10)$$

It follows from equations (14.3) and (14.10) that:

$$dr_1 \simeq \frac{n_2 h_1}{M} \cos\theta_2 \, d\theta_2 \qquad (14.11)$$

Equation (14.6) then takes the form:

$$\Psi'_0(r_0) = \Psi_0 \frac{k_2 a}{M} \int_0^{\theta_0} J_1\left(\frac{k_2 a \sin\theta_2}{M}\right) J_0\left(\frac{k_2 r_0 \sin\theta_2}{M}\right)$$

$$\times \exp(-2jk_2 z \cos\theta_2) \sin\theta_2 \, d\theta_2 \qquad (14.12)$$

where $\sin\theta_0$ is the numerical aperture of the lens. The value of $I(z)$ may be found from the result of equation (14.12) by writing:

$$I(z) = \frac{2}{\pi a^2 \Psi_0^2} \int_0^a |\Psi'_0|^2 r_0 dr_0 \qquad (14.13)$$

An important case of interest is that for a pinhole infinitesimal in size. Now equation (14.12) takes the form:

$$\Psi'_0 = \frac{1}{2} \left(\frac{k_2 a}{M}\right)^2 \Psi_0 \int_0^{\theta_0} \exp(-2jk_2 z \cos\theta_2) \sin\theta_2 \cos\theta_2 \, d\theta_2 \qquad (14.14)$$

In this case Ψ'_0 is uniform over the radius of the pinhole and, with $z = 0$, i.e. when the reflector is at the focus, is of the value:

$$\Psi'_0 = \left(\frac{k_2 a \sin\theta_0}{2M}\right)^2 \Psi_0 \qquad (14.15)$$

It is often convenient to work with the normalized form of equation (14.14) and write:

$$V(z) = \frac{2}{\sin^2 \theta_0} \int_0^{\theta_0} \exp(-2jk_2 z \cos \theta_2) \sin \theta_2 \cos \theta_2 \, d\theta_2 \qquad (14.16)$$

where the quantity $V(0)$ has a value of unity. We note that if the system is paraxial, i.e., if θ_2 is not too large, we may put $\cos \theta_2 \simeq 1$. equation (14.16) then takes the simple form:

$$V(z) = \frac{\sin[k_2 z(1 - \cos \theta_0)]}{k_2 z(1 - \cos \theta_0)} \exp[-jk_2 z(1 + \cos \theta_0)] \qquad (14.17)$$

We may derive from equation (14.17) a very useful formula for the spacing of the 3dB points

$$d_z(3\text{dB}) = \frac{0.45\lambda}{n_2(1 - \cos \theta_0)} \qquad (14.18)$$

The magnitude squared of $V(z)$, or normalized intensity $I_{norm}(z)$, given by equations (14.16) and (14.17) for a numerical aperture of 0.95 has been determined. It is found that even with a 0.95 aperture, the difference between the two results for the 3dB resolution $d_z(3\text{dB})$ is small. For an infinitesimal pinhole size, $\lambda = 546$nm and a numerical aperture of 0.95, the depth resolution given by equation (14.18) is $d_z(3\text{dB}) = 353$nm, while a more exact calculation from equation (14.16) yields $d_z(3\text{dB}) = 375$nm. Thus there is a 6% error in the use of equation (14.18) which reduces to a 2% error for N.A.=0.9. The analytic form of equation (14.18) is, therefore, a very convenient one.

Plots of the efficiency $I(0)$ with respect to the parameter $n_2 a/\lambda M$ are given for different numerical apertures in Figure 14.7 for $n_2 = 1$. For a typical real-time scanning optical microscope with $n_2 = 1$, $M = 65$X, $\lambda = 546$nm, $\sin \theta_0 = 0.9$ and $a = 10\mu$m, $n_2 a/\lambda M = 0.28$ and the reflected power passing back through the pinhole is 22% of the incident power. Plots of the depth resolution $d_z(3\text{dB})$ as a function of $n_2 a/\lambda M$ for different numerical apertures are given in Figure 14.8. It will be observed that, as might be expected, the resolution becomes worse as the pinhole size is increased. It will be noted that the case treated by Wilson et al. (1987) for the confocal microscope is that for a collimated beam illuminating the objective, which is equivalent to a transmitting pinhole of infinitesimal size but a receiving pinhole of finite size. Thus, their results, as shown in Figure 14.9, are naturally different from ours.

Efficiency vs. Pinhole Radius

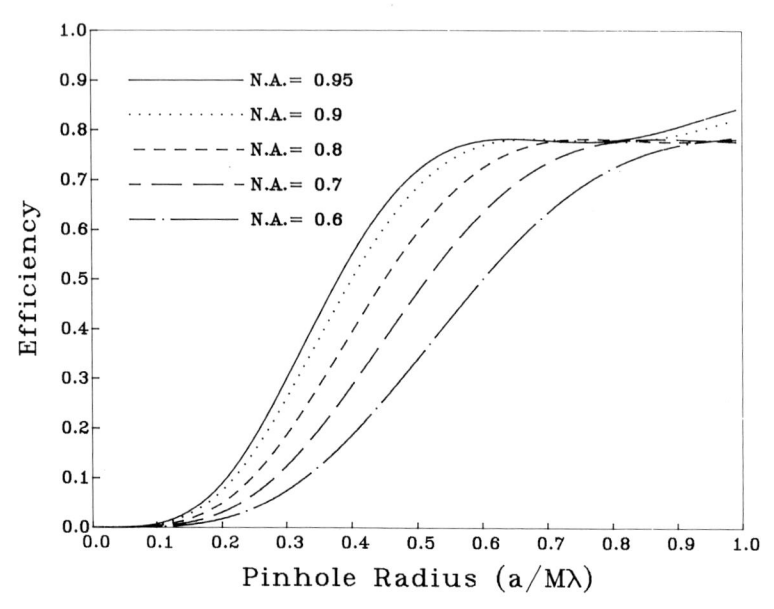

Figure 14.7. Plots of the efficiency as a function of $a/\lambda M$ for different numerical apertures.

Depth Resolution vs. Pinhole Radius

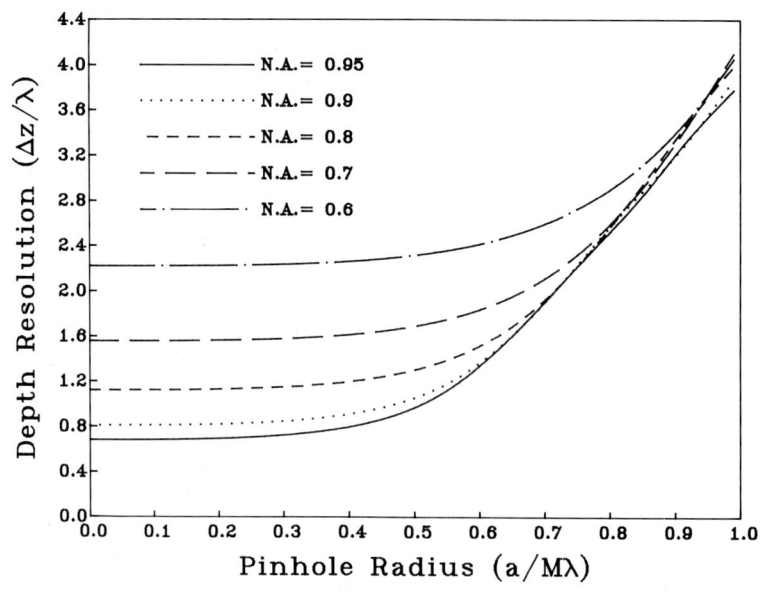

Figure 14.8. Plots of the depth resolution as a function of $a/\lambda M$ for different numerical apertures.

Depth Resolution Comparison

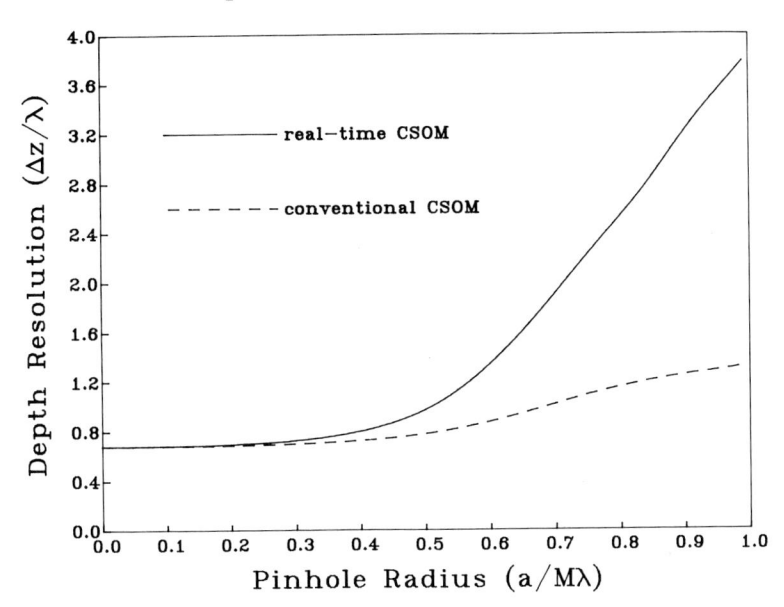

Figure 14.9. A comparison of the depth resolution of an RSOM with a standard CSOM for N.A.=0.95.

IV. The transverse response of a confocal microscope

We will determine the transverse response of a confocal microscope without using the paraxial approximation. We will first calculate the fields at the pupil plane D_3 where $z = 0$. Then we will determine the excitation at the pinhole by the waves reflected from a point reflector located at the plane D_3 a distance x from the axis. Since we need to determine the image seen by a stationary observer through a rotating set of pinholes, the analysis cannot be identical in form to either the case of a beam entering and leaving through a stationary pinhole, or that of a collimated transmitting beam and a stationary receiving pinhole of finite size.

We therefore need to define a *time-averaged point spread function*. We assume that an observer looks at one point in space and determine the signal arriving at that point, as the point reflecting object is moved a distance x from the axis. The point object is illuminated by a set of moving pinholes, and the image is observed through this same set of rotating pinholes. Therefore, we first determine how the illumination of the point object varies as the position of the illuminating hole varies;

then we determine the average signal received at a point of observation on the axis at the plane of the pinholes. This is the *time-averaged point spread function*.

We first assume that the pinhole is located with its center on the axis and use equations (14.1) and (14.3) to determine the field $\Psi_2(r_2, \theta_2)$ at the point r_2, θ_2 on the exit plane D_2 corresponding to the position of a ray which passed through r_1, θ_1 on the pupil plane D_1 and had a field $\Psi_1(r_1, \theta_1)$. We can then determine the value of $\Psi_2(r_2, \theta_2)$ from conservation of power, $P = (1/2)\text{Im} \int \Psi^* \nabla \Psi \cdot n \, dS$ or by use of the reciprocity theorem [Kino, 1987].

Taking the elemental areas in a plane perpendicular to the z-axis which include the rays passing through the corresponding points r_1, θ_1 and r_2, θ_2 to be dS_1 and dS_2, respectively, it then follows from ray optics that:

$$\Psi_1 \Psi_1^* \cos \theta_1 dS_1 = n_2 \Psi_2 \Psi_2^* \cos \theta_2 dS_2 \qquad (14.19)$$

We write:

$$dS_1 = h_1^2 \sec^3 \theta_1 \sin \theta_1 d\theta_1 \, d\phi_1 \qquad (14.20)$$

and

$$dS_2 = h_2^2 \sec^3 \theta_2 \sin \theta_2 d\theta_2 \, d\phi_2 \qquad (14.21)$$

respectively. It follows from equations (14.19)—(14.21), with $d\phi_1 = d\phi_2$ that:

$$\left|\frac{\Psi_2}{\Psi_1}\right|^2 = \frac{\sin \theta_1}{n_2 \sin \theta_2} \left(\frac{h_1 \cos \theta_2}{h_2 \cos \theta_1}\right)^2 \frac{d\theta_1}{d\theta_2} \qquad (14.22)$$

However, by differentiating equation (14.3), we can show that:

$$n_2 \cos \theta_2 \, d\theta_2 = M \cos \theta_1 \, d\theta_1 \qquad (14.23)$$

So that it follows from equations (14.3), (14.22) and (14.23) that:

$$\left|\frac{\Psi_2}{\Psi_1}\right| = \frac{\sqrt{n_2} h_1}{M h_2} \left(\frac{\cos \theta_2}{\cos \theta_1}\right)^{3/2} \qquad (14.24)$$

Thus we can write from equations (14.1) and (14.24):

$$\Psi_2(\theta_2) = \frac{\sqrt{n_2} k a^2 \Psi_0}{M h_2} \frac{J_1(ka \sin \theta_1)}{ka \sin \theta_1} \exp(jkR_{20}) \cos^{3/2} \theta_2 \qquad (14.25)$$

where

$$R_{20} = \sqrt{h_2^2 + r_2^2} \qquad (14.26)$$

In equation (14.25) we have written $\cos\theta_1 \simeq 1$ to be consistent with the paraxial assumption on the pinhole side of the lens. We have neglected phase terms that are uniform over the cross-section, and we have chosen the phase term $\exp(jkR_{20})$ to correspond to focusing on a point on the axis at the plane D_3.

We may determine the value of $\Psi_3(r_3)$ at the plane D_3 by writing the distance between the point $r_3 = x$ on the plane D_3 and the point r_2, ϕ_2 on the plane D_2 as:

$$R_2 = \sqrt{h_2^2 + (r_2\cos\phi_2 - x)^2 + r_2^2\sin\phi_2} \qquad (14.27)$$

We will assume that $x^2 \ll h_2^2$ but will make no approximations to the r_2 terms. We will write, from equation (14.26):

$$r_2 = R_{20}\sin\theta_2 = h_2\tan\theta_2 \qquad (14.28)$$

It therefore follows from equation (14.27) that:

$$R_2 \simeq R_{20} - x\sin\theta_2\cos\phi_2 \qquad (14.29)$$

The field at the plane D_3 at the point x from the axis is therefore of the form:

$$\Psi_3(x) = \frac{n_2}{j\lambda} \int_{\theta_2=0}^{\theta_0} \int_{\phi_2=0}^{2\pi} \frac{\exp(-jk_2R_{20})}{R_{20}} \Psi_2(r_2)$$

$$\times \exp(jk_2 x\sin\theta_2\cos\phi_2)\cos\theta_2\, d\phi_2\, r_2\, dr_2 \qquad (14.30)$$

This may be integrated to yield the relation:

$$\Psi_3(x) = -jk_2 \int_{\theta_2=0}^{\theta_0} \frac{\exp(-jk_2R_{20})}{R_{20}} \Psi_2(r_2) J_0(k_2 x\sin\theta_2)\cos\theta_2\, r_2\, dr_2 \qquad (14.31)$$

Substitution of equation (14.25) into equation (14.31) yields the result:

$$\Psi_3(x) = A\Psi_0 \int_0^{\theta_0} J_0(k_2 x\sin\theta_2) J_1\left(\frac{k_2 a\sin\theta_2}{M}\right) \cos^{1/2}\theta_2\, d\theta_2 \qquad (14.32)$$

where A is a constant.

We now consider the situation when the center of the pinhole has moved to a point R_0, α. In this case the radial distance of the center of the demagnified image of the pinhole from the point a distance x from the axis is:

$$\rho = \sqrt{\left(\frac{R_0}{M}\right)^2 + x^2 - 2\frac{R_0 x}{M}\cos\alpha} \qquad (14.33)$$

So the more general form of equation (14.32) becomes:

$$\Psi_3(\rho) = A\Psi_0 \int_0^{\theta_0} J_0(k_2\rho\sin\theta_2) J_1\left(\frac{k_2 a \sin\theta_2}{M}\right) \cos^{1/2}\theta_2 \, d\theta_2 \qquad (14.34)$$

We now assume that there is a reflector of infinitesimal size at the point $x, 0$ on the object plane D_3 and that this point reflector reradiates a signal proportional to the incident amplitude $\Psi_3(x)$. The point reflector will therefore give rise to a reflected wave at the exit plane D_2 of the form:

$$\Psi_2'(r_2, \phi_2) = \frac{B\exp(-jk_2 R_2)}{R_2}\Psi_3(\rho) \qquad (14.35)$$

where B is a constant. It follows from equations (14.29) and (14.35) that:

$$\Psi_2'(r_2, \phi_2) = \frac{B\exp[-jk_2(R_{20} - x\sin\theta_2\cos\phi_2)]}{R_{20}}\Psi_3(\rho) \qquad (14.36)$$

Following the procedures leading to equation (14.25), it can be shown that:

$$\Psi_1'(r_1, \phi_1) = C\exp[j(kR_{10} + k_2 x\sin\theta_2\cos\phi_1)]\cos^{1/2}(\theta_2)\,\Psi_3(\rho) \qquad (14.37)$$

where C is a constant and it is assumed that a point on the plane D_3 at $x = 0$ is focused on-axis at the plane of the pinhole. The field on-axis at the plane of the pinhole is given by the relation:

$$\Psi_0'(0) = \frac{1}{j\lambda}\int_0^b \int_0^{2\pi} \Psi_1'(r_1, \phi_1)\frac{\exp(-jkR_{10})}{R_{10}} d\phi_1 \, r_1 \, dr_1 \qquad (14.38)$$

Substituting equation (14.37) into equation (14.38), using equation (14.11) and writing the integrand in terms of the angles θ_2 and ϕ_2, equation (14.38) takes the form:

$$\Psi_0'(0, x) = D\Psi_3(\rho)\int_0^{\theta_0}\int_0^{2\pi} \exp(jk_2 x\sin\theta_2\cos\phi_2)$$

$$\times \sin\theta_2 \cos^{1/2}\theta_2 \, d\phi_2 \, d\theta_2 \qquad (14.39)$$

where D is a constant.
Equation (14.39) can be integrated to yield the result:

$$\Psi'_0(0,x) = D\Psi_3(\rho)U(0,x) \qquad (14.40)$$

where

$$U(0,x) = \int_0^{\theta_0} J_0(k_2 x \sin\theta_2) \sin\theta_2 \cos^{1/2}\theta_2 \, d\theta_2 \qquad (14.41)$$

This integral cannot be directly evaluated in terms of tabulated functions, unlike the equivalent paraxial form which can be expressed in terms of the familiar Airy function.

We must now determine the average signal intensity seen by an observer at a point on-axis at the plane of the disk. As the disk rotates, the average signal received on-axis is of the form:

$$I(0,x) = U^2(0,x)Q(x) \qquad (14.42)$$

where

$$Q(x) = <\Psi_3^2(\rho)> \qquad (14.43)$$

Transverse Resolution vs. Pinhole Radius

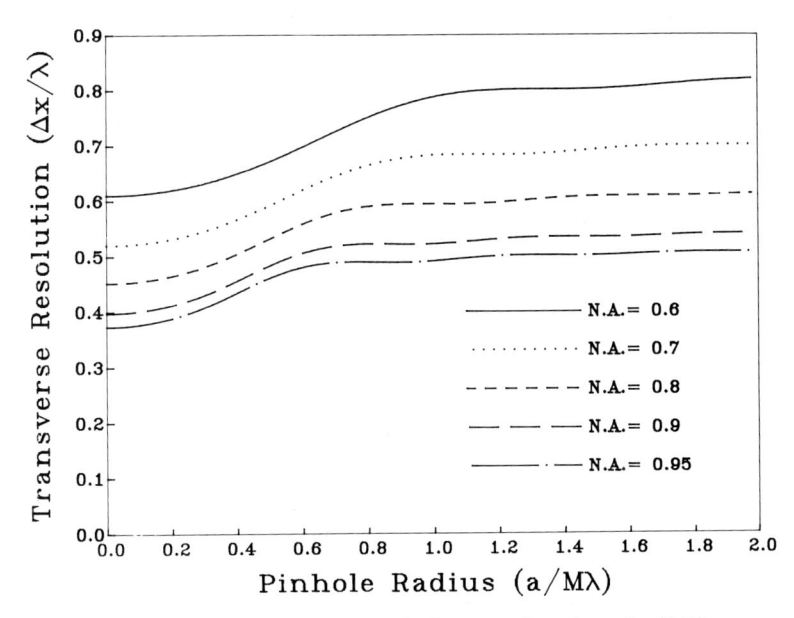

Figure 14.10. Plots of the transverse resolution as a function of $a/\lambda M$.

Assuming that the sample is uniformly scanned by the pinhole, we can convert the time average into an area integral, which is expresssed as:

$$Q(x) = E \int_0^{2\pi} \int_{R_0 \leq a} \Psi_3^2 \left(\sqrt{\left(\frac{R_0}{M}\right)^2 + x^2 - 2\frac{R_0}{M} x \cos \alpha} \right) R_0 \, dR_0 \, d\alpha$$

(14.44)

where E is a constant. The integration limit is determined by the fact that when the center of the pinhole is away from the axis by a distance greater than a, the signal at the observation point on the axis is blocked by the pinhole.

A plot of the 3dB resolution $d_{RS}(3\text{dB})$ as a function of the parameter $n_2 a/\lambda M$ for different numerical apertures is given in Figure 14.10 for $n_2 = 1$. It will be observed that the 3dB resolution increased by a factor of approximately 1.4 as the pinhole size is increased from infinitesimal to very large values. When the pinhole size is large, the system becomes a standard microscope. When it is infinitesimal, the point spread response becomes that of a perfect confocal microscope.

V. Experimental results

A series of black and white images of an integrated circuit taken with the RSOM using a lens with a numerical aperture of 0.9 are shown in Figure 14.11. Because of the chromatic aberration of the lens, different planes at different levels of the integrated circuit would show up in different colors on a color reproduction. This can be useful for inspection purposes; however, if a filter is used, the light is most intense on the plane on which the beam is focused. Typically planes less than 500 nanometers apart can easily be observed separately, as can be seen from the photos.

A plot of the intensity response as a function of distance z is shown in Figure 14.12 for our microscope with a N.A. of 0.9 and an objective lens magnification of 65X. It is seen that theory and experiment are in good agreement. The $d_z(3\text{dB})$ width is theoretically 460nm for a $20\mu\text{m}$ pinhole diameter and $\lambda = 546\text{nm}$. This compares to the value of 440nm for an infinitesimal size pinhole. The measured value of $d_z(3\text{dB})$ is 501nm. Results taken with other lenses are in similarly fairly good agreement with theory; for instance, with N.A. = 0.8, the theoretical value is 621nm while the experimental value is 651nm. The sidelobes of the response do tend to be somewhat different from the theoretical levels. This is due to spherical aberration in the lens and can be accounted for [Corle et al., 1986]. More careful studies with mechanically-scanned confocal microscopes lead to the same conclusion.

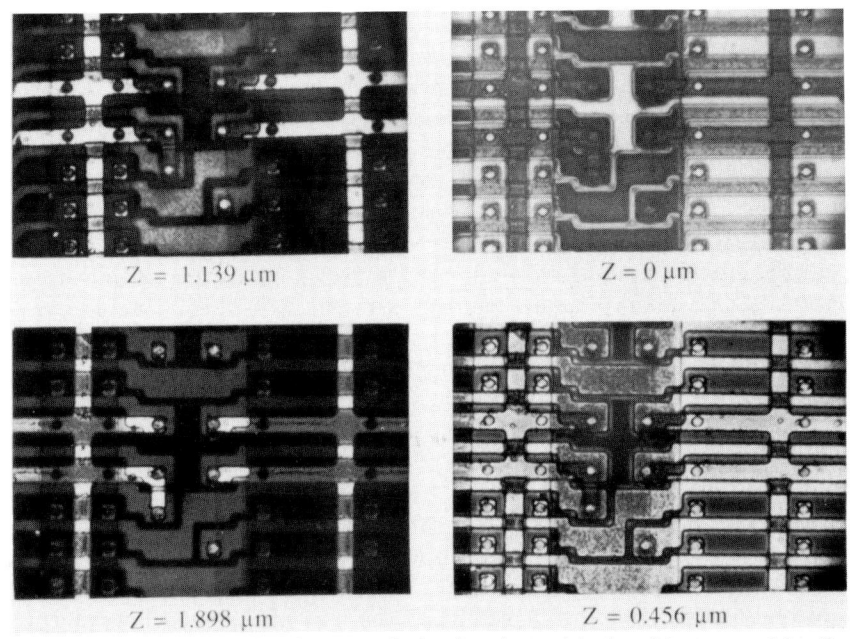

Figure 14.11. Images of an integrated circuit taken with the object moved by the distances shown in the figure.

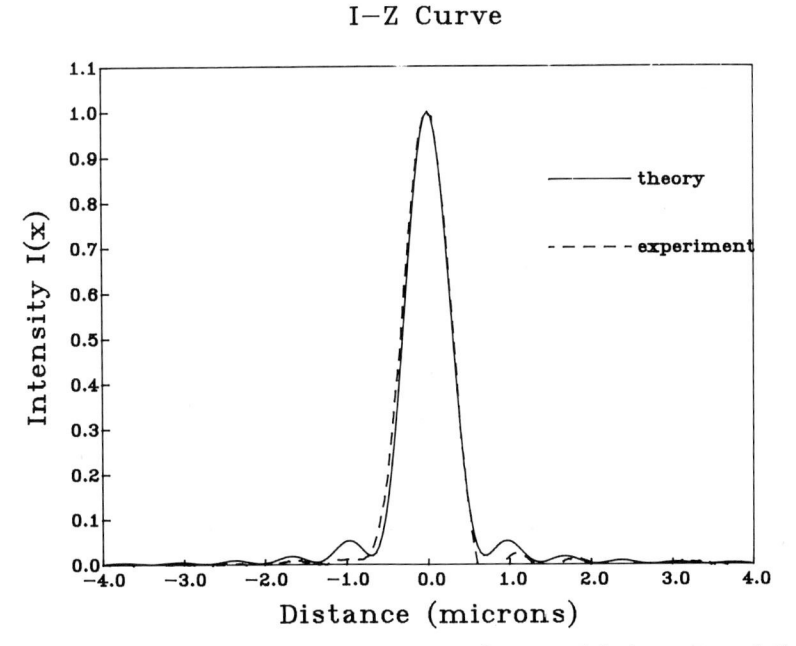

I−Z Curve

Figure 14.12. Plots of theoretical and experimental curves of the intensity variation as a function of z for a numerical aperture of N.A.$=0.9$.

With the help of Barry Masters of the Emory Eye Center of Emory University in Atlanta, we have taken several optical micrographs of the cornea and lens of a rabbit eye. These results show the real-time cross-sectioning ability of this type of microscope to great advantage. Since the reflectivity of these features of the eye is only 0.01 percent, the image can barely be seen in real time with the naked eye, and is best displayed in real time on a video screen with the use of an image intensifier. The images shown here have also been averaged for 5 frames, and the small amount of glare caused by weak scattering of light from various optical components has been removed by subtracting digitally a constant value from the signal output of the vidicon. However, it should be emphasized that the images can be seen directly by eye and on the output of the image intensifier. All images are taken with a 25X, 0.6 N.A. water immersion lens.

Figure 14.13 shows an image of the corneal endothelium. A second picture in Figure 14.14 shows the surface of the outer epithelial surface of the cornea; cell nuclei can be clearly seen. The depth resolution from section to section appears to be less than 1μm and the transverse resolution is of the order of 0.6μm. A third picture in Figure 14.15 shows the fibrous nature of the lens of the eye. This picture shows considerable detail along the fibers. In all of these pictures the magnification is approximately 600.

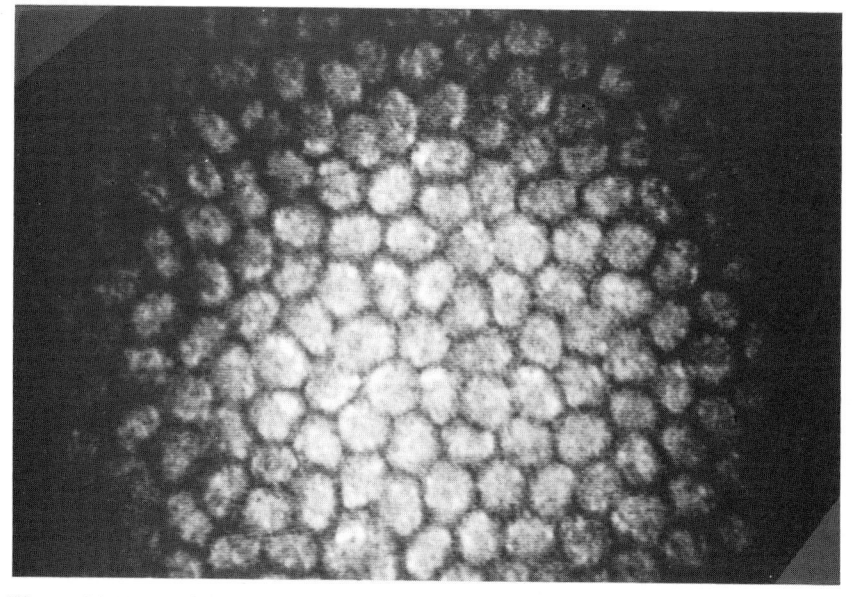

Figure 14.13. An image of the surface of the corneal endothelium of a rabbit eye taken by the RSOM.

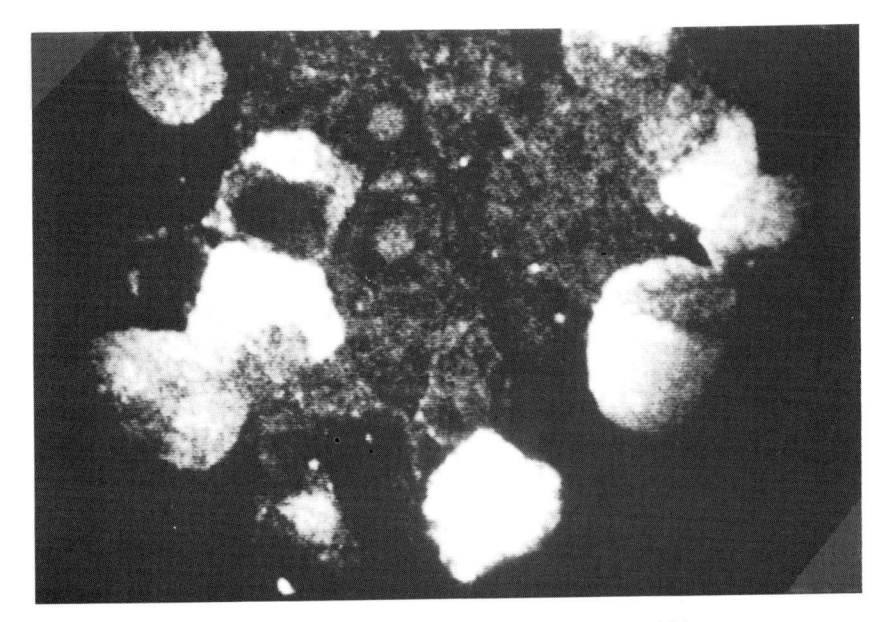

Figure 14.14. An image of the outer epithelial surface of a rabbit eye.

Figure 14.15. An image of the fiber of the lens of a rabbit eye.

One important example of the use of this microscope, which we have demonstrated for the first time, is to look at a deep trench in silicon. In this case, with the beam focused below the surface of the silicon sample, much of the beam is cut off and the beam passing into the optical waveguide formed by the trench behaves as a cylindrically focused strip beam. Thus focusing action still occurs, and the depth of the bottom of the trench may be determined by determining the point where the intensity is maximum. With the microscope, we have been able to visually observe the bottom of a 6μm deep trench only 1μm wide and determine its depth accurately. Images with the beam focused on the top and bottom surfaces of the trench are shown in Figures 14.16(a) and 14.16(b) respectively. Line scans of these images for a 2μm wide trench are shown in Figure 14.17 and compared to a theoretical analysis developed in this laboratory by Kino and Chim. Some results of these trench measurements are summarized in Table 1.

Parameter	Cross sectional Measurement	Depth Measurement
length	5.88μm	5.95μm
width	1.03μm	
length	6.01μm	6.25μm
width	1.54μm	
length	5.97μm	6.30μm
width	1.99μm	
length	6.03μm	6.25μm
width	2.40μm	

Table 10.1. Trench measurements.

(a)

(b)

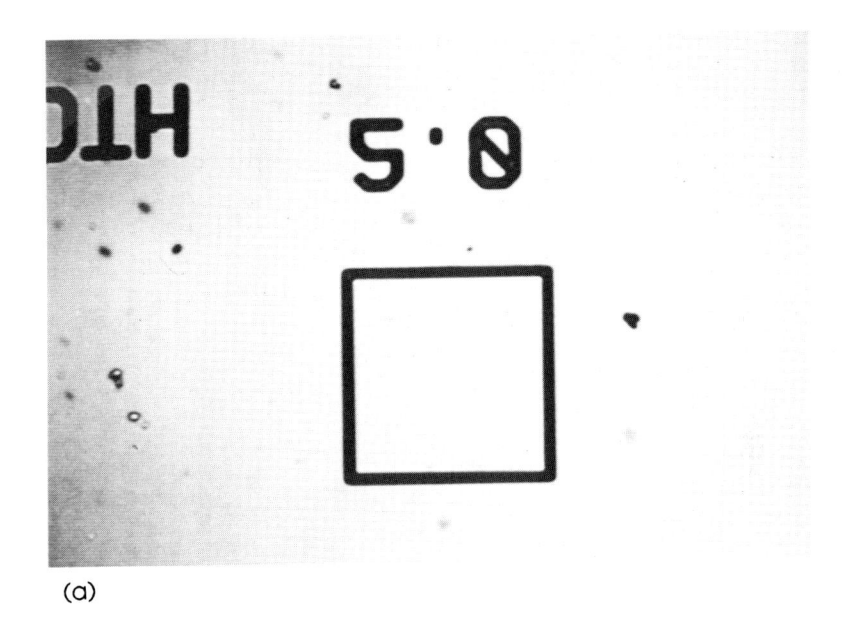

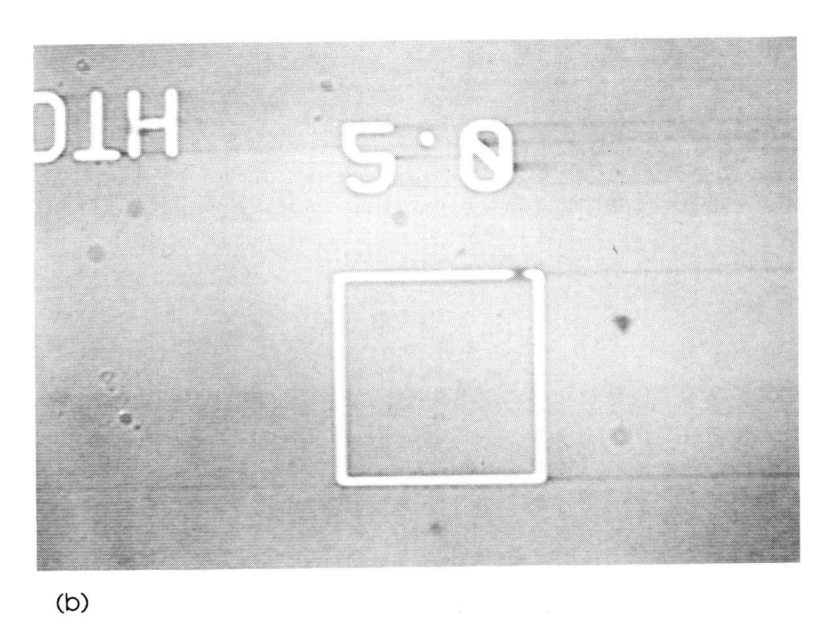

Figure 14.16. (a) A picture of the top surface of a 2μm wide, 6μm deep silicon trench. (b) A picture of the bottom of the trench.

Linescan over Trench

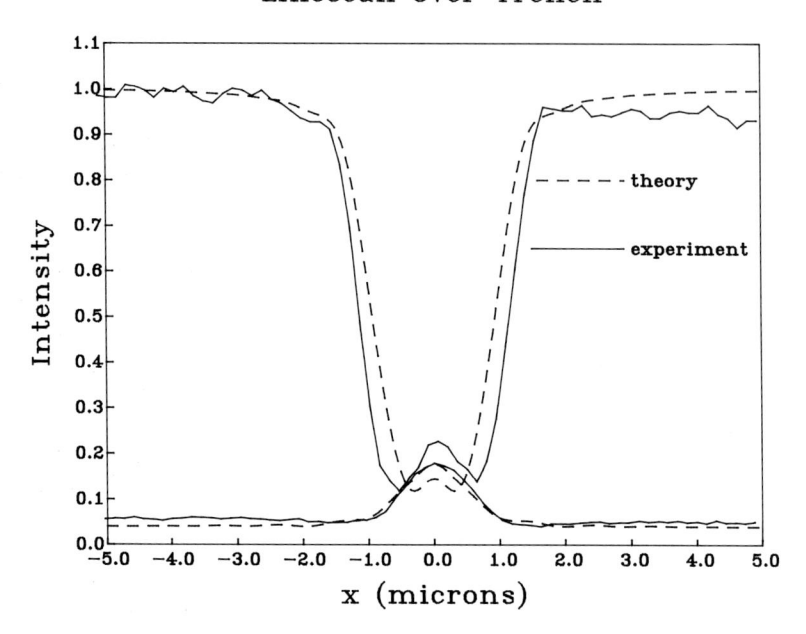

Figure 14.17. Line scans of a 2μm wide, 6μm deep trench.

The transverse resolution of a microscope can be defined in several different ways. On the basis of paraxial theory, if a point reflector is observed with a standard microscope, the distance between half power points of the point spread function is $d_s(3\text{dB})$, where

$$d_s(3\text{dB}) = \frac{0.51\lambda}{\sin\theta_0} \tag{14.45}$$

For a confocal microscope with infinitesimal pinhole size, the point spread function is the square of the point spread function of the standard microscope, and the distance between half power points $d_{sc}(3\text{dB})$ is:

$$d_{sc}(3\text{dB}) = \frac{1.37\lambda}{\sin\theta_0} \tag{14.46}$$

An alternative definition, the Abbé criterion, is based on the maximum spatial frequency which can be detected. For a standard microscope illuminated through a condenser with the same aperture as the objective lens, the maximum spatial frequency which can be detected is the same as for the confocal microscope using a lens of the same aperture. In this case, the minimum size of grating which can be observed has a period p_A where

$$p_A = \frac{0.5\lambda}{\sin \theta_0} \qquad (14.47)$$

Thus, it might be expected that for a grating with equal size strips and gaps, the minimum size of observable strip width would be $d_A = 0.5p_A$.

Other criteria based on the two-point definition can also be used. Such criteria lead to an 8% better definition for the confocal microscope compared to that of the incoherently illuminated standard microscope. The main reason for the better performance of the real-time confocal microscope is that its point spread function is the square of the point spread function of the standard microscope. This yields better contrast with the very low sidelobe levels of the point spread function, along with easy adjustment for optimum performance. These advantages make it, in practice, a better instrument for metrology. As an example, in the image shown in Figure 14.18, there are 300nm wide strips, 300nm apart. We can can observe visually, with this microscope, gratings with strips only 200nm wide. The Abbé criterion yields the result $d_A = 150$nm, with $d_{sc}(3\text{dB}) = 225$nm for a lens with a numerical aperture of 0.9. Since the higher spatial frequency terms in the spatial frequency response of the lens are very weak, the 3dB criterion appears to be a better one to use for this example.

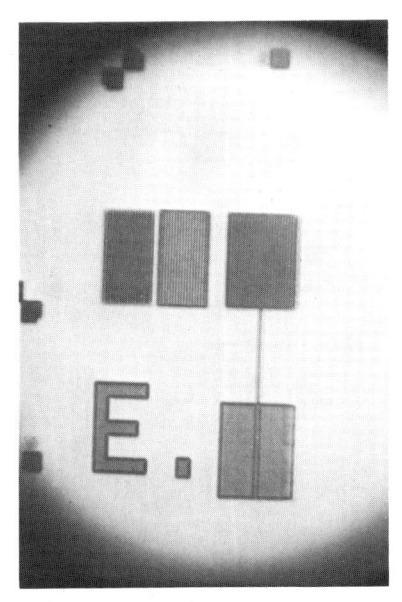

Figure 14.18. A picture of a 0.3μm grating.

VI. Conclusions

The tandem scanning and real-time optical microscopes provide real-time images. The RSOM gives results fully equivalent to the standard confocal scanning microscope, but with a much faster scan time and with the use of broad-band light, typically a simple mercury arc lamp. The tandem scanning microscope has given excellent results on biological samples, while the real-time scanning optical microscope is presently being adapted for use on integrated circuits and other structures where precision metrology is required. These microscopes are adaptable to new modes of phase contrast imaging, differential imaging, and fluorescence imaging. Research is presently being carried out on the use of these modes for determining the slopes of structures, and therefore obtaining better precision in defining edges of structures in integrated circuits. It is expected that since these microscopes are relatively simple to construct, they will become standard tools in optical microscopy.

References

Ash, E. A. (Editor), (1980). *Scanned image microscopy.* Academic Press.
Born, M. and Wolf, E., (1975). *Principles of optics.* Pergamon Press.
Corle, T. R., Chou, C.-H. and Kino, G. S., (1986). Depth response of confocal optical microscopes. Opt. Lett., **11**, 770-772.
Corle, T. R. and Kino, G. S., Phase imaging in scanning optical microscopes. Proc. S.P.I.E., **1028**, 114-121.
Goodman, J. W., (1968). *Introduction to Fourier optics.* McGraw-Hill.
Keller, J. B., (1957). Diffraction by an aperture. J. Appl. Phys., **28**, 426-443.
Kino, G. S., (1987). *Acoustic waves: devices, imaging, and analog signal processing.* Ch. 3: Wave propagation with finite exciting sources, 154-318. Prentice-Hall, New Jersey.
Lemons, R. and Quate, C. F., (1973). A scanning acoustic microscope. Proc. I.E.E.E. Ultrasonics Symp.
Petráň, M., Hadravský, M., Egger, M. D. and Galambos, R., (1968). Tandem-scanning reflected-light microscope. J. Opt. Soc. Am., **58**, 661-664.
Petráň, M., Hadravský, M. and Boyde, A., (1985). The tandem scanning reflected light microscope. Scanning, **7**, 97-108.
Wilson, T. and Sheppard, C. J. R., (1984). *Theory and practice of scanning optical microscopy.* Academic Press, London.
Wilson, T. and Carlini, A. R., (1987). Size of the detector in confocal imaging systems. Opt. Lett., **12**, 227-229.

Xiao, G. Q. and Kino, G. S., (1987). A real-time confocal scanning optical microscope. In: *Scanning imaging technology.* Wilson, T. and Balk, L. (Editors), Proc. S.P.I.E., **809**, 107-113.

Xiao, G. Q., Corle, T. R. and Kino, G. S., (1988). Real-time confocal scanning optical microscope. Appl. Phys. Lett., **53**, 716-718.

15. Confocal Interference Microscopy

C. J. R. SHEPPARD

I. Introduction

Confocal microscopy is a coherent imaging method [Sheppard and Choudhury, 1977] which suggests that if an interference method is employed, the phase of the image signal as well as its modulus may be recovered. This is useful for subsequent processing of the images for restoration, measurement of surface height variations or investigation of phase objects. Confocal methods have the major advantage for interference microscopy that the shape of the reference beam wavefront is immaterial: only its phase and amplitude at the detector pinhole is important [Sheppard and Wilson, 1980a], removing the requirement for matched optics and making alignment much less critical. Furthermore, it is possible to combine the system with multiple detectors and real-time processing to separate imaging of particular forms. The microscope may be operated with a collector lens of high aperture, thus improving resolution whilst retaining coherent illumination. Laser illumination also simplifies the optical arrangement because of the long coherence length. A transmission-mode confocal interference microscope of the Mach-Zehnder type has been described by Brakenhoff (1979). Hamilton and Sheppard (1982) described a reflection-mode confocal interference system of the Michelson geometry, which is illustrated in Figure 15.1. Figure 15.2 shows an interference image produced by displaying the signal from a single detector of Figure 15.1 using HeNe light (λ=633nm), and an objective of numerical aperture 0.5. The image exhibits a series of fringes representing the surface height of a TEM grid, where local deformations of the fringes correspond to surface detail. A deep scratch is also visible. If the object beam has a complex amplitude t and the reference beam a complex amplitude r, the intensity from detector A, according to paraxial theory, is:

$$I_A = |t|^2 + |r|^2 + 2\text{Re}\{tr^*\} \qquad (15.1)$$

so that the fringes are superimposed on a confocal image, with intensity $|t|^2$. The confocal image, produced with no reference beam, is shown in Figure 15.3.

It can be shown from reciprocity considerations that when two beams are mixed in a beam-splitter, the two resulting outputs correspond to the sum and difference of the original beams. The two detectors of Figure 15.1 thus give signals:

$$I_{A,B} = |t|^2 + |r|^2 \pm 2\mathrm{Re}\{tr^*\} \tag{15.2}$$

The photodiode outputs may be amplified to a standard level and their sum and difference extracted electronically to give signals proportional to:

$$I_S = \frac{1}{2}(I_A + I_B) = |t|^2 + |r|^2 \tag{15.3}$$

and

$$I_D = \frac{1}{2}(I_A - I_B) = 2\mathrm{Re}\{tr^*\} \tag{15.4}$$

It is apparent that the sum signal consists of the normal confocal image $|t|^2$ superimposed on a constant background $|r|^2$ whereas, in the difference signal, these two components cancel to leave a pure interference image. Because this signal can be bidirectional, an electronic offset is added to it, producing for display purposes a signal which is always positive-going; alternatively the signal may be squared. In the former case the fringes are modulated by the amplitude of the object beam a and centred about mid-grey, whereas in the latter case the fringes vary in intensity between black and a^2, the fringes now having half the spacing.

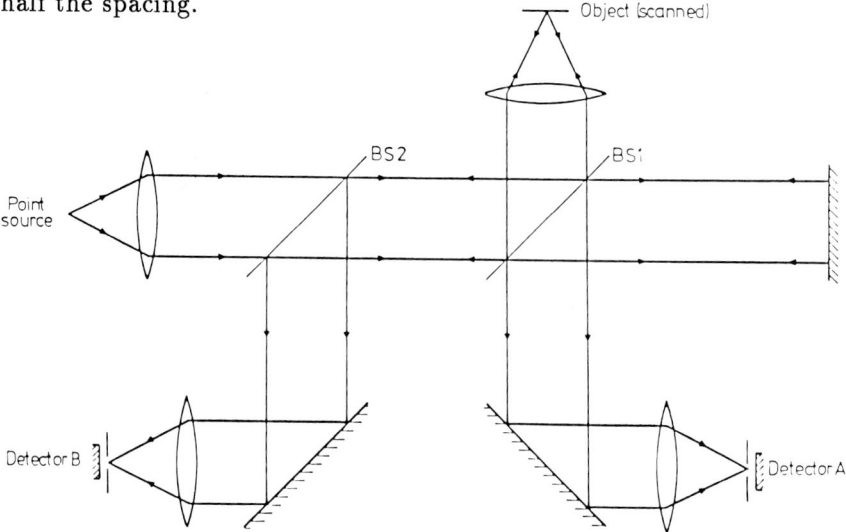

Figure 15.1. A reflection-mode confocal interference microscope.

Confocal Interference Microscopy 391

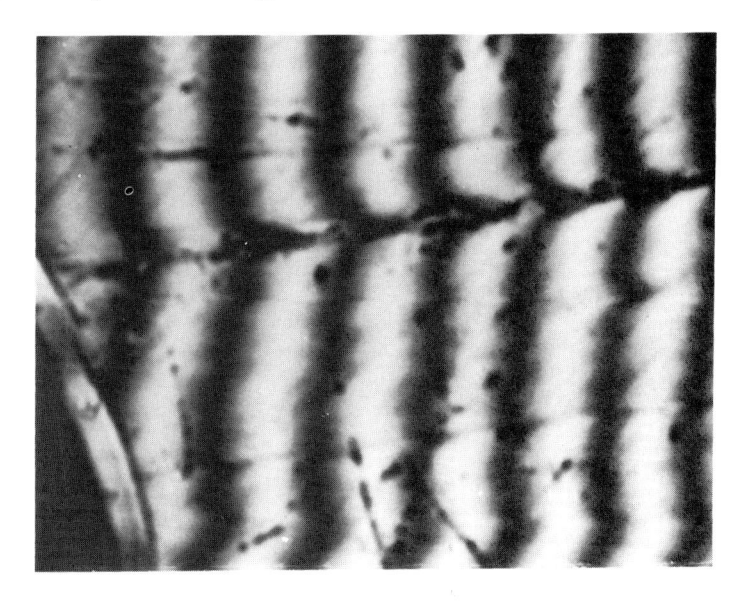

Figure 15.2. A confocal interference image of a TEM grid, produced using a single detector, showing a fringe pattern superimposed on a confocal image.

Figure 15.3. A confocal (non-interference) image of the TEM grid.

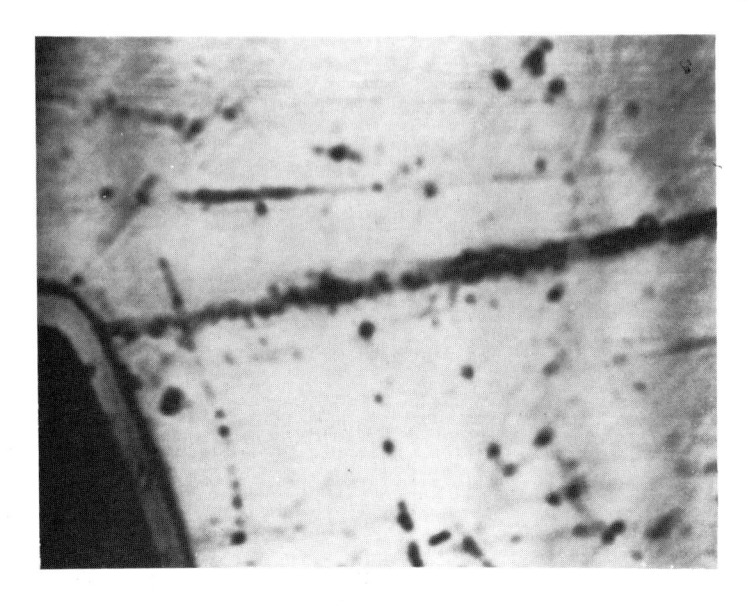

Figure 15.4. The image formed from the sum of the two detector signals in an interference microscope. The fringes cancel, leaving a confocal image.

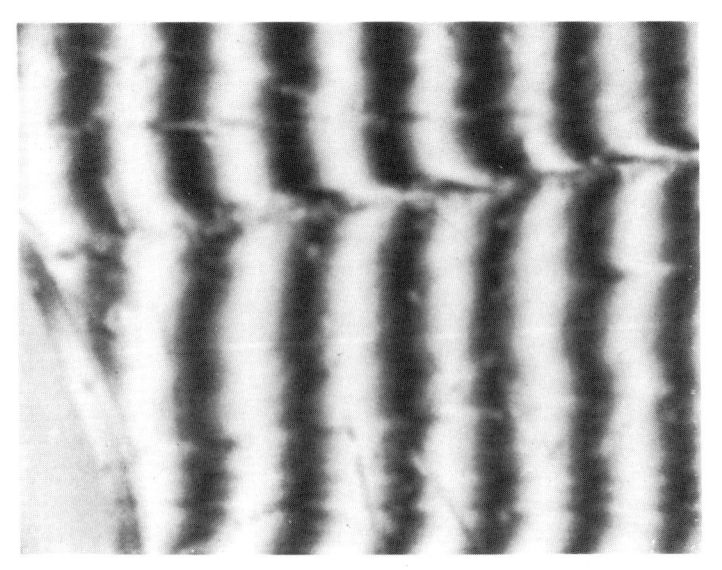

Figure 15.5. The image formed from the difference of the two detector signals, showing the fringe pattern with no superimposed confocal image.

For I_S and I_D to be produced accurately, it is important that the ratio $|t|/|r|$ is approximately the same at each detector, which relies on the beam-splitter BS1 having equal transmittance and reflectance. These properties depend on the polarisation of the incident beam, and fine adjustment can be made by rotating the laser until the two ratios are equal. Alternatively, and perhaps preferably, the laser polarisation can be set vertically and the signals equalised electronically. The image I_B, from the second detector, is similar to I_A except that the fringes are displaced by a half-period, which means that the signal I_S produced by addition should exhibit no fringes. In practice, small optimisations of the relative gains and pinhole alignments are needed to achieve the optimum result as shown in Figure 15.4, which is, as predicted, substantially similar to the confocal image of Figure 15.3. Figure 15.5 shows the difference image I_D with no further adjustments made to the system, showing the fringe pattern without the superimposed confocal image. Even though the brightness of the fringes is modulated by the reflectivity of the object, their positions are now much more easily observed, allowing the surface topography to be deduced more precisely. The interference image I_D gives the component of the object beam in phase with the reference beam. If necessary the phase and amplitude of the object beam can be separated, either from I_D and I_S, or by using two reference beams in order to image the real and imaginary parts of the object simultaneously. The latter method is preferable from the point of view of sensitivity and also for fringe counting and interpolation. By further electronic processing we could, for example, display an image in which intensity changes in direct proportion to phase changes in the object. Furthermore, if the object may be thought of as a dielectric slab of varying thickness, absorption and refractive index, by simultaneous examination with interference microscopy in reflection and transmission, it is possible to separate these material properties [Sheppard and Wilson, 1980a].

So far we have ignored diffraction effects but if we incorporate diffraction the interference term may be written:

$$I_D(x) = 2\text{Re}\left\{r^* \int_{-\infty}^{\infty} c(m)T(m) \exp 2\pi jmx\, dm\right\} \qquad (15.5)$$

Imaging is in principle identical in conventional and confocal systems with circular aberration-free pupils, so it appears we have lost the resolution advantage of confocal microscopy. Actually, by using interference techniques, we have improved the imaging of conventional microscopy to the level of confocal imaging, but the advantage of confocal microscopy — that the shape of the reference beam wavefront is not important — still remains. In addition, reflection confocal interference microscopy exhibits an optical sectioning property.

The space-invariant nature of scanning microscopes with object-scanning also has advantages for multiple-beam interferometry, as the reflecting element may be situated before the objective lens. This is unlike the conventional multiple-beam method, in which the resonator must be placed after the objective lens, thereby restricting the available numerical aperture. It is the coherence of the laser beam which allows the optical path to be made so large. Furthermore, the high power of the laser allows high-reflectivity mirrors to be used, thus giving great sensitivity. Various optical arrangements of this resonant microscope have been discussed by Sheppard and Kompfner (1978).

The resonant microscope is very sensitive to small changes in specimen phase and also in specimen absorption, resulting in contrast enhancement [Sheppard and Wilson, 1980a]. The sensitivity of the microscope is increased for decreasing resonator losses, and these losses will be smaller if the system is focused so that the objective pupil transmits the lowest-order mode without appreciable loss. The loss can be reduced still further by placing the reflecting element between the objective and object. Use of a spherical mirror is possible in object-scanning systems, so that the numerical aperture need not be restricted. Furthermore, the aberrations of the objective are not of great importance, because the spatial resolution is related to the lowest-order resonator mode. The higher modes are rapidly attenuated by higher diffraction losses, and contribute little to the energy density at the focus.

The techniques of scanning interference microscopy are extremely powerful: the confocal interference microscope allowing simultaneous measurement of signal phase and amplitude. The limitation to their usefulness is, however, set by the fact that in a reflection system the signal is sensitive to stray vibrations and axial movement of the specimen stage. The images in Figures 15.2 and 15.5 show the type of performance which can be achieved in practice. If the sensitivity is to be increased greatly, say to the value of better than 1nm, then it is necessary to measure or stabilise the object axial position, which, of course, must be done with even greater sensitivity. An alternative is to derive the reference beam from the object, as in differential interference microscopy.

In the Nomarski (1955) method of differential interference contrast, the object is imaged at two spatially separated points, the beam being split into two and recombined using birefringent elements. If the spatial separation between the observation points is 2Δ and the beams are recombined with a relative phase $\pi - 2\delta$, the effective transfer function $C'(m;p)$ for the system can be shown to be [Sheppard and Wilson, 1980b]:

$$C'(m;p) = C(m;p)\sin(2\pi m\Delta + \delta)\sin(2\pi p\Delta + \delta) \qquad (15.6)$$

The quantity δ is adjusted by the compensator setting. If $\delta = 0$ then differential contrast results, which is non-linear and somewhat similar to darkfield. For weak objects, the transfer function $C'(m;0)$ can be separated into its even part, responsible for amplitude imaging, and its odd part:

$$C'_o(m;0) = \frac{1}{2}C(m;0)\sin(2\pi m\Delta)\sin(2\delta) \qquad (15.7)$$

responsible for differential phase contrast. The shape of this curve is independent of the value of δ. If δ is small, as is usually the case in practice, the contrast:

$$|C'_o(m;0)/C'(0;0)| = C(m;0)\sin(2\pi m\Delta)\cot\delta \qquad (15.8)$$

is maximised, and so too is the strength of the differential phase imaging relative to the amplitude imaging. However, how small δ may be made depends on the strength of the object modulation, for as δ is decreased the strength of the non-linear term increases. In a scanning microscope the contrast is not so important, as it can be restored by electronic enhancement. The absolute strength of the differential phase term can then be maximised by increasing δ to a value of $\pi/4$, although this is not possible for phase objects, where the amplitude contrast term is not negligible. This is the approach followed by Allen et al. (1981a, 1981b) in TV microscopy, where, again, the contrast may be enhanced electronically. The Nomarski method can be performed with a confocal microscope, an advantage being the simpler form of the image expressions for this coherent system.

A flying spot differential interference system was described by Roberts and Causley (1954). Instead of using polarisation to stop the beams interfering in the specimen region, an alternative is to use heterodyning to produce a pair of beams frequency-shifted relative to one another [Laub, 1972].

II. Surface profiling by confocal interferometry

Contacting methods of surface profilometry, in which a fine-tipped diamond stylus is traversed across a surface, run the risk of causing damage to delicate specimens. There are a large number of non-contacting methods available, with various corresponding advantages and disadvantages. For high-sensitivity measurements it is often advantageous to use a common-path interferometer, in which the phase at a given point is measured relative to a neighbouring point or to an average over some region so that the effects of vibrations and air currents cancel out. Downs et al. (1985) have described a confocal profiling system in which the phase of the signal beam reflected from the specimen surface is compared with that of a defocussed reference beam. However, we have chosen to use a confocal interference microscope, as in the previous section, because the image is then simpler to analyse theoretically, and it is easier to provide quantitative results. For surface profiling, a sensitive method is to use a feedback system to lock on to the condition of phase quadrature by altering either the specimen axial position [Hamilton and Matthews, 1985] or the phase of the reference beam [Matthews, Hamilton and Sheppard, 1986] as the object is scanned. The two experimental systems are illustrated in Figures 15.6 and 15.7 respectively. In the former, the object is mounted on a piezoelectric bimorph, so the method is rather slow: a single line scan takes about two minutes. In the alternative scheme, the phase of the reference beam is altered using an electro-optic phase modulator, allowing much higher scanning speeds so that a complete frame can be scanned in only 2 seconds. In each case the difference signal I_D, equation (15.4), is fed to an operational amplifier integrator which is fed either to the piezoelectric bimorph or to the drive amplifier of the electro-optic modulator. This system then forms a phase-locked loop with error signal I_D. Integral feedback was chosen, as it has zero steady-state error. The output of the integrator is the height signal, which can be dislayed using either a pen recorder or an oscilloscope. For a frame scan, the oscilloscope brightness can be modulated as it is scanned in synchronism with the specimen, or an isometric view of the surface, as illustrated in Figure 15.8 which shows the bonding pad of a surface acoustic wave device, can be observed using the second scheme. This clearly shows small pits in the deposited aluminium of 5nm depth along with other smaller surface details. The height sensitivity for this scheme was in the region of 0.5nm, which compared with a sensitivity of about 10nm for the former approach. The sensitivity attainable is determined primarily by noise, both external, such as that due to air currents and vibrations, and internal, such as laser and electrical noise. Since this is a closed-loop method, internal noise is reduced by the loop gain and can be neglected. In contrast, external noise will be interpreted as object height variations, but such disturbances are reduced at slower scan speeds as the integration time per picture point is increased. The resolution is noise-limited because a null detection method, which has inherent advantages over maximum detection methods, is used. A further difference between the two schemes is that if the object is moved axially to lock on to phase quadrature, large total height variations can be tracked, whereas if the reference beam is altered, the surface will go out of focus. However, the latter scheme could be extended to general phase imaging, and could be used in a Mach-Zehnder interferometer for imaging in transmission [Hansen et al., 1981].

Confocal Interference Microscopy

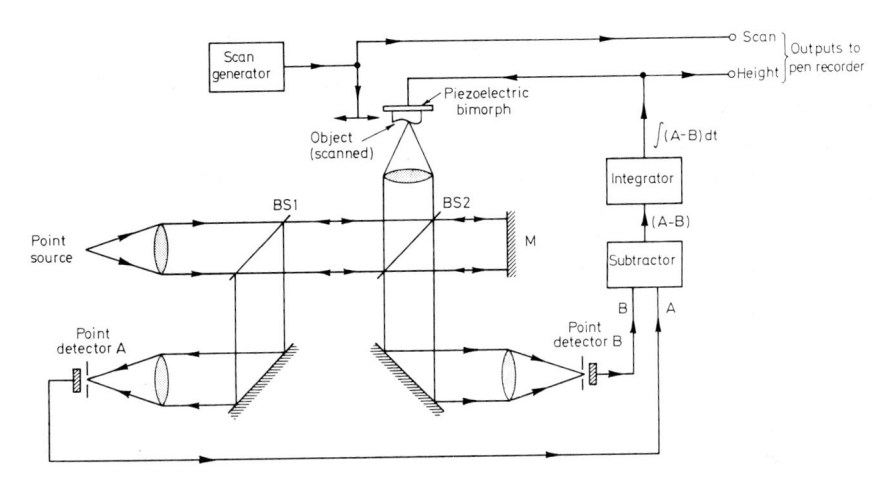

Figure 15.6. A surface profiling system using piezoelectric focusing.

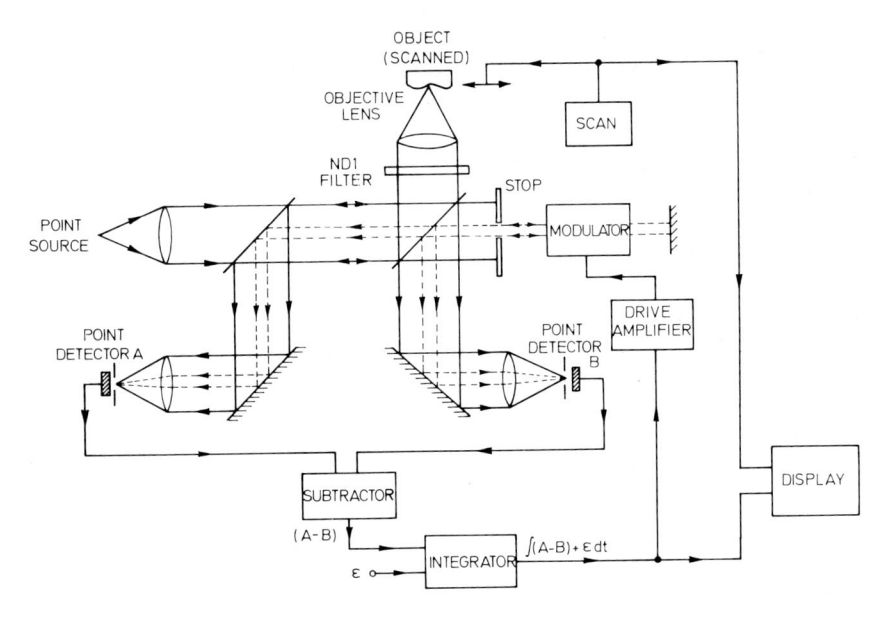

Figure 15.7. A surface profiling system using electro-optic phase-shifting.

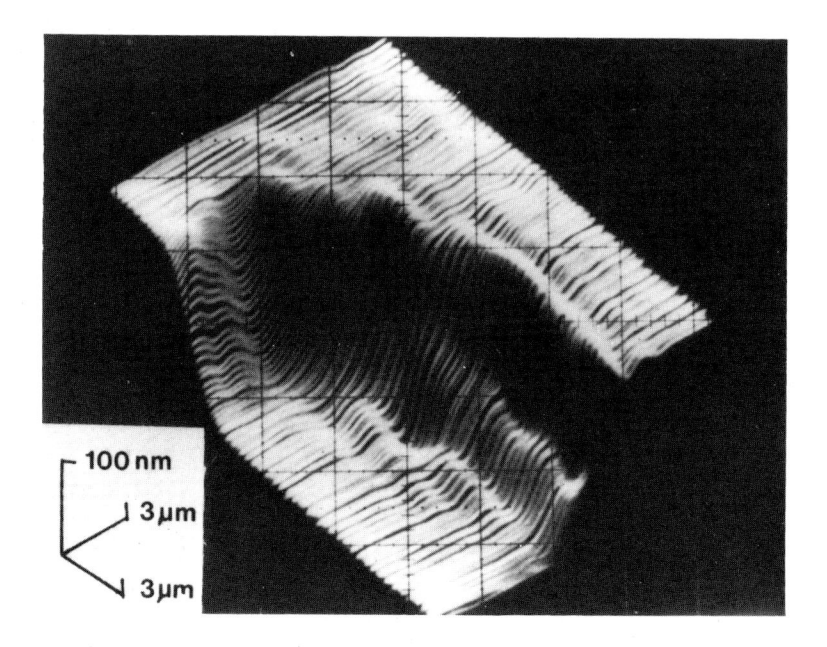

Figure 15.8. The surface profile of the bonding pad of a surface acoustic wave device.

III. Heterodyne interferometry

The well-defined frequency of the laser radiation makes possible the use of optical heterodyne methods, in which the laser beam is split into two components. The frequency of one component is changed by a small amount $\Delta\omega$ by reflection from a mirror vibrating in the direction of its normal or by an acousto-optic interaction. After the other beam is transmitted through or reflected from the object, the two beams are combined on a photodetector, and, from the output of the photodetector, the component at the difference frequency:

$$I = a \cos(\Delta\omega t + \phi) \qquad (15.9)$$

is selected.

The heterodyne method has a number of interesting features which have been applied in microscopical systems. Firstly, the signal may be used to obtain the phase of the object beam. In this respect, it has some similarities with the normal interference methods which are, in electronics terminology, called homodyne techniques to distinguish them from heterodyne techniques. A scanning differential interference microscope using heterodyning in which both beams pass through the object, one being laterally displaced, was reported by Laub (1972). By using a differential technique, vibrations and microphonics in the system are not imaged. In this microscope the frequency difference is used to stop the beams interfering in the specimen region, rather than polarisation, as is used in the ordinary differential interference microscope. Such a system could also be used to obtain an image showing variations in the object's optical dispersion. An advantage of heterodyne detection is that it is extremely sensitive and enables very small phase changes to be measured. A confocal heterodyne microscope has been described by Jungerman, Hobbs and Kino (1984).

A further interesting feature of heterodyne detection is its imaging property: a significant signal is detected only if the phase front of the object beam coincides with that of the reference beam [Corcoran, 1965; Siegman, 1966]. This principle was used by Fujii and Takimoto (1976) to construct a scanning microscope in which a divergent reference beam renders a lens unnecessary. This has possible applications in microscopy with radiation for which lenses are not available, and may also prove useful in focusing deep into a specimen, which would be physically impossible with a real lens. It has been shown by Fujii et al. (1981) that the detector acts exactly as a lens with numerical aperture equal to that of the detected portion of the reference beam. In Fujii's system, the object is illuminated approximately uniformly, so that it behaves similarly to a point-detector microscope of our terminology. Thus, in some ways the sensitivity advantage of heterodyne detection is counteracted by the poor detection efficiency of point-detection. The divergent reference beam also needs to be created in some way: if it is formed by focussing with a lens we have lost the advantage of lensless imaging, and, furthermore, the aberrations of the lens appear on the reference beam phase-front and result in analogous aberrations in the heterodyne image. If, on the other hand, a small pinhole is used to form a divergent beam, the efficiency of utilisation of the laser power is low. Sawatari (1973) described a microscope in which the object was illuminated by the focussed beam from a lens, and this combined with heterodyne imaging to form a confocal microscope. Now the detection efficiency is improved and the sensitivity of heterodyne detection appreciated. If the object and reference beam are focussed on the detector, no detector pinhole is necessary to achieve confocal imaging because the reference beam acts as a *synthetic* pinhole.

IV. Aberration measurement by confocal interferometry

If a perfectly reflecting surface is observed in a confocal microscope, it is found that the signal intensity falls off as the surface is displaced from focus. The form of this defocus signal depends on the aberrations (and apodization) of the microscope objective [Corle et al., 1986] and also on the angular variation in reflectance of the surface [Cox et al., 1982] which is in turn a function of the optical properties of the object material. By comparison, in a conventional system the defocus signal is simply constant.

Confocal imaging is very sensitive to the presence of small amounts of aberration, which suggests that observation of the defocus signal may be a basis for measurement of the system aberrations. Confocal imaging is a coherent process, and it is therefore necessary to use interference techniques to extract the modulus and phase of the defocus signal [Hamilton and Sheppard, 1986]: otherwise only the intensity of the signal is observed, the phase information being lost.

An analogous effect occurs in acoustic microscopy [Atalar, 1978] when, however, the defocus signal exhibits strong ringing as a result of the generation of surface-waves, and hence the quality of three-dimensional image formation is restricted. In acoustic microscopy, the low frequency of the measured signal ensures that the complex amplitude of the resulting signal can be measured directly [Hildebrand et al., 1983]. The defocus signal is commonly referred to as $V(z)$, signifying the variation in transducer output-voltage with defocus distance.

A confocal interference technique can be used to measure the complex amplitude of the defocus signal in a confocal optical system. From this signal the modulus and phase of the pupil function of the objective lens may be extracted. Quantitative knowledge of the pupil function is useful in several ways:

(1) As a guide to the quality of imaging of an objective obtained *in situ* in the complete optical system.

(2) For quality control in an optical manufacturing facility.

(3) To allow correction of aberrations by variation in tube length, spacing of lens elements or insertion of correction lenses.

(4) The point spread function of the lens is given by the Hankel transform of the pupil function.

(5) The pupil funciton may be treated as *a priori* knowledge for subsequent image-processing operations such as deconvolution by Wiener filtering or calculating the eigenfunctions of the imaging operator.

The defocus signal is essentially a high-angle property which implies that the paraxial approximation may not be used and that a full high-angle analysis must be employed instead. Such an analysis has been described by Sheppard and Wilson (1981) for a rotationally-symmetric aplanatic system that satisfies the sine condition. The defocus signal is given by:

$$V(z) = \int_0^\alpha P^2(\theta) R(\theta) \exp(2jkz\cos\theta) \sin\theta \, d\theta \qquad (15.10)$$

where θ is the angle of convergence of a ray and lies in the range $0 \le \theta \le \alpha$, $P(\theta)$ is the pupil function of the lens including a factor $\cos^{1/2}\theta$ for a high aperture aplanatic lens, $R(\theta)$ is the angular reflectivity of the object, k is the wavenumber of the illumination given by $2\pi/\lambda$ and z is the distance of the object from the focal plane.

The physical interpretation of the integral in equation (15.10) is that the pupil function $P(\theta)$ is multiplied by a phase term $\exp(jkz\cos\theta)$ because of the movement of the object away from the focal plane. This is then multiplied by the angular reflectivity of the object to give $P(\theta)R(\theta)\exp(jkz\cos\theta)$. Since the light then passes back through the lens, it is again multiplied by the phase-shifted pupil function to give $P^2(\theta)R(\theta)\exp(2jkz\cos\theta)$. This is then integrated over all angles of incidence.

Two alternative substitutions may be made in equation (15.10), each of which is appropriate for a particular physical situation. First, consider the substitution:

$$t = \frac{2}{\lambda}\cos\theta \qquad (15.11)$$

equation (15.10) then becomes:

$$V(z) \doteq \int_{(2/\lambda)\cos\alpha}^{2/\lambda} P^2(t) R(t) \exp(2\pi jzt) \, dt \qquad (15.12)$$

Since $P(t)$ is non-zero only in the range $(2/\lambda)\cos\alpha \le t \le 2/\lambda$, the limits of integration in equation (15.12) may be extended to $\pm\infty$. Equation (15.12) may then be interpreted as stating that $V(z)$ is the inverse Fourier transform of $P^2(t)R(t)$. Therefore, $P(t)$ may be found by taking the Fourier transform of $V(z)$, dividing by $R(t)$ and taking the square root. This is the case that is normally studied [Liang et al., 1985].

An alternative substitution may be made in equation (15.10). Putting:

$$s = \frac{\sin^2(\theta/2)}{\sin^2(\alpha/2)} \qquad (15.13)$$

we get:

$$V(z) \doteq \int_0^1 P^2(s) R(s) \exp\left\{2jkz\left[1 - 2s\sin^2(\alpha/2)\right]\right\} ds \qquad (15.14)$$

Rearranging, and again extending the limits of integration to $\pm\infty$, gives:

$$V(z)\exp(-2jkz) = \int_{-\infty}^{+\infty} P^2(s)R(s)\exp\left[-4kjzs\sin^2(\alpha/2)\right]ds \quad (15.15)$$

which can be rewritten as:

$$V(z)\exp(-2jkz) = \int_{-\infty}^{\infty} P^2(s)R(s)\exp(-jus)\,ds \quad (15.16)$$

where

$$u = 4kz\sin^2(\alpha/2) \quad (15.17)$$

This is another Fourier transform relationship, this time between u and s.

For the case where the object is moved and the lens remains stationary, there is a change in the total optical path of $2z$, owing to the movement of the object. This corresponds to a phase change of $\exp(2jkz)$ in $V(z)$. This quickly varying linear phase term is normally considered part of $V(z)$ and is not written out explicitly. The t substitution of equation (15.11) is best suited to this method, as equation (15.12) relates $V(z)$ and $P(t)$ directly. In contrast, if the lens is moved and the object remains stationary [Hamilton and Sheppard, 1986] the total optical path length does not change, as the lens is moved and so there is no $\exp(2jkz)$ term implicitly present in $V(z)$ when measured in this way. Hence the function that is measured by this method is not $V(z)$, but $V(z)\exp(-2jkz)$ instead, so the s substitution of equation (15.13) is the most appropriate as equation (15.16) relates $V(z)\exp(-2jkz)$ and $P(s)$ directly. This method of measuring $V(z)$ by moving the lens, suppresses the rapid phase variation of the normal method and constrains the phase of $V(z)$ to lie within approximately $\pm\pi$ with no phase discontinuities, unlike the object movement method where these jumps in the phase of $V(z)$ are not suppressed. It is this feature of the lens-movement system which makes it so attractive compared to the more common object-movement method.

By definition, $P(s)$ is identicallly zero for $s < 0$. Great use can be made of the single-sided nature of $P(s)$, as it allows a simplification of the experimental apparatus and reduces the amount of data required by a factor of two. To see why this is so, let the Fourier transform of $p(z)$ be $P(s)$:

$$F[p(z)] = P(s) \quad (15.18)$$

Now consider just the real part of $p(z)$:

$$\operatorname{Re}[p(z)] = \frac{1}{2}[p(z) + p^*(z)] \qquad (15.19)$$

Taking the Fourier transform of both sides gives:

$$F\{\operatorname{Re}[p(z)]\} = \frac{1}{2}[P(s) + P^*(-s)] \qquad (15.20)$$

For a function that is non-zero for both positive and negative values of s, the two terms on the right-hand side of equation (15.20) cannot be separated but, since $P(s)$ is single-sided by definition, $P(s)$ and $P^*(-s)$ are on opposite sides of the origin and are thus completely separable. This is an important result: it comes from the fact that the real and imaginary parts of $p(z)$ are connected by a Hilbert-transform relationship.

A diagram of a suitable apparatus is shown in Figure 15.9. This system is based on a confocal interference microscope with an electro-optic phase-modulator in the reference beam. Only one detector is used. The principle of the method involves using the modulator to switch the phase of the reference beam between two distinct phases. An electronic demultiplexer, driven in synchronism with the modulator, is then used to separate the single alternating detector-signal from a reflecting surface into two non-multiplexed signals corresponding to the two states of the reference beam. These two signals are then subtracted to remove the terms in each corresponding to the object-beam and the reference-beam powers, leaving just the interference term, equation (15.10). This interference signal is then sampled as a function of defocus and stored for subsequent processing.

The use of one detector and two reference-beam phases, rather than the standard confocal interference microscope employing two detectors

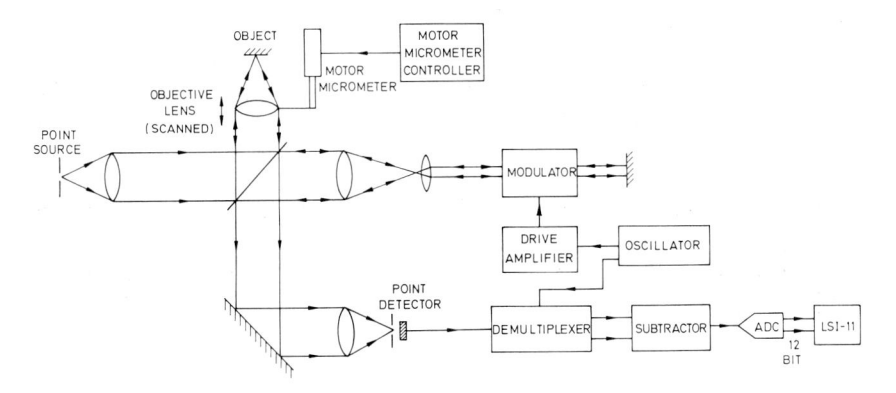

Figure 15.9. A confocal interference system for measuring *in situ* aberrations.

and one reference-beam phase, has the advantage that the problems of alignment of the system are much reduced. With two detectors, it is difficult to achieve alignment such that both detectors give the same signal from both the object beam and reference beam over a depth of field of more than 400 wavelengths.

Hamilton and Sheppard (1982) have shown that for an object-beam amplitude t and reference-beam amplitude r, the signal at the detector is:

$$I_A = |t|^2 + |r_1|^2 + 2\operatorname{Re}\{tr_1^*\} \tag{15.21}$$

for one reference beam state and

$$I_A = |t|^2 + |r_2|^2 + 2\operatorname{Re}\{tr_2^*\} \tag{15.22}$$

for the other. Subtracting equation (15.22) from equation (15.21) gives the difference signal I_D:

$$I_D = \left(|r_1|^2 - |r_2|^2\right) + 2\operatorname{Re}\left[t\left(r_1^* - r_2^*\right)\right] \tag{15.23}$$

The first bracketed term is a constant representing the difference in the intensities of the two states of the reference beam, and the second term is a pure interference term. Since both r_1 and r_2 are complex quantities they may be represented on a phasor diagram, where their difference will be a phasor r_3. Equation (15.23) may hence be written as:

$$I_D = A + 2\operatorname{Re}\left[tr_3^*\right] \tag{15.24}$$

where A is a constant. Since r is also a constant and t is equivalent to $V(z)$, if the phase origin is chosen to lie along r_3^* (this is entirely valid as the choice of phase origin is immaterial, its choice here being merely for mathematical convenience) equation (15.24) becomes:

$$I_D = A + 2r\operatorname{Re}\left[V(z)\right] \tag{15.25}$$

If the constant A is removed and the Fourier transform of both sides is taken, the resulting equation is identical to equation (15.20) to within a multiplicative constant. For dry objectives, a glass surface is a suitable object for measuring the defocus signal. It is preferable to use anisotropic materials or metals which have a complex refractive index. Also, glass has a reflectivity of about 4%, making the object beam and reference beam almost equal in intensity, thereby maximising the visibility of the fringes. The reflectivity $R(\theta)$ is almost constant and affects only the modulus of $P(s)$ and not its phase. For oil immersion lenses it is necessary to use a reflective mirror.

The discrete Fourier transform of the data points can be calculated and the pupil function $P(s)$ can thus be recovered from the Fourier transform of only one component of $V(z)$ by taking only positive values of s. This is equivalent to zeroing all points outside the range $0 \leq s \leq 1$. Real and imaginary parts of $V(z)$ can then be calculated by taking the inverse discrete Fourier transform of $P(s)$. From this the modulus and phase of $V(z)$ can be derived and compared to the theoretical aberration-free curves. Figure 15.10 shows the theoretical and the experimental $|V(z)|^2$ for a 0.5 N.A. lens [Matthews et al., 1989]. These two curves agree closely, the calculated curve being slightly wider than the theoretical curve and also having larger sidelobes caused by slight spherical aberration. The corresponding phase curves [Matthews et al., 1989] are shown in Figure 15.11. These curves do not agree as well as the modulus curves and it is here that, in general, the differences between ideal and real lenses are most apparent.

The pupil function, as just calculated, contains a large linear phase term since the zero of z, and hence u, is not at the peak of $|V(z)|^2$. This linear phase-term was removed by shifting the calculated $V(z)$, so that the peak of the modulus curve became the first sample, and transforming again. This operation affects only the phase of $P(s)$ and not its modulus.

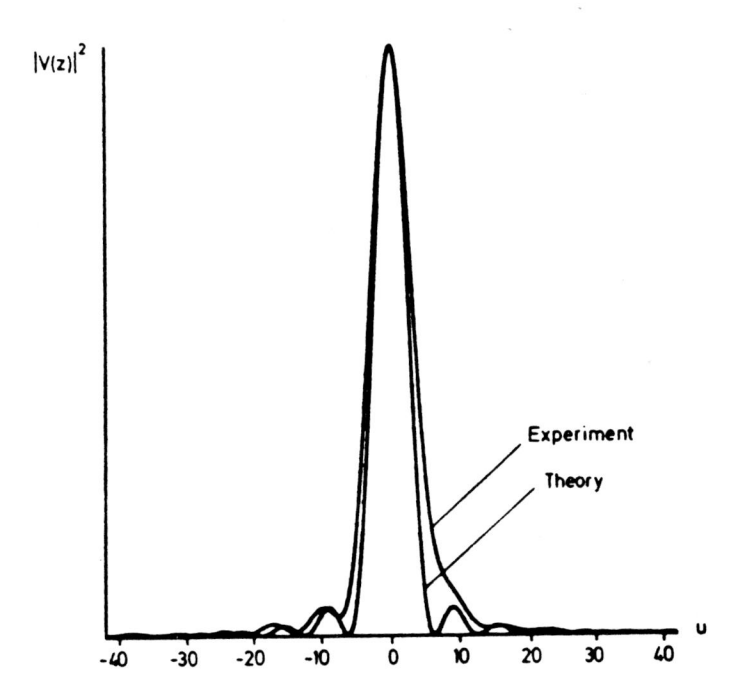

Figure 15.10. Experimentally measured intensity of the defocus signal for a 0.5 N.A. lens, compared with the theoretical prediction for an aberration-free lens.

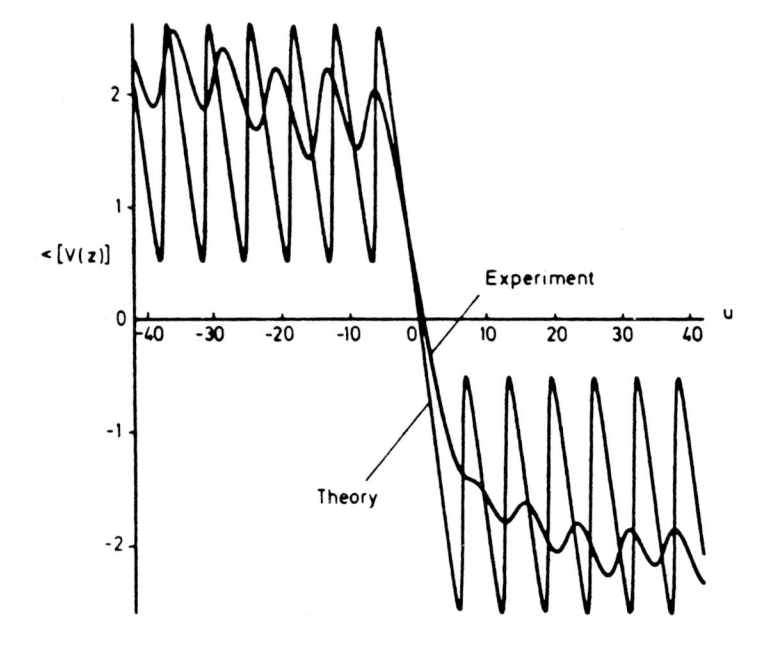

Figure 15.11. Experimentally measured phase of the defocus signal for a 0.5 N.A. lens, compared with the theoretical prediction for an aberration-free lens.

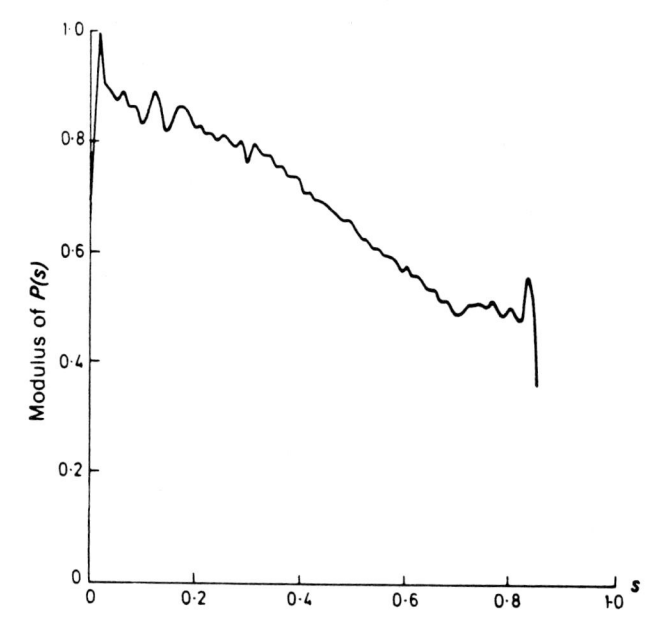

Figure 15.12. The modulus of $P(\theta)/\cos^{1/2}\theta$ for a 0.5 N.A. lens, plotted as a function of s.

The modulus of $P(s)$ for the 0.5 N.A. lens is shown in Figure 15.12, the aplanatic $\cos^{1/2}\theta$ being suppressed. All our measured moduli share the same characteristic shape: falling from unity at the centre of the lens as s increases and cutting off to zero as s tends to unity. The spikes at the beginning and end of the curves are artefacts from the sampling process. $|P(s)|$ falls sharply towards zero when s is just greater than 0.8 for the 0.5 N.A. lens. This corresponds to the outer boundary of the pupil and, as s is defined by equation (15.13) with $\sin\alpha$ taken as the quoted numerical aperture of the lens, the results imply that the true numerical aperture is 0.46. The fall-off in transmissivity is caused by Fresnel loss at each of the curved surfaces of the glass elements comprising the lens. The Fresnel loss affects only the modulus of $P(s)$ and not its phase.

The phase of $P(s)$ may be expressed as a polynominal in s:

$$\arg[P(s)] = a_0 + a_1 s + a_2 s^2 + a_3 s^3 + \ldots + a_n s^n \qquad (15.26)$$

In this expression the constant coefficient a_0 represents the choice of the phase origin and can be subtracted from the calculated phase so that the phase at the origin is zero. The linear coefficient a_1 represents the distance of the diffraction focus of the lens away from the peak of $|V(z)|^2$. All the higher-order coefficients, a_2, a_3 etc., represent aberrations. Since the pupil is assumed to be radially symmetric, aberrations such as distortion and coma will average out to zero when integrated around the whole pupil. Of the three remaining primary Seidel aberrations, both astigmatism and curvature of field are proportional to the square of the distance of the object point from the axis. But the system is imaging purely on-axis, which means that these two aberrations should be unimportant, leaving spherical aberration as the only significant primary aberration.

The separate coefficients $a_2, a_3, \ldots$ could be obtained by fitting a polynomial through the measured phase of $P(s)$, but this method has the problem that if a polynomial of a sufficiently high order to give a good fit to the data is used, then the coefficients of the higher-order terms alternate between large positive and large negative values. These large terms imply that the polynomial becomes large as s tends to unity. Such terms do not affect the fit near the origin, but they introduce large errors in the phase at the edges of the lens where a significant proportion of the light passes. Therefore, this course was not taken, and the measured phase curve was used to determine the average wavefront deformation of the lens instead.

The ratio of the maximum intensity in the three-dimensional point spread function in the presence of aberrations to the maximum in their absence is called the Strehl intensity (more properly called the

Strehl ratio). Since the lens is assumed to be radially symmetric, the maximum in the point spread function will lie along the optic axis. From Richards and Wolf (1959) the intensity along the axis is $|h(z)|^2$, where $|h(z)|$ is given by:

$$h(z) = \int_0^\alpha (1 + \cos\theta) P(\theta) \exp(jkz\cos\theta) \sin\theta \, d\theta \qquad (15.27)$$

or

$$h(u) \doteq \int_0^1 |P(s)| \left[1 - s\sin^2(\alpha/2)\right] \exp[j\phi(s)] \exp(-jus/2) \, ds \qquad (15.28)$$

The value u_0 of u which maximises the intensity can thus be calculated, giving the distance of the diffraction focus from the geometric focus. The Strehl intensity can also be found and, from this, the wavefront deformation. For the 0.5 N.A. lens, the measured Strehl intensity is 0.989, corresponding to an rms wavefront deformation of $\lambda/59$.

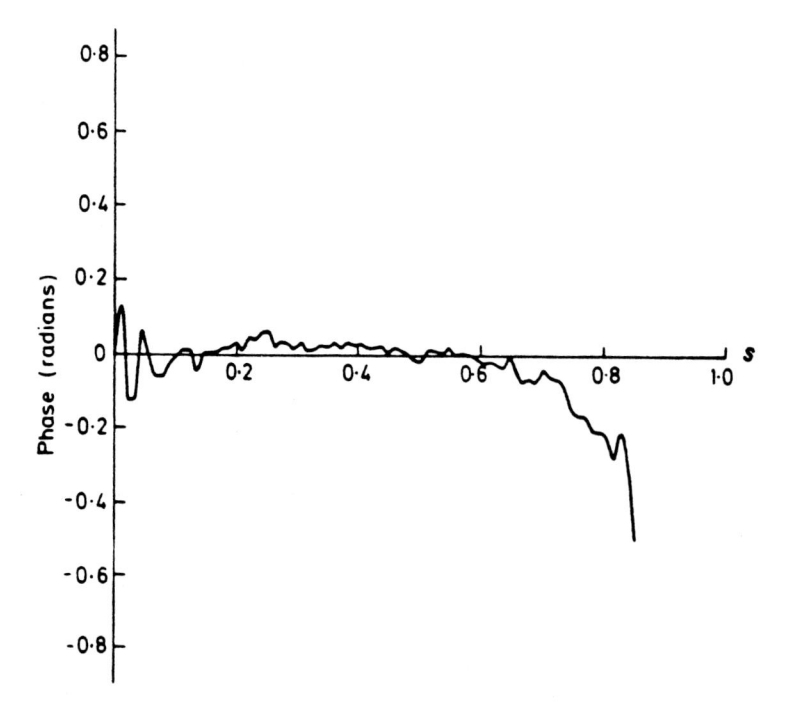

Figure 15.13. The phase of the pupil function for a 0.5 lens N.A., plotted as a function of s.

Confocal Interference Microscopy

Once u_0 is known, the first two terms of the polynomial expansion of $\arg[P(s)]$, equation (15.26), can be subtracted, leaving only the aberration terms. This phase plot [Matthews et al., 1989] for the 0.5 N.A. lens is shown in Figure 15.13. It is plotted against s, but also shown against $\sin\theta$ (corresponding to the radius of the rear element) in Figure 15.14. It should be noticed that, when plotted against s, the curves for amplitude and phase include odd powers of s, but when plotted against θ, symmetry dictates that they must be even functions. These curves have been truncated at the value of radius for which the modulus of the pupil function fell to zero. It is apparent that the pupil-function phase of the 0.5 N.A. lens is very well-behaved, showing a significant error only at the edge of the pupil where the modulus decreases sharply.

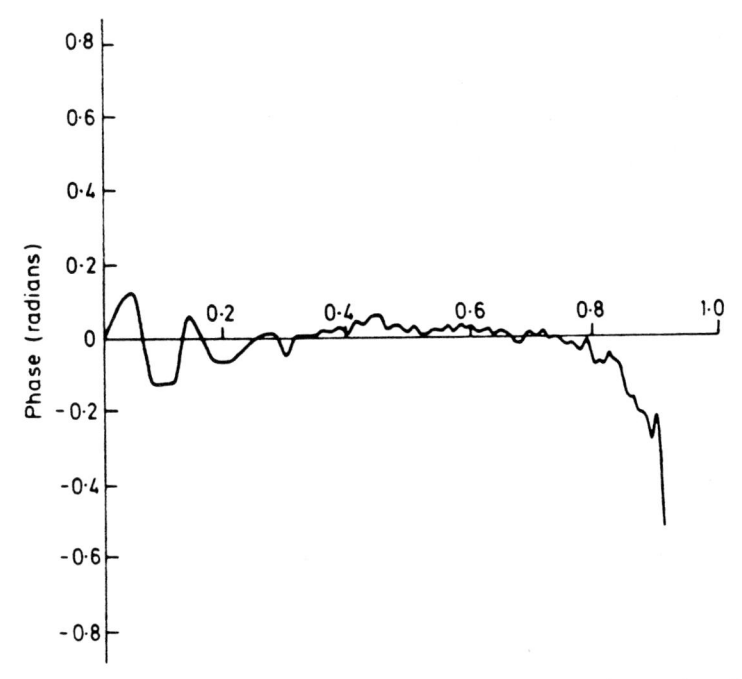

Figure 15.14. The phase of the pupil function for a 0.5 lens N.A., plotted as a function of r.

V. Conclusions

Confocal interference methods can be used to extract the amplitude and phase of the object transmittance. Indeed, it could be argued that as confocal microscopy in reflection or transmission is a coherent process, interference is necessary in order to render the image signal suitable for restoration or object reconstruction. Interference methods can be used to measure surface profiles with high sensitivity, and also to investigate *in situ* objective lens aberrations and apodization.

References

Allen, R. D., Allen, N. S. and Travis, J. L., (1981a). Video-enhanced contrast, differential interference contrast (AVEC-DIC) microscopy: a new method capable of analysing microtubule related mobility in the reticulopodial network of *allogromia laticollaris*. Cell Motility, **1**, 291-302.

Allen, R. D., Travis, J. L., Allen, N. S. and Yilmaz, H., (1981b). Video-enhanced polarization (AVEC-POL) microscopy: a new method applied to the detection of birefringence in the motile reticulopodial network of *allogromia laticollaris*. Cell Motility, **1**, 275-287.

Atalar, A., (1978). An angular spectrum approach to contrast in reflection acoustic microscopy. J. Appl. Phys., **49**, 5130-4139.

Brakenhoff, G. J., (1979). Imaging modes of confocal scanning light microscopy. J. Microsc., **117**, 233.

Corcoran, V. J., (1965). Directional characteristics in optical heterodyne detection processes. J. Appl. Phys., **36**, 1819-1825.

Corle, T. R., Chou, C-H. and Kino, G. S., (1986). Depth response of confocal optical microscopes. Opt. Lett., **11**, 770-772.

Cox, I. J., Hamilton, D. K. and Sheppard, C. J. R., (1982). Observation of optical signatures of materials. Appl. Phys. Lett., **41**, 604-606.

Downs, M. J., McGivern, W. H. and Ferguson, H. J., (1985). Optical system for measuring the profiles of super-smooth surfaces. Precision Engineering, **7**, 211-215.

Fujii, Y. and Takimoto, H., (1976). Imaging properties due to the optical heterodyne and its application to laser microscopy. Opt. Comm., **18**, 45-47.

Fujii, Y., Takimoto, H. and Igarashi, T., (1981). Optimum resolution of laser microscope by using optical heterodyne detection. Opt. Comm., **38**, 85-90.

Hamilton, D. K. and Matthews, H. J., (1985). The confocal interference microscope as a surface profilometer. Optik, **71**, 31-34.

Hamilton, D. K. and Sheppard, C. J. R., (1982). A confocal interference microscope. Opt. Acta, **29**, 1573-1577.

Hamilton, D. K. and Sheppard, C. J. R., (1986). Interferometric measurments of the complex amplitude of the defocus signal $V(z)$ in the confocal scanning microscope. J. Appl. Phys., **60**, 2708-2712.

Hansen, E. W., Allen, R. D., Strohbehn, J. W., Gold, C., Chaffee, M., Riley, M. F. and Pillsbury, T. A., (1981). Laser scanning phase modulation microscope. J. Opt. Soc. Am., **71**, 1557 (abstract).

Hildebrand, J. A., Liang, K. and Bennett, S. D., (1983). Fourier transform approach to materials characterisation with the acoustic microscope. J. Appl. Phys., **54**, 7016-7020.

Jungerman, R. L., Hobbs, P. C. D. and Kino, G. S., (1984). Phase sensitive scanning optical microscope. Appl. Phys. Lett., **45**, 846-849.

Laub, L. J., (1972). A.C. differential interferometry. J. Opt. Soc. Am., **62**, 737 (abstract).

Liang, K. K., Kino, G. S. and Khuri-Yakub, B. T., (1985). Material characterisation by the inversion of $V(z)$. I.E.E.E. Trans. Sonics Ultrasonics, **32**, 213-224.

Matthews, H. J., Hamilton, D. K. and Sheppard, C. J. R., (1986). Surface profiling by phase-locked interferometry. Appl. Opt., **25**, 2372-2374.

Matthews, H. J., Hamilton, D. K. and Sheppard, C. J. R., (1989). Aberration measurement by confocal interferometry. J. Mod. Opt., **36**, 233-250.

Nomarski, G., (1955). Microintefermetrie differential à ondes polarisees. J. Phys. Radium, **16**, 9.

Richards, B. and Wolf, E., (1959). Electromagnetic diffraction in optical systems. II. Structure of the image field in an aplanatic system. Proc. Roy. Soc. Lond. **A253**, 358-378.

Roberts, F. and Causley, D., (1954). A flying spot interference microscope. Research VII, No.6 (June).

Sawatari, T., (1973). Optical heterodyne scanning microscope. Appl. Opt., **12**, 2768-2772.

Sheppard, C. J. R. and Choudhury, A., (1977). Image formation in the scanning microscope. Opt. Acta, **24**, 1051-1073.

Sheppard, C. J. R. and Kompfner, R., (1978). Resonant scanning optical microscope. Appl. Opt., **17**, 2879-2882.

Sheppard, C. J. R. and Matthews, H. J., (1987). Imaging in high aperture optical systems. J. Opt. Soc. Am., **A4**, 1354-1360.

Sheppard, C. J. R. and Wilson, T., (1980a). Fourier imaging of phase information in scanning and conventional microscopes. Phil. Trans. Roy. Soc. Lond. **A295**, 513-536.

Sheppard, C. J. R. and Wilson, T., (1980b). Image formation in confocal optical systems. Proc. Soc. Photo-opt. Instrum. Eng., **232**, 197-202.

Sheppard, C. J. R. and Wilson, T., (1981). Effects of high angles of convergence on $V(z)$ in the scanning acoustic microscope. Appl. Phys. Lett., **38**, 858-859.

Siegman, A. E., (1966). The antenna properties of optical heterodyne receivers. Appl. Opt., **5**, 1588-1594.

INDEX

Abbé criterion, 384, 385
Aberrations, 102–108
  correction of, 227–229
  effect of annular pupil plane filter, 124
  effects of, 187
  measurement by confocal interferometry, 399–409
  measurement of, 59
  see also Astigmatism; Chromatic...; Coma; Spherical...;Wavelength aberration
Absorption effects, correction of, 191
Acousto-optic deflector system, 135, 136
Afocal 4-*f* optics, 135, 351
Airy disc, 24, 173, 174
Alginate, immobilization by, 332
Alignment
  importance of, 139, 187
  tandem scanning microscopy, 260–261, 363
  test of, 139
  transmission confocal microscopy, 237
Allium triquetium pollen cells, 191, 192
Aluminium coating
  reflectance scans of, 181
  reflection images of, 180
Alzheimer's disease plaques, 225–226
Amphipleura pellucida (diatom), reflected-light image of, 136
Amplitude point spread functions
  composite, 172
  modified pupils, 174–176
Amplitude transmittance, 47
Annular filters, 121–122
  aberrations affected by, 124
Annular lenses
  image affected by, 122
  single point object imaged using, 149, 150, 151
Annular pupils
  axial response of, 166

single point object imaging affected by, 152–155
Ant's leg, hairs on, 16, 17
Aperture discs
  choice of, 248–249
  design information, 259
  driving of, 259–260
  materials used, 258
  pinhole layout used, 256–257
  pinhole size used, 257–258
  size of, 255–256
Aperture size, spatial frequencies affected by, 158, 159
Aperture spacing effects, direct view microscope, 133–134
Apodisation, meaning of term, 174
Apodising pupil filters
  amplitude point spread functions affected by, 175, 176
  diffraction affected by, 173, 174
  pupil functions affected by, 177, 178, 179
Applanating objectives, use in ophthalmology, 320, 321
Arc lamps, 254
Archimedean spirals (pinhole pattern), 256
Auto-focus technique, 16–17, 143, 144–145, 221
  compared with extended-focus technique, 147
  images produced using, 17, 146
  images using, 17, 220, 221, 239
Autocollimation tandem scanning microscope, 253
Autofluorescence, orchid ovules, 67
Autofocus technique, 102
Autoradiography, 327
Axial resolution, coverglass thickness effects, 228
Axial response, 161–164
  definition of, 162
  as indicator of three-dimensional imaging capabilities, 167
  low-aperture system, 164–167

Beam-scanning microscopes, 5–6
    advantage of, 5–6
    axial response of, 167
    compared with on-axis scanning, 214–216
    drawbacks of, 5
    edge response of, 347–348
    image formation in, 95, 97
    limitations of, 216–217
    metrology use of, 350–351
    scanning mechanisms for, 135
    scattered light in, 39
    z-scanning method for, 203
Beam-scanning reflection microscope
    Heidelberg Instruments version, 350–351
    schematic diagram of, 345
Biological applications, 325–333
    confocal fluorescence microscopy used, 205–210
        MDCK cells studied, 205, 206–209
        PC-12 cells studied, 209–210
    objectives for, 263
BioRad confocal microscope, images produced using, 215, 314, 315–317, 319
Bone-marrow, fluorescent confocal images of, 117
Bowman's membrane, 307
Brightfield microscopy
    comparison with fluorescence microscopy, 41, 42
    design considerations, 214–218
    image formation in, 94–113
    image interpretation problems, 59
    imaging techniques, 213–242
    optimum design configuration for optics, 214–217
    reflection-mode, 218–234
        advantage of, 188
        electronic differentiation used, 233–234
        imaging in, 187
        resolution improved for, 227–232
        sample characteristics for, 218–227
        visual quality improved for, 227–234
    scanning speed for, 217–218
    transmission-mode, 327–328
    see also Reflection brightfield microscopy

C6-NBD-ceramide, metabolism in MDCK cells, 205, 206–209
$C(m;p)$ function, 36–37
    symmetry of, 48
Cavalieri's principle, 289, 292
Cell counting, 68
Ceramics applications, objectives for, 263
Chinese hamster V-79 cells, 68
Chodanthus puberulus, woody stem, 224, 225
Chromatic aberration, 108
    colour-coding by, 272
Chromosome positions, graphic techniques used to image, 195
Closely spaced points, imaging of, 29
Coherent detectors, 118–120
Coherent imaging, 25
Coherent slit detectors, 119–120
    depth discrimination strength of, 120
Coherent transfer function, 31
    confocal compared with convential microscope, 32–33
Colour coding, 84
Colour information, confocal brightfield microscopy, 224–225
Colour-coding, depth indicated by, 221, 272, 274
Coma (aberration), 105, 106
    detection of, 166
Communication bus, 82–83
Comparison (of microscope systems), 37
Composite amplitude point spread functions, pupil combinations, 175–176
Composite image
    electronic processing of, 129–130
    optical sectioning strength improved by, 125–126
    resolution improved by, 126–128
Composite point spread functions, 172
Composite pupil functions, modified pupils affecting, 177–179
Computer Aided Tomography systems, 76–77, 196
    memory size requirements, 77
Computer considerations
    data processing, 83–85
    instrument control, 81–82

Computers
  PHOIBOS instrument controlled via, 70, 81–82
  TSM images processed by, 268–269
Confocal 2002 tandem scanning microscope, 252
  optical components in, 250
Confocal differential interference contrast method, 237, 239–241
Confocal imaging
  first discovered, 171
  pupils in, 174–181
  requirement for, 171
Confocal interference microscopy, 389–410
Confocal microscopy
  first developed, 185
  image formation in, 8
  introduction to, 1–60
  optical aspects of, 93–139
Confocal principle, first described, 10
Confocal scanning light microscopy (CSLM), comparison with conventional microscopy, 93
Connectivity, study of, 299
Contrast enhancement, 3
Control processor, selection of, 81–82
Conventional fluorescence microscopy, limitations of, 199–201
Conventional microscope
  combined with scanning optical microscope, 6
  image formation in, 7, 8, 25–30
  parallel processing in, 1
  power variation with defocus, 38–39
Correction lenses, 228, 229
Coverglass thickness, resolution affected by, 107, 108, 228
Crepis capillaris
  chromosomal structure, 195
  root top cell, 193, 194
Cross-sections, representation of, 190–191

Data
  collection of, 188–189
  filtering of, 78
  information available, 80
  post-processing of, 70
  reduction of, 78
  storage of, 79–81
  size requirements, 80, 84
Dawsonia superba (moss), spores, 326
Defect inspection, VLSI circuits, 343
Defocus
  confocal fluorescent transfer function affected by, 43, 44
  effects of, 26–28
  in Fourier imaging, 33–35
  power variation with, 38–39
  single point object imaging affected by, 152–155
Defocus response, definition of, 162
Delesse's principle, 287, 289
Depth coded projections, 72, 73
  example of, 73
Depth discrimination
  applications of, 13–22
  confocal fluorescence microscope, 200, 201–202
  meaning of term, 201
  pinhole size affecting, 95–96, 99
  principle of, 9, 10–12
  pupil filters affecting, 181
  see also Optical sectioning
Depth map, surface normal calculation from, 75, 76
Depth resolution, real-time scanning optical microscope, 372, 373
Descemet's membrane, 307
Detector, alternative geometries, 108–113
Detector arrays, 112–113
  resolution improved by, 231–232
Detector signal enhancement, 90–91
Detector size, image intensity affected by, 94–95
Diagnostic ophthalmology, 320–322
Differential amplitude contrast
  electronic production of, 50
  optical arrangement to produce, 51
  optical production of, 50
  production of, 50–52
  transfer function for, 48, 49, 51
Differential contrast
  meaning of term, 48
  production of, 49
  transfer function for, 49
Differential interference contrast (DIC), 237, 239–241, 276
  optics to achieve, 239

Differential phase contrast, 234–236
  optical sectioning system with, 55, 58
  production of, 53–58
  stereo pairs produced, 235, 236
  transfer function for, 48, 49, 53
Diffraction grating, minimum size observable, 385
Diffuse surface models, 73–74
Dimensions, 286–289
Dirac delta functions, 23
Direct view confocal microscopy, 245–278
  see also Tandem scanning microscope (TSM)
Direct-view fluorescence microscope, resolution of, 132
Direct-view microscope, 130–134
  aperture size effects, 131
  aperture spacing effects, 133–134
  fluorescence microscopy, 132
  see also Tandem scanning microscope (TSM)
Disc scanning, 247–248
  advantages of, 248
  choice of aperture disc, 248–249
  design information for, 259
  disc materials used, 258
  driving of discs, 259–260, 366
  implementation methods, 249–254
  one-sided configuration for, 253, 268
  pinhole layout for, 256–257
  pinhole size for, 257–258, 366, 368
  size of aperture disc, 255–256
Disector counting, 292
Display techniques, 71–78
Double scanning, 246–247
Double-pass system, 59
Double-pass transmission confocal microscopy, 237
  images produced by, 238
Drosophila melanogaster (fruit fly), compound eye, 219, 220, 221

Edge enhancement
  example of, 52
  techniques used, 51, 233, 236, 238
Edge-response, metrology use of, 347–348
Effective point spread function, 93, 174
8-bit microprocessor systems, 82, 83

Electrical contrast enhancement, 3
  example of use, 4
Electron microscopy, compared with confocal microscopy, 185, 354, 358–359
Electronic differentiation, visibility of image features improved by, 233–234
Electro-optic phase shifting, surface profiling technique using, 396, 397
Encircled energy, 150, 151
EPROM, confocal images of, 100–101
Escherichia coli, 328
Ethernet, 83
Ewald sphere construction, 157–158
  application to confocal microscopy, 158–159
Extended objects, pupil filters used to image, 179–180
Extended-focus images, 265
Extended-focus imaging, 351, 352
Extended-focus method, 14–16, 102, 145–147
  compared with auto-focus technique, 147
  confocal microscopy used, 15, 16, 17
  high-resolution images using, 17
  images using, 15, 17, 265, 351, 352
  mathematical basis of, 15
  principle of, 14
  slit-detector image using, 113
Eye
  movements of, 321
  optical properties of, 305–306

Filtering, of data, 78–79
Finite-sized detector
  composite image with, 127
  images with, 99–102
Fixation techniques, effects of, 270
Flare, 39–40
Fluorescence confocal microscopy, 328–331
  advantages of, 330–331
  biological applications of, 205–210, 329–330
  depth discrimination of, 200, 201–202
  MDCK cells studied using, 205, 206–209

Fluorescence confocal microscopy, *cont.*
  PC-12 cells studied by, 209–210
  staining methods for, 200
  x–z image generation in, 202–205
Fluorescence generation process, model of, 41–42
Fluorescence microscopy, 41–45
  advantages for biological specimens, 188
  compared with brightfield microscopy, 41, 42
  composite-image method, 127, 128
  conventional, limitations of, 199–201
  direct-view response for, 132
  high-aperture objectives used, 4–5
  image formation in, 42, 45, 114–118
  limitations of, 328–329
  ocular tissue imaged by, 310–312
  optical sectioning in, 44, 114, 118
  resolution of, 41
  scanning mechanisms for, 135, 136
  slit detectors used, 115–117
  spatial distribution determined by, 199
  spatial frequency response for, 161
Fluorescence tandem scanning microscopy, 266, 271
Fluorescent monoclonal antibodies, use in ocular tissue, 322
Flying-spot differential interference system, 395
Fourier imaging, 31–37
Fourier optics
  coherent detectors, 118–119
  incoherent detectors, 94
Fourier series, three-dimensional generalisation of, 157
Frame buffers, 84
Frame store, 274
Frame store interface, 2, 3
Fraunhofer approximation, 367
Fresnel loss, 407
Fringe-scan flow cytometry, 332
Fruit fly (*Drosophila*), compound eye, 219, 220, 221
Full-colour reflected brightfield imaging, 224–225
Full-width-half-maximum (FWHM), resolution measured by, 58
Fundus camera, 306

Gaussian kernels, 79
Generalised least squares techniques, 45
Geology, objectives for, 263
Geometrical-optics approximation, 97
Glass industry, objectives for, 263
Gold, scattering from, 327
Gold-covered silicon, 354
Gold-labelling, 225–227, 230, 232, 235
Golgi complex
  MDCK cells, 205, 206, 207
  PC-12 cells, 209, 210
Gordon–Reynolds algorithm, 75
Gouraud (shading) method, 74
GPIB communication bus, 83
Gradient filters, 79
Graphical techniques
  image representation by, 194–195
  rotations using, 77–78
  surface shading by, 75
Grillotia erinaceus neurons, 298

Haemanthus katherinae (African blood lily), endosperm, 226, 227, 229, 230, 231–232
Hafnium carbide lamp, 254
Hankel transform, 148, 400
Heidelberg Instruments LPM, 350–351
  linewidth measurements, 340
  optics of, 350
Height image technique, 102
  slit-detector image using, 113
Height measurements, semiconductors, 355
Heterodyne interferometry, 59, 398–399
Heterodyne techniques, imaging by, 399
High-speed scanning, 86
Hilbert-transform relationship, 403
Histochemistry, 331
Homodyne techniques, 398–399
Human jaw bone, 287, 291, 294, 300
Human skull bone, 266

Image collection times, 87, 189, 217
Image deconvolution, 277
Image formation, 7–12, 186–188
  confocal compared with convential microscopes, 25–30

Image formation, *cont.*
  geometry of, 23
  polarization affecting, 186
  theory of, 22–24
Image intensity, 11
  auto-focus, 16
  confocal microscopy, 95
  effect of pinhole size, 96
  extended-focus, 15
  fluorescence microscopy, 41
  Fourier optics, 31
  point detector, 109
  point objects, 26, 28, 29, 94, 346
  single point, 148–149, 150, 155
  slit detector, 109
  stereo pairs, 20
  straight edge, 29
  three-dimensional images, 159, 196
Image-processing algorithms, 299, 300
Immersion, effect of, 262
Immunogold labelling, 225–227, 230, 232, 235, 327
Incoherent imaging, 25
Information theory, resolution in terms of, 275–276
Infrared images, 21, 22
Instrument control requirements, 81–83, 85–87
Integrated circuit
  brightfield and DIC images compared, 239
  real-time scanning images, 379
Integrated intensity, point image, 38
Integrator
  maximum integration time used, 90
  noise-suppression using, 90
Interference microscopy, 389–410
  aberration measurement by, 399–409
  heterodyne techniques used, 398–399
  reflection-mode confocal microscope, 389, 390
    images produced, 391–392
  surface profiling by, 19, 20, 395–398
Introduction (to confocal microscopy), 1–60
Isotropic uniform random (IUR) sections, 293

Keeler–Konan wide-field specular microscope, 309

Lambert's law of reflection, 74
Lamprey spinal chord neuron cell, 65, 66
Large-area detectors
  advantage of, 50–51
  differential contrast obtained using, 50, 53
  quadrant detectors, 55
    images using, 56
  split-half detectors, 50, 53
Laser based confocal microscope
  display techniques for, 71–78
  tandem scanning microscope compared with, 46–47
Laser scanning ophthalmoscope, 313
  images produced by, 315–317, 319
Lateral resolution, pinhole size effects, 229–231
Length estimation, 295
Lens-less imaging, 399
Line objects, fluorescence microscopy image of, 116
Linewidth measurements
  semiconductors, 341, 354–355
  VLSI circuits, 341
Living matter, study of, 247, 278
Look-through projection, 72
  example of, 66
Low-aperture system, axial response for, 164–167
Lukosz's principle, 9

Mach–Zehnder type transmission-mode confocal interference microscope, 389
Madin-Darby Canine Kidney (MDCK) cells, investigation of, 205, 206–209
Magnification, confocality affected by, 97
Masks, critical dimensions of, 343–344
Mechanical rastering, 2–4
Mechanical scanning (of specimen), 188
  see also Object...; Specimen scanning...
Mechanical stylus measurement system, 355
Medical applications, objectives for, 263

INDEX 419

Michelson-type reflection-mode confocal interference microscope, 389
Microcircuit
 confocal image of, 9
 confocal images of, 13
 edge-enhanced image of, 52
 extended-focus images of, 15
 low-magnification image of, 3
 profiling of, 18
Microstructural geometry, measurement of, 285–301
Microtomoscopy, 199–211
Mirror objectives, 262, 263, 278
Modified pupils
 amplitude point spread functions affected by, 175–176
 imaging with, 173–174
 transfer functions affected by, 176–179
Morphological studies, ocular tissue, 321–322
Mouse cerebellum, Purkyne cells, 267
MS-DOS, 84
Multiple-wavelength images, 86

Networking, computer, 83
Nipkow disc, 255
 schematic diagram of, 362
Nomarski method (for differential phase contrast), 53, 55, 237, 239–241, 276, 394–395
 image produced by, 57, 239, 240
Non-contacting surface profilometry, 395–398
Non-visible radiation, images using, 21, 22, 399
Normalised radial coordinate, definition of, 148
Nucleator sampling, 297–298
Number estimation, 290–292
Number-weighted distribution, 295–296

Object function, three-dimensional imaging, 156
Object scanning microscopes, 2–4
 axial response of, 167
 drawbacks of, 349–350
 edge response of, 347–348
 image formation in, 95
 scattered light in, 39
 z-scanning method for, 203–204
 see also Specimen-scanning...
Object table, precision of, 86
Objective lenses
 determination of quality of, 12
 further development of, 277–278
 tandem scanning microscopy, 261–263
Objective scanning microscopes, 6–7
 scattered light in, 39
 z-scanning for, 203–204
Ocular tissue
 confocal microscopy of, 305–322
 fluorescence techniques used, 310–312, 322
 morphological studies of, 321–322
Off-axis scanning, 188, 214
 see also Beam-scanning...
On-axis scanning microscope
 brightfield imaging using, 213–242
 see also Objective-scanning...; Specimen-scanning...
One-dimensional object, Fourier imaging of, 34–36
One-sided tandem scanning microscope (OTSM), 253, 268
Operating system (OS)
 function of, 84
 types of, 84, 85
Ophthalmology
 clinical use of real-time confocal system, 320–322
 confocal microscopy applied to, 309–319
 development of optically sectioning microscopes for, 306–309
Optical aspects, 93, 139
Optical beam induced-current technique, image affected by, 6
Optical fibre, image of, 3, 4
Optical sectioning
 differential phase contrast with, 55, 58
 enhancement of, 125–130
 fluorescence microscopy, 44, 114, 118, 203, 204
 principle of, 9, 10
 pupil filters affecting, 181
 slit detector, 110

Optical sectioning, *cont.*
  theory of, 38–39
  width as function of numerical aperture, 11
  see also Depth discrimination
Optical sectioning microscopes, ophthalmology applications, 306–309
Orchid seeds, embryo cells from, 66, 67
Orientation, probes affected by, 286–287
Orthogonal triplet probes (ORTRIPS), 295
Osteocyte lacunae, 287, 291, 294, 300

Paleontology, objectives for, 263
Parallel projection geometry, 274
Paranthropus (Australopithecus) Boisei, fossil tooth, 265
Partially coherent imaging, 25
Partially coherent transfer function, 159
Particle counting, 275
Particle size, estimation of, 295–298
PC-12 cells, colocalization studies, 209–210
Perfect amplitude imaging, 47
Petráň–Hadravský microscope, 247–278
  compared with Xiao–Kino microscope, 310
  images produced by, 265–267, 364
  see also Tandem scanning microscope
Phase boundaries
  reflections at, 222–223
  resolution affected by, 327
Phase specimens, reflection from, 326–327
PHOIBOS instrument
  communication bus for, 82–83
  computer considerations for, 81–82, 83–85
  control processor for, 81–82
  data amount generated, 84
  detector signal enhancement in, 90–91
  display techniques used, 71–78
  electronics for, 87–91
  examples of use, 66–68
  instrument control requirements, 81–83, 85–87
  microprocessor system for, 82, 87
  operator interactions, 70, 81–82
  opto-mechanical principle of, 69–70
  position-encoding of scanning mirror in, 87–89
  scanning speed of, 87
  system concepts, 68–71
  system configuration for, 81–82
  time-critical parts of, 85
Phong (shading) method, 74
Photoacoustic effect, image formation by, 59
Photoresist residue, detection of, 54
Photothermal effect, image formation by, 59
Piezoelectric focusing, surface profiling technique using, 396, 397
Pinhole detectors
  compared with slit detectors, 110, 112
  images with, 99–102, 112
Pinhole layout, tandem scanning microscopy, 256–257
Pinhole offset, axial response affected by, 166
Pinhole size
  depth-discrimination affected by, 95–96
  disc-scanning, 257–258
  efficiency affected by, 372
  image quality varying with, 100–101
  real-time scanning optical microscope, 368
  resolution affected by, 95, 229–231, 372, 373, 377
  tandem scanning microscope, 131
Pinholes, effect of, 39
Pitch measurements, 353
Pixels, meaning of term, 71, 188
Planapochromat objective
  performance of, 215, 216
  spherical aberrations in, 106, 107
Planar reflectors
  fluorescence microscopy response, 117
  image formation of, 95–99
  optical sectioning demonstrated using, 9, 37
  performance of microscope tested using, 12

Plzeň discs, 251, 257, 258, 259
Point counting methods, 289
Point detector, size of, 93
Point objects
  fluorescence microscopy image of, 116
  imaging of, 26–29, 94–99
    contour plots for, 27
    isometric surface projections for, 28
    reflection mode, 28
  optical sectioning of, 38
Point spread function
  definition of, 121
  effective, 93, 174
  real-time confocal microscopy, 385
Polarization contrast, 356–357
Polarization imaging, 332
Polygon representation, 77
Polysilicon
  etched, 355
  resist on, 356, 357
Polytrichum commune (moss), spores, 326
Position-encoding, 87–89
  digital signal method, 88, 89
  pulse-burst method, 87
  transmission grating method, 87–88, 89
Power, point image, 38
Preparation methods, conventional, 200, 245
Probes, metrology, 286–289
Profiling, *see also* Surface profiling
Profiling technique, 17–18
  interferometry used, 19, 20
  isometric image for, 19
  typical trace using, 18
Pseudo-colouring, 221
Pupil filters, production of, 182
Pupil functions
  derivation for confocal system, 177
  form of, 24
  meaning of term, 23
  non-uniformity of, 99
  phase of, 408, 409
  uses of, 400
Pupil modification
  amplitude point spread functions affected by, 175–176
  coherent imaging systems, affected by, 173–174
  transfer functions affected by, 176–179
Pupil plane filters, 120–124, 171–182

Quadrant detector, 55
  images using, 56
Quasi-incoherent contrast (QIC), 357

Rabbit cornea
  confocal microscopy images of, 314–317
  optical section through, 312
Rabbit eye
  corneal endothelium, 380
  epithelial surface of, 381
  lens fiber of, 381
Raman effect, image formation by, 59
Rat cerebellum, 144, 145
Rat lung tissue, 67
Rat tibia cortical bone, 266
Ray-tracing, 76
Rayleigh–Sommerfeld theory, 369
Real-time imaging, 331–332
Real-time scanning, 247, 277, 361
Real-time scanning optical microscope (RSOM), 365–366
  disc scanning for, 366
  experimental results for, 378–385
  light sources used, 366, 386
  pinhole size for, 368
  schematic diagram of, 365
  theory of, 367–373
  see also Xiao–Kino confocal scanning optical microscope
Redox fluorometry, 311
Reflected brightfield microscopy, 218–234
  advantage of, 188
  electronic differentiation used, 233–234
  imaging in, 187
  resolution improved for, 227–232
  sample characteristics, 218–227
  visual quality improved for, 227–234
Reflection confocal microscopy
  biological applications of, 326–327
  defocussed transfer function for, 33
  optical sectioning in, 9, 10
Reflection interference contrast, 275

Reflective coatings, 271–272
Reflective stains, 271
Reflective surfaces, confocal images from, 219–221
Refractive index boundaries (within samples), 222–223
Registration measurements
 semiconductors, 342, 354
 VLSI circuits, 342
Resolution
 brightfield microscopy, improvement of, 227–234
 confocal microscopy, 58
 detector arrays affecting, 231–232
 enhancement of, 9
 fluorescence microscopy, 330–331
 improvement by composite imaging, 126–128
 information-theory approach, 275–276
 pinhole size affecting, 229–231
Resonant microscope, 394
Reticles, critical dimensions of, 343–344
Rhoeo spathacea leaf, 222, 223
Rotations, ways of achieving, 76–78

Scanning
 advantages of, 3–4
 disc-based, 247–248
  advantages of, 248
  choice of aperture disc, 248–249
  design information for, 259
  disc materials used, 258
  driving of discs, 259–260, 366
  pinhole layout for, 256–257
  pinhole size for, 257–258, 366, 368
  size of aperture disc, 255–256
 speed of
  image affected by, 6, 86–87
  specimen-scanning microscope, 217–218
Scanning electron microscopy (SEM), compared with confocal microscopy, 354, 358–359
Scanning mechanisms, 135–139
Scanning mirror, position encoding of, 87–89
Scanning optical microscopes
 arrangement of, 2–3
 combined with convential microscope, 6
 disadvantages of, 361
 image formation in, 7–12
 non-confocal imaging in, 47–58
 theory of imaging in, 22–40
Scanning slit microscope (SSM), 307–308
 image taken with, 308
Scattered light, 39–40
Scattering function, 156
Scheimpflug camera, 306
Schlieren system, 160
Secretogranin-I, colocalization in PC-12 cells, 209–210
Sectional images, 190–191
Sectioning
 dimensional reduction caused by, 288
 optical *see* Optical sectioning
Seidel aberrations, 407
Semiconductor industry, objectives for, 263
Semiconductor metrology, 339–359
 contrast modes used, 348–349
Sieving distribution, 297
Signal-to-noise ratio
 detector size affecting, 40
 effect of filtering, 79
 improvement of, 78, 90–91, 189
 limitation of, 90, 91
Silicon
 aluminium lines on, 352
 calibration standard for, 267
 trench in
  bottom of, 383
  linescans of, 384
  measurements for, 382
  top surface of, 383
Silicon discs, 258
Silicon nitride, resist on, 359
Silicon semiconductor devices, imaging of, 21, 22
Silver-enhanced gold, 225–226
Simple objects, imaging of, 25–30
Simulated fluorescence processing (SFP), image representation by, 193–194
Single aperture confocal microscope, 137, 139
 slit width effects, 138, 139

Single point object, image formation for, 147–155
Singular value decomposition techniques, 45
16-bit microprocessor systems, 82, 83, 84
Slit aperture scanning
 equipment, 137
 slit width effects, 138
Slit detectors
 asymmetry of image, 108, 109
 coherent, 119–120
 compared with pin-hole detectors, 110, 112
 fluorescence imaging with, 115–117
 image intensity from, 109
 images with, 112, 113
 optical sectioning affected by, 110
Slit lamp (biomicroscope), 306
Slit lens system, depth discrimination strength of, 120
Slit tandem scanning microscopy, 259
Smoothing, 79
Snake muscle motor nerve terminal, 138
Space-invariant imaging, 4
Spatial filters, fluorescence microscopy use of, 331
Spatial frequency, 157
 range affected by pupil filters, 180
 range for coherent imaging systems, 158, 159
 stereo pair in terms of, 158
Specimen preparation, TSM samples, 271
Specimen-scanning confocal microscope
 compared with beam-scanning microscope, 214–216
 scanning speed for, 217–218
 see also Object-scanning…
Spectroscopic grating, differential phase contrast image of, 54
Specular microscope, 307
Sperm whale dentine, chick osteoclast, 269
Spherical aberration, 105, 106, 407
 axial response affected by, 163, 164, 165
 correction for, 328
 effect of cover-slip thickness on, 107

Stage-scanning microscope, 349–350
 see also Object-scanning…;
 Specimen-scanning microscope
Step height, theory dealing with, 30
Stereo images, TV display of, 71
Stereo pairs
 brightfield microscopy, 235, 236
 differential interference contrast used, 239, 240
 differential phase contrast used, 237, 238
 double-pass transmission confocal microscopy used, 237, 238
 generation of, 147
 spatial frequencies needed to form, 158
 TSM-generated, 273, 274
Stereo viewing, 20
 mathematical basis of, 20
Stereological measurement, 285–301
 future developments in, 299–301
Stereology, 275, 285
Stereology applications, 331
Stereoscopic image recording, TSM used for, 272–274
Stereoscopic images, 192, 193
 TSM, 268
Stigmatic imaging theory, 368
Straight edge, imaging of, 29–30
Strehl intensity/ratio, 407–408
Sub-wavelength scattering objects, 225–227
Superposition images, 192, 193
Superresolving pupil filters
 amplitude point spread functions affected by, 175, 176
 diffraction affected by, 173, 174
 extended objects imaged using, 179–180
 production of, 182
 pupil functions affected by, 177, 178, 179
Surface estimation, 287, 292–294
Surface gradient, calculation of, 76
Surface normal, calculation from depth map, 75, 76
Surface profile technique, 17–18, 143–144, 219, 221
 images produced using, 146, 221
 interferometry used, 19, 20
 isometric image for, 19

Surface profile technique, *cont.*
  typical trace using, 18
Surface profiling
  confocal interferometry for, 19, 20, 395–398
    electro-optic phase shifting used, 396, 397
    piezoelectric focusing used, 396, 397
Surface shading, 73–76
  graphical systems for, 75
  voxel-based systems for, 75
Surface slope, axial response affected by, 167
Surface texture, imaging of, 237, 241
System concepts, 68–71
System configurations, 81–82

Talysurf, 355
Tandem scanning microscope (TSM), 7, 45–47, 361–364
  advantages of, 47
  alignment aids in, 260–261
  alternatives to, 276–277
  aperture spacing effects, 133–134
  applications of, 265–267, 270
  beam-splitters in, 260
  compared with other CSLMs, 276, 277
  comparison with laser-based confocal microscope, 46–47
  disadvantages of, 363
  disc scanning used, 247–249, 361–362
  disc size used, 255–256
  drawbacks of, 47
  eyepieces for, 263–264
  filters in, 260
  first experiments with live animals, 247
  first patented, 246
  fluorescence microscopy, 132, 266, 271
  further development of, 277–278
  illumination source for, 254–255, 361
  image intensifiers for, 264
  image processing for, 268–269
  image-formation properties of, 46–47
  limitations of, 213
  mirrors in, 260
  non-confocal modes combined with, 268
  objective lenses in, 261–263
  photographic recording equipment for, 264
  pinhole layout for, 256–257
  pinhole size for, 131, 257–258
  principle of, 45–46, 130, 361–362, 363
  reflection mode
    autocollimation variants, 253
    one disc with central symmetry used, 250–253
  reflective coatings used, 271–272
  reflective stains used, 271
  relay optics for, 263–264
  resolution of, 247, 275–277
  schematic diagram of, 363
  slits in, 259
  specimen preparation for, 271
  specimens suitable for, 270
  stage automation for, 265, 268
  stands for, 264–265
  transmission mode
    one disc with central symmetry used, 254
    requirements for, 249, 254
    two discs used, 254
  TV cameras for, 264
  see also Direct-view ...; Petráň–Hadravský microscope
TEM grid, confocal interference image of, 391
Thick specimens
  imaging of, 14
  study of, 201, 211
  theory dealing with, 30
Thick translucent objects, imaging of, 275
32-bit microprocessor systems, 82, 84
Three-dimensional data sets
  representation of, 190–193
  software processing of, 70–79, 190–193
Three-dimensional image formation, 155–161
  metrology use of, 351–352
Three-dimensional image representation, 185–196
Three-dimensional imaging, 143–167
  data collection for, 188–189
  display technique used, 71–78
  method used, 69

INDEX

Through-focus series, 14
Tichonov regularisation, 45
Tiling algorithms, 299
Time-averaged point spread function, 373–374
Tracor Northern tandem scanning microscope, 252
　discs used, 257, 258
Tradescantia stamen hair cell, 146, 147, 233, 236, 238, 326
Transfer functions, 31
　aberration effects on, 105, 106
　annular lenses, 123
　coherent slit detector, 119
　confocal fluorescence microscopy, 42–44
　differential amplitude contrast, 48, 49, 51
　differential contrast, 48, 49
　differential phase contrast, 48, 49, 53
　general form of, 36–37
　modified pupils affecting, 176–179
　symmetry of, 32
　three-dimensional, 159–160
Transfer lens system, 135
Transistor bond pad, metallisation on, 122
Transmission confocal microscopy, 236–237, 254
　biological applications of, 327–328
　defocussed transfer function for, 33–34
　drawback of, 188
　imaging in, 187
Transmission cross-coefficient, 36
Transmission tandem scanning microscopy, 254
Transparent microstructures, size measurement of, 355–357
Transverse resolution, real-time scanning optical microscope, 377
Transverse response, confocal microscope, 373–378
Tube length
　adjustment of, 107
　tandem scanning microscopy, 261
Two-dimensional probes, 287
Type I scanning microscope, 245
Type II scanning microscope, 247, 276
　see also Tandem scanning microscope (TSM)

Type III microscope, 113
　see also Detector arrays

Ultraviolet microscopy, 358
UNIX operating system, 80, 84, 85

$V(z)$ response, 11, 400–401
　aberration determined from, 103–107
　alignment tested by, 139
　annular filters, 121
　beam-scanning compared with specimen scanning, 214–215
　composite images, 127, 128
　performance of objective tested by, 228
Video-enhanced contrast method, 277
Video-enhanced fluorescence microscopy, 201–202
　compared with confocal fluorescence microscopy, 205
VLSI circuits
　linewidth measurements of, 341
　metrology of, 341–344
　　confocal microscopy used, 345
　registration measurements of, 342, 354
Volume data
　filtering of, 78–79
　reduction of, 78
　structure of, 80
Volume estimation, 287, 289–290
Volume-weighted distribution, 297
Voxel, meaning of term, 71, 188
Voxel-based systems, surface shading by, 75

Wavefront aberration, 105, 187
Waveguide techniques, 30
Wavelength, absorption/reflection varying with, 224–225
Weak object transfer function, 35
　real and imaginary parts of, 35
White-light confocal microscopy, 332
Wide-field scanning specular microscope, 309
Wiener filters, 45

x–z images, generation in fluorescence microscopy, 202–205
Xiao–Kino confocal scanning optical microscope, 310, 365–366
  compared with Petráň–Hadravský microscope, 310
  experimental results, 378–386
  theory of, 367–373
  see also Real-time scanning optical microscope (RSOM)

z-response, 346–347
z-scanning, 203–204
Zero-dimensional probes, 286
Zero-dimensional quantities, 298–299